Current Topics in Microbiology
and Immunology

151

Bacterial Adhesins

Edited by
K. Jann and B. Jann

With 23 Figures

Springer-Verlag
Berlin Heidelberg New York
London Paris Tokyo Hong Kong

Prof. Dr. Klaus Jann
Dr. Barbara Jann

Max-Planck-Institut für Immunbiologie
Stübeweg 51, 7800 Freiburg, FRG

ISBN 3-540-51052-4 Springer-Verlag Berlin Heidelberg New York
ISBN 0-387-51052-4 Springer-Verlag New York Berlin Heidelberg

© Springer-Verlag Berlin Heidelberg 1990
Library of Congress Catalog Card Number 15-12910
Printed in Germany

The use of registered names, trademarks, etc. in this publication does not imply, even in
the absence of a specific statement, that such names are exempt from the relevant
protective laws and regulations and therefore free for general use.

Product Liability: The publishers can give no guarantee for information about drug
dosage and application thereof contained in this book. In every individual case the
respective user must check its accuracy by consulting other pharmaceutical literature.

Phototypesetting by Thomson Press (India) Ltd, New Delhi
Offsetprinting: Saladruck, Berlin; Bookbinding: B. Helm, Berlin
2123/3020-543210 – Printed on acid-free paper

Preface

The great majority of bacterial infections are initiated by the adhesion of pathogenic bacteria to cells and mucosal surfaces of the host. The sequela of adhesion may range from the action of toxins outside target cells to their penetration into or through tissue. Besides the consequences of bacterial adhesion related in infection, the result may be colonization of mucosal surfaces with normally harmless bacteria, which in stress situations may become virulent, a phenomenon known as nosocomial infections.

With very few exceptions, adhesion is carbohydrate specific. It is mediated by bacterial recognition proteins that are, according to the phenomenon studied, termed adhesins or hemagglutinins; the term "lectin" is sometimes also used. The chemical nature of the adhesins and their organization on the bacterial surface have been studied intensively in many laboratories. The application of genetic and biochemical techniques has led to substantial progress in the molecular characterization of adhesins in recent years. We now know that adhesins may occur as structural subunits of fimbriae and that they may form fimbriae which can be considered as mono- or multifunctional linear adhesin polymers. Other adhesins do not form recognizable structures and are tentatively called nonfimbrial. Adhesins may even be components of bacterial cell walls. Adhesin-receptor specificities have been unravelled. The study of the distribution of receptors in tissue has created implications about the possible susceptibility to infections. The genetics of adhesins provide insight into the biogenesis of adhesins and into the relationships between structure and function. Thus, a comprehensive view of bacterial adhesins is now available, revealing vistas about the molecular details of adhesins, both about pathogenicity in general, and on possible applications of the findings to medicine and the abatement of infections. The observer of the field will, however, also find many hitherto unanswered questions, for instance regarding protein–protein and protein–carbohydrate interactions. This will certainly be a stimulus for further research.

In this volume important aspects of adhesin research are presented, covering the most pertinent areas of genetics, serology, and biochemistry, and the approaches to medical applications. Since these areas overlap, we have encouraged the authors to present not only their most pertinent data but also more general outlooks on their field and to expound principles behind the facts. Although the book is restricted to the adhesins of *E. coli*, one of the genera on which the most information is available today, the principle findings will doubtlessly also be applicable to other systems. We feel that focussing on one system in this way will ensure a more concise and coherent presentation of general principles and their molecular basis.

Klaus Jann
Barbara Jann

July 20, 1989

Table of Contents

Indexed in Current Contents

List of Contributors

You will find the adresses at the beginning of the respective contribution

ANDERSSON, B.
ANIANSSON, G.
DE GRAAF, F. K.
EVANS, D. G.
EVANS, JR., D. J.
HACKER, J.
HOLTHÖFER, H.
HOSCHÜTZKY, H.
JANN, K.
KORHONEN, T. K.
LEFFLER, H.
LINDSTEDT, R.
DE MAN, P.

MOON, H.
NIELSEN, A.
OFEK, I.
ØRSKOV, F.
ØRSKOV, I.
PARKKINEN, J.
SCHMIDT, M. A.
SHARON, N.
SVANBORG EDÉN, C.
VIRKOLA, R.
WESTERLUND, B.
WOLD, A.

Genetic Determinants Coding for Fimbriae and Adhesins of Extraintestinal *Escherichia coli*

J. HACKER

1 Introduction

Extraintestinal *Escherichia coli* are a frequent cause of urinary tract infection (UTI), sepsis, and newborn meningitis (NBM) (ØRSKOV and ØRSKOV 1985; SVANBORG EDEN and DE MAN 1987). Uropathogenic *E. coli* produce different factors involved in pathogenicity, i.e., O antigens (e.g., O4, O6, O18, and O75), capsule antigens (e.g., K5, K12, and K13; ORSKOV et al. 1977), and cytotoxic proteins, termed hemolysins (Hly, HUGHES et al. 1983). It has often been shown that strains causing meningitis and sepsis belong to particular serotypes, e.g., O1:K1, O2:K1, O7:K1, O83:K1, or O75:K1 (KORHONEN et al. 1985; ACHTMAN and PLUSCHKE 1986; HACKER et al. 1989). In addition, extraintestinal *E. coli* strains produce adhesins which enable the isolates to attach to eukaryotic cells (SVANBORG EDEN et al. 1977; OFEK and SILVERBLATT 1982). The majority of adhesins produced by extraintestinal *E. coli* are associated

Institut für Genetik und Mikrobiologie, Röntgenring 11, D-8700 Würzburg, FRG

Current Topics in Microbiology and Immunology, Vol. 151
© Springer-Verlag Berlin·Heidelberg 1990

with fimbriae, structures of the cell envelope which are easily identified electron microscopically and which consist of about 1000 identical major protein subunits. Their binding capacity can be determined by hemagglutination, a simple assay in which *E. coli* strains and erythrocytes of different species are mixed on slides or in microtiter plates (DUGUID et al. 1955; JONES and RUTTER 1974; EVANS and EVANS 1983).

Fimbrial adhesins are distinguished by their receptor specificity. Type 1 fimbriae recognize a mannose-containing receptor, and were therefore termed mannose-sensitive (MS) fimbrial adhesins (ORSKOV and ORSKOV 1983). The type 1 fimbriae are widely distributed among pathogenic and nonpathogenic *E. coli* strains (ORSKOV and ORSKOV 1985). However, recent data suggest that MS adhesins contribute to extraintestinal infections, especially those of the bladder (HAGBERG et al. 1983; O'HANLEY et al. 1985; MARRE and HACKER 1987; KEITH et al. 1986).

Mannose-resistant (MR) adhesins bind to eukaryotic receptor substances other than mannose. P fimbrial adhesins, which represent the main group of MR adhesins associated with UTI, are able to bind to the α-D-Gal-(1-4)-β-D-Gal part of glycolipid receptor structures (KÄLLENIUS et al. 1980; VÄISÄNEN-RHEN et al. 1984). Another, important group of MR adhesins, the S fimbrial adhesins, recognize an α-sialyl-(2-3)-β-Gal-containing receptor structure (KORHONEN et al. 1984; PARKKINEN et al. 1986). S fimbriae are very often found among *E. coli* strains causing sepsis or NBM (KORHONEN et al. 1985; HACKER et al. 1989a). Apart from their binding property, fimbrial adhesins can be separated into different F groups according to their serologic specificity, e.g., type 1 fimbriae form group F1A whereas P fimbriae may belong to groups F7–F14 (ØRSKOV and ØRSKOV 1983; ABE et al. 1987).

In addition to fimbrial adhesins, adhesive proteins of extraintestinal *E. coli* which do not exhibit fimbrial structures have been described. The M agglutinin, the afimbrial adhesins, and the nonfimbrial adhesins belong to this group of adhesive cell structures (RHEN et al. 1986b; GOLDHAR et al. 1987; LABIGNE ROUSSELL and FALKOW 1988). The occurrence of fimbriae (pseudotype I) which are devoid of any detectable hemagglutinating activity has also been reported for *E. coli* strains in UTI (PERE et al. 1985). These fimbriae belong to serogroup F1C. Recent observations, however, indicate that F1C fimbriae mediate adherence to human kidney cells (VIRKOLA et al. 1988).

2 Genetic Approaches for Analyzing Adherence

2.1 Chromosomal Mapping of Adhesin-Specific Genes

Several attempts have been made to find out whether the gene clusters coding for fimbrial adhesins are located on the chromosome or on plasmids. It was shown that the great majority of, if not all, adhesin determinants of extraintestinal *E. coli* are chromosomally encoded. Transposition of adhesin determinants or parts thereof to plasmids rarely occurs (HALES and AMYES 1986a, b; J. HACKER and M. SCHMITTROTH, unpublished results). In addition it is generally accepted that an *E. coli* strain may carry more than one adhesin determinant on its chromosome (RHEN 1985; HULL et al. 1985; HACKER et al. 1989b).

Using conjugative bacterial transfer experiments it was demonstrated that gene clusters coding for type 1 fimbriae are located in the vicinity of the gene loci *thr* and *leu* at map position 98 (see BACHMANN 1987). This was shown for both K-12 strains and wild-type isolates (see Table 1). The genetic determinants which code for P or P-related fimbriae (serotypes $F7_1$, $F7_2$, and F13) have been mapped in regions next to the *serA* gene (map position 57-63), at position 85-90 (in the vicinity of the *ilv* gene cluster), and at position 95 on the chromosomes of O4 and O6 wild-type strains (HULL et al. 1986; HOEKSTRA et al. 1986). In contrast the P determinant of serotype F8 (*fei*) has been mapped at position 17 on the chromosome of an O18:K5 strain (KRALLMANN-WETZEL et al. 1989). It therefore seems that the gene cluster coding for type 1 fimbriae are located at fixed positions on the chromosomes of *E. coli* strains. P fimbrial determinants may be located at different regions on the chromosomes of extraintestinal *E. coli* isolates.

2.2 Cloning of Fimbrial and Adhesin Determinants

As summarized in Table 2, the genetic determinants coding for various types of fimbriae and adhesins have been cloned. In most cases the chromosomal DNAs of the *E. coli* parental strains were isolated and partially cleaved with the enzyme *Sau*3A, and DNA fragments of about 30 kb were ligated into cosmid vector molecules (for references see Table 2). The recombinant cosmid DNAs were transduced into *E. coli* K-12 strains and fimbrial-positive clones were selected by hemagglutination, by using specific antisera prepared against the fimbrial structures or by a subsequent inspection of clones in electron microscopes.

The type 1 fimbrial determinants of an *E. coli* K-12 strain (termed *fim*) as well as of different wild-type strains (termed *pil*) and the genetic determinants coding for

Table 1. Chromosomal map positions of adhesin determinants

Strain	Adhesin determinant[a]	Coselective markers[b]	Map position[b]	Reference
K-12	Type 1 (*fim*)	*thr-leu*	98	SWANEY et al. 1977
J96	Type 1 (*pil*)	*thr-leu*	98	HULL et al. 1986
J96	P (F13-*pap*)	*serA-thyA*	61–63	HULL et al. 1986
J96	P related (F13-*prs*)	*ilv*	85	HULL et al. 1986
AD110	P ($F7_1$-*fso*)	*purA-thr*	95–0	HOEKSTRA et al. 1986
AD110	P ($F7_2$-*fst*)	*thyA-pheA*	61–62	HOEKSTRA et al. 1986
2980	Type 1 (*pil*)	*thr-leu*	98	KRALLMANN-WETZEL et al. 1989
2980	P (*fei*)	*gal-pyrD*	17–21	KRALLMANN-WETZEL et al. 1989

[a]The gene clusters *fim* and *pil* both code for type 1 fimbriae; *pap* refers to pili associated with pyelonephritis and codes for P fimbriae of serotype F13; *prs* refers to P-related sequences and codes for fimbrial adhesins of serotype F13 with a GalNAc-(1-3)GalNAc specificity (see LUND et al. 1989b);
[b]Map positions and marker genes according to BACHMAN (1987)

Table 2. Cloned fimmbrial and adhesin determinants of extraintestinal *E. coli* strains

Adhesin/fimbria type (gene abbreviation)	Receptor specificity	F-type	Subunit (kd)[a]	Parental strain	Serotype	References
Fimbrial adhesins						
Type 1 fimbriae (*fim*)	α-D-Mannose	F1A	17	PC31	K-12	Klemm et al. 1985
(*pil*)	α-D-Mannose	F1A	17	J96	O4:K6	Hull et al. 1981; Orndorff and Falkow 1984a
(KS71D)	α-D-Mannose	F1A	17	KS71[b]	O6:K2	Rhen 1985
(*pil*)		Non-F1A	17	536	O6:K15	Hacker (unpublished)
P fimbriae (*fso*)	α-D-Gal-(1-4)-β-D-Gal	F7₁	20	AD110[b]	O6:K2	Van Die et al. 1985
(*fst*)	α-D-Gal-(1-4)-β-D-Gal	F7₂	17	AD110	O6:K2	Van Die et al. 1983; Van Die et al. 1984b
(*fei*)	α-D-Gal-(1-4)-β-D-Gal	F8	19.5	2980	O18:K5	Hacker et al. 1986b
	α-D-Gal-(1-4)-β-D-Gal	F9	21	C1018	O2:K5	De Ree et al. 1985c
(*fel*)	α-D-Gal-(1-4)-β-D-Gal	F11	18	C1976	O1:K1	De Ree et al. 1985a
	α-D-Gal-(1-4)-β-D-Gal	F12$_{rel}$	16.8	20025	O4:K12	High et al. 1988
(*pap*)	α-D-Gal-(1-4)-β-D-Gal	F13	19.5	J96	O4:K6	Hull et al. 1981; Normark et al. 1983
	α-D-Gal-(1-4)-β-D-Gal	F13	18.8	20025	O4:K6	High et al. 1988
	α-D-Gal-(1-4)-β-D-Gal	F14	20.5	20025	O4:K6	High et al. 1988
(KS71A)	α-D-Gal-(1-4)-β-D-Gal	F7₁	22	KS71	O6:K2	Rhen et al. 1983a; Rhen et al. 1985b
(KS71B)	α-D-Gal-(1-4)-β-D-Gal	F7₂	20	KS71	O6:K2	Rhen et al. 1983; Rhen et al. 1985b
(pDC1)	α-D-Gal-(1-4)-β-D-Gal	F11		IA2	O6	Clegg 1982; Clegg and Pierce 1983
S fimbrial adhesin (*sfa*)	α-Sialyl-(2-3)-β-Gal		16	536	O6:K15	Berger et al. 1982; Hacker et al. 1985
G fimbriae	GlcNac		19.5	IH11165	O2	Rhen 1986a
P-related sequence (*prs*)	GalNac-α-(1-3)-GalNac	F13	19.5	J96	O4:K6	Lund et al. 1988b

A (non-)fimbrial adhesins						
M agglutinin (*bma*)	Glycophorin A^M		21	IH11165	O2	Rhen et al. 1986a
D hemagglutinin (previously O75X)	IFC blood group complex		21	IH11128	O75:K5	Nowicki et al. 1987, 1988
A fimbrial adhesin (*afaI*)	Unknown		16	KS52	O2	Labigne et al. 1984
(*afaII*)	Unknown		?	A22	O15	Labigne and Falkow 1988
(*afaIII*)	Like D hemaggl.?		?	A13	O75	Labigne and Falkow 1988
Nonfimbrial adhesin (*nfaI*)	Unknown		21	827	O83:K1	Hales et al. 1988
Nonadhesive fimbriae						
Pseudo type 1 fimbriae (*foc*)	?	F1C	17	AD110	O6:K2	Van Die et al. 1984a
(KS71C)		F1C	17	KS71	O6:K2	Rhen 1985
(*foc*)		F1C	16.5	20025	O4:K6	High et al. 1988
S/F1C-related fimbriae (*sfr*)	?		17	BK568	O75:K1	Pawelzik et al. 1988
P-related fimbriae (*prf*)	?		22	536	O6:K15	Hacker et al. 1989b

[a] The molecular masses of fimbrial subunit proteins or, in the case of A-fimbrial adhesins, of the adhesive proteins are given on the basis of SDS–PAGE experiments;
[b] Strain AD110 is identical to strains C1212 and KS71 (see ØRSKOV 1983; I. VAN DIE, personal communication)

eight of the ten serologic variants of P fimbriae were ligated into suitable vector molecules. The same was done for the gene clusters coding for S fimbriae and G fimbriae and for the determinants coding for different nonfimbrial adhesins and for nonhemagglutinating fimbriae. It is interesting to note that more than two adhesins or fimbrial determinants were cloned from the chromosomes of wild-type strains J96 (O4:K6), AD110 (O6:K2), KS71 (O6:K2), 20025 (O4:K6), and 536 (O6:K15).

2.3 Molecular Characterization of Cloned Adhesin Determinants

In order to characterize the different cloned adhesin and fimbrial determinants more accurately the coding regions were subcloned and physical maps were established. In addition, their gene products were elucidated by the expression of the proteins in minicell and maxicell systems and by DNA sequence studies. With the help of subclones and transposon insertional mutants the gene organization of several gene clusters was determined. The generation of operon and protein fusions together with the detection of promoter regions and the characterization of mRNA species yielded new insights into the regulation of expression of adhesin determinants (see UHLIN et al. 1985b; BAGA et al. 1988).

It has been shown that the genetic composition of different adhesin determinants is similar. Different regions of the determinants which code for the major fimbrillin (pilin) subunits and for minor fimbrillin proteins, including the corresponding adhesin proteins, can be distinguished. In addition, sequences necessary for transport and assembly functions as well as regions involved in the regulation of transcription can be observed (KLEMM 1985; HACKER et al. 1985; NORMARK et al. 1986). These different regions of adhesin determinants are described below.

3 Genes Coding for Major Fimbrial Subunits

The intact fiber-like structures, consisting of about 500–1000 identical subunit proteins, seem to be necessary for the presentation of the adhesive molecules since they bridge the distances between adhesins and eukaryotic receptor structures. In addition, at least in particular E. coli wild-type strains fimbriae may function as carriers to help the adhesins overcome the shielding effect of the LPS-O side chains (VAN DIE et al. 1986c).

The genes coding for the major fimbrial subunit proteins have been mapped at the proximal (5′) end of the determinants analyzed. The nucleic acid sequences and the predicted amino acid sequences have been established from nine different fimbrial structural genes (BAGA et al. 1984; VAN DIE and BERGMANS 1984; VAN DIE et al. 1984a, 1986a; ORNDORFF and FALKOW 1985; RHEN et al. 1985a; KLEMM 1984; SCHMOLL et al. 1987). On the basis of the sequence data summarized in Fig. 1 the degree of homology between the mature fimbrial proteins has been calculated (Fig. 2). Five P fimbriae show a sequence homology of about 60% to each other. The

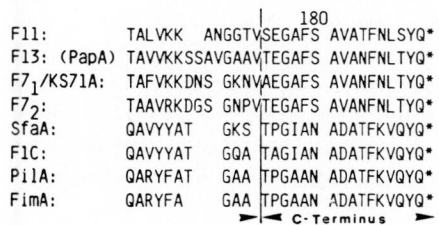

```
                        20        |          40        |          60          |  80
F11:        MIKSVIAGAVAMAVVSFGVN  AAPTIPQGQGKVTFNGTVV  DAPCSISQKSADQSIDFGQL  SKSFLEAGGTSKPNDL  DIE
F13: (PapA) MIKSVIAGAVAMAVVSFGVN  NAAPTIPQGQGKVTFNGTVV DAPCSISQKSADQSIDFGQL  SKSFLEAGGVSKPMDL  DIE
F7₁/KS71A:  MIKSVIAGAVAMAVVSFGAN  AAASIPQGQGEVSFKGTVV  DAPCGIETQSAKQEIDFGQI  SKSFLQEGGETQPKDL  NIK
F7₂:        MIKSVIAGAVAMAVVSFGAY  AAPTIPQGQGKVTFNGTVV  DAPCGIEAQSAGQSIDFGQV  SKLFLENTGESQPKSL  DIK
SfaA:       MKLKFISMAVFSALTLGVAT  NASAVTTVNGGTVHFKGEVV DAACAVNTNSANQTFS GQV RSAKLANTGEKSSPVGFSIE
F1C:        MKLKFISMAVFSALTLGVAT  NASAVTTVNGGTVHFKGEVV NAACAVNTNSFTQTVNLGQV RSERLKVTGAKSNPVGFTIE
PilA:       MKIKTLAIVVLSALSLSSTT  ALAAATTVNGGTVHFKGEVV NAACAVDAGSVDQTVQLGQV RTASLAQEGATSSAVGFNIQ
FimA:       MKIKTLAIVVLSALSLSSTA  ALAAATTVNGGTVHFKGEVV NAACAVEAGSVDQTVQLGQV RTASLAQEGATSSAVGFNIQ
                 ◄      Leader     ►|◄    N-Terminus     ►|◄         C-C-loop
```

```
            |          100       |          120       |          140       |        160
F11:        LVNCDITAFK    QGQAAKN GKVQLSFTGPQVTGQAEELA TNGGTGTAIVVQAAGKNVSF DG TAGDAYPLKDGDNVLHY
F13: (PapA) LVNCDITAFK    GGNGAKK GTVKLAFTGPIVNGHSDELD TNGGTGTAIVVQGAGNNVVF DG SEGEANTLKDGENVLHY
F7₁/KS71A:  LVNCDITNLKQLQGGA AKK GTVSLTESGVPAENADDMLQ TVGDTNTAIVVTSSGKRVKF DGATETGASNLINGDNTIHF
F7₂:        LINCDITNFKKAAGGGGAKT GTVSLLTFSGVPSGPQSDMLQ TVGATNTAIVVTPHGKRVKF DGATATGVSYLVDGDNTIHF
SfaA:       LNDCSSATAG    HASIIFA GNVIAT HNDVLSLQNSAAG SATNVGIQILDH TGTAVQF DGVTASTQFTLTDGTNKIPF
F1C:        LNDCDSQVSA    GAGIVFS GPAV TGKTDVLALQSSAAG SATNVGVQITDH TGKVVPL DG TASSTFTLTDGTNKIPF
PilA:       LNDCDTNVAS    KAAVAFL GTAIDAGHTNVLALQSSAAG SATNVGVQITDR TGAALTL DGATFSSETTLNNGTNTIPF
FimA:       LNDCDTNVAS    KAAVAFL GTAIDAGHTNVLALQSSAAG SATNVGVQITDR TGAALTL DGATFSSETTLNNGTNTIPF
                 ►|◄                    Hydrophilic region
```

```
            |          180
F11:        TALVKK   ANGGTVSEGAFS AVATFNLSYQ*
F13: (PapA) TAVVKKSSAVGAAVTEGAFS AVANFNLTYQ*
F7₁/KS71A:  TAFVKKDNS GKNVAEGAFS AVANFNLTYQ*
F7₂:        TAAVRKDGS GNPVTEGAFS AVANFNLTYQ*
SfaA:       QAVYYAT   GKS TPGIAN ADATFKVQYQ*
F1C:        QAVYYAT   GQA TAGIAN ADATFKVQYQ*
PilA:       QARYFAT   GAA TPGAAN ADATFKVQYQ*
FimA:       QARYFA    GAA TPGAAN ADATFKVQYQ*
                              ►|◄ C-Terminus ►
```

Fig. 1. Comparison of the amino acid sequences of the fimbrillin subunit proteins of P fimbriae (serotypes F11, F13 Pap, F7₁/KS71, F7₂), S fimbriae, F1C fimbriae, and type 1 fimbriae (PilA, FimA). The amino acids sequences are given in single letter code. The five main regions of the proteins are indicated (for references see text). The amino acids of the premature protein counted

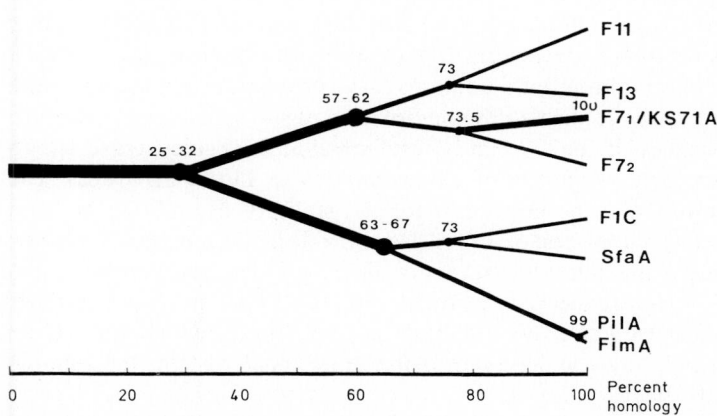

Fig. 2. Diagram showing the degrees of homology among the mature fimbrillin subunit proteins presented in Fig. 1

fimbriae of the serotypes F13 and F11 on the one hand and those belonging to serogroups $F7_1$ and $F7_2$ on the other form two "subgroups" (homology of about 73%) within the P family.

The two type 1 fimbrial proteins, FimA and pilA (see above), exhibit 99% homology. The type 1 proteins in general are more related to pseudotype 1 (F1C) and S fimbrial adhesins (homology of 63% and 67%) than to the P fimbrial proteins (homology 25%-32%). The F1C- and Sfa-specific proteins, which show a degree of homology of 73%, form another related group of fimbrial antigens (OTT et al. 1988). Thus on the basis of the sequence homologies three groups of fimbriae can be determined: P fimbriae, type 1 fimbriae, and a group consisting of Sfa and F1C fimbrillin proteins.

An analysis of the fimbrillin protein sequences shows that five different regions can be discriminated (for details see VAN DIE et al. 1987). These parts differ with respect to sequence homologies observed and to their putative functions.

1. The leader sequences necessary for the transfer of the proteins across the inner membrane show a homology of up to 90% within each of the three groups of fimbriae mentioned above. As demonstrated in Fig. 1, the length of the leader peptides also differs between the three putative groups.

2. The N-terminal sequences are the most homologous among all nine proteins. It is interesting to note here that a synthetic peptide consisting of seven amino acids of the N-terminus of the F13 fimbrillin protects animals against infection by P fimbriated strains of different F-types (SCHMIDT et al. 1988). These data confirm the view that the N-terminal sequences represent well-conserved parts of the major fimbrial subunits.

3. The two cysteine residues of the molecules which form a Cys-Cys bridge in native proteins (JANN et al. 1981) are located exactly at the same place in all nine molecules. Some of the differences in the primary protein structure of the Cys-Cys loops have no relevance to the secondary structure of the proteins because they reveal conservative amino acid exchanges (VAN DIE et al. 1987).

4. The regions between the second cysteine residue and the C-terminal end of the proteins consist of a remarkably high amount of hydrophilic amino acids (therefore they have been termed "hydrophilic regions") and show several differences in their amino acid composition. In these regions five to six "hypervariable parts," which should represent the immunodominant domains of the proteins, can be found (HOPP and WOODS 1981). This was confirmed by the isolation of site-specific mutants with altered protein sequences of the $F7_1$ and F11 fimbrillin molecules; these led to differences in the serologic properties of the proteins (VAN DIE et al. 1988a). The hypervariable parts also may be important for the stability of fimbriae because deletions of corresponding pieces of DNA led to a total absence of fimbriae. Furthermore, foreign antigenic determinants like peptides specific for foot-and-mouth disease virus can be incorporated into the hypervariable region of the fimbrial subunits (VAN DIE et al. 1988b). It is also suggested that the hypervariable regions may play a role in the escape of fimbriated bacteria from the immune system (DE REE et al. 1985b).

5. The C-terminal regions of the proteins which carry some hydrophobic and aromatic amino acids exhibit a high degree of homology. In all cases the two amino acids at the C-terminal end of the proteins (tyrosine and glutamine) are identical (see

Fig. 1). It is suggested that the C-termini are involved in the interaction of the fimbrial protein subunits and the formation of an intact fiber-like structure (KLEMM 1985).

4 Genes Coding for Minor Fimbrial Subunits and Adhesins

It is generally accepted that fimbriae of extraintestinal *E.coli* isolates do not represent the adhesive proteins per se. Rather it has been shown that "minor fimbrial proteins" are incorporated into the fiber-like structures. Evidence revealed that one of these minor proteins represents the actual adhesin of P fimbriae (LINDBERG et al. 1984; UHLIN et al. 1985b; VAN DIE et al. 1985; HOSCHÜTZKY et al. 1989), S fimbriae (HACKER et al. 1985), and type 1 fimbriae (MAURER and ORNDORFF 1985; MINION et al. 1986), whereas another two minor proteins are involved in anchoring and/or presentation of the adhesins and in initiation of fimbrial polymerization (RIEGMAN et al. 1988; NORMARK et al. 1988; see also sect. 5). The major and minor fimbrial proteins form a structure termed "fimbria–adhesin complex" (see HOSCHÜTZKY et al. 1989).

4.1 P Type

Three minor components are detected for P fimbriae of serotype F13 which are encoded by the *pap* (pili associated with pyelonephritis) operon (NORMARK et al. 1983; LUND et al. 1985) and for fimbriae of the serotypes F7$_1$ (*fso* determinant; VAN DIE et al. 1985) and F7$_2$ (*fst* determinant; VAN DIE et al. 1984b). The genes are termed *papE*, *papF*, and *papG* and are analogous to *fsoE, F, G* and *fstE, F, G* (RIEGMANN et al. 1988). As indicated in Fig. 3a, these proteins have comparable molecular weights and in each case the corresponding genes are located at the distal (3′) end of the determinants. Genetic complementation data argue that the PapG protein represents the adhesin protein of the *pap* determinant (LUND et al. 1987, 1988a). It has also been suggested that the proteins FsoG, FstG, and FelG (for the F11 determinant) mediate adherence to the Gal-Gal receptor.

DNA sequence studies have shown that the putative adhesin proteins are nearly 32 kd in size. The genes *fstG* and *felG* are rather similar to each other, whereas the *papG* gene differs from the other two cistrons (only 45% sequence homology). The corresponding protein sequences, however, show similarities to the C-terminal part of the major P fimbrial subunits and they all carry four cysteine residues at well-defined positions. Very recently Hoschützky et al. (1989) isolated the adhesin proteins of three different P fimbrial adhesin complexes. The identity of the isolated proteins to the proteins extrapolated from DNA sequences was demonstrated by these authors.

It is interesting to note that the adhesin-specific genes may vary among fimbriae of the same F serogroup. LUND et al. (1988b) described two different binding specificities, Pap [α-D-Gal-(1-4)-β-D-Gal specificity] and Prs [P-related sequences, GalNAc-α-(1-3)-GalNAc specificity; see also Table 2] which are associated with

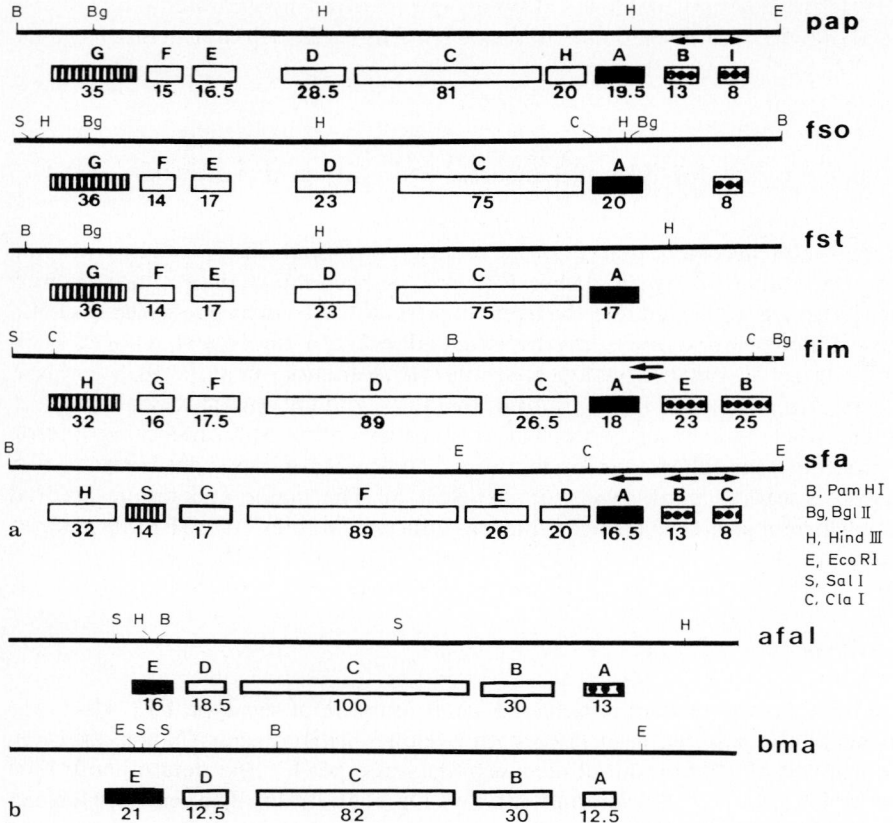

Fig. 3. a Comparison of the physical maps and the gene products of different fimbrial adhesin determinants. Gene clusters coding for P fimbriae (F13-*pap*, F71-*fso*, F72-*fst*), type 1 fimbriae (*fim*), and S fimbriae (*sfa*) are given. The *black boxes* represent fimbrial structural proteins; the *shaded boxes* represent adhesin proteins and the *boxes with closed circles* represent proteins involved in regulation of expression. The *arrows* indicate putative promoter regions and transcriptional orientations. The main direction of transcription is from right to left. The two *arrows* in the 5′ region of the *fim* determinant reflect the phase variation process (for further explanations and references see text). **b** Comparison of the physical maps and the gene products of two nonfimbrial adhesin determinants. The *black boxes* represent adhesin proteins; the *box with closed circles* represents a protein which seems to be involved in regulation of transcription. The main direction of transcription is from right to left (for references see text)

fimbriae of serotype F13. A similar observation was made with fimbriae of serotype F12, which may be associated with P-specific adhesins as well as with adhesins which recognize a different receptor located on canine erythrocytes (GARCIA et al. 1988a,b). In addition, it was shown that the genes coding for the nonadhesive minor proteins E and F but not those coding for the G proteins are *trans*-complementable between *pap* and *prs* determinants, a fact which also argues for a specific heterogeneity of the adhesin-specific genes of these two gene clusters.

While mutations in the *papG* gene (and also in the *fsoG* and *fstG* cistrons) abolished hemagglutination, *papE* mutants showed a reduced adherence (NORGREN et al. 1984; LINDBERG et al. 1984, 1986, 1987). Incubation of F_{71} and F_{72} fimbrial adhesins with anti-FsoE as well as anti-FstE antibodies also inhibited hemagglutination (RIEGMAN et al. 1988). It is therefore suggested that the E gene products directly anchor the adhesins (G proteins) to the fimbriae. Mutants in the F genes of the three determinants, however, show a reduced number of fimbriae per cell. The corresponding proteins (and also the FstG product) seem to act by initiating the assembly of new fimbriae and may be necessary for the determination of the degree of fimbriation (LINDBERG et al. 1987; RIEGMAN et al. 1988).

Electron micrographs which show Pap-positive cells incubated with G-, F-, and E- specific polycolnal antisera indicate the presence of this complex at the tip of the fimbriae (LINDBERG et al. 1987). On the other hand data on the use of FstE- specific monoclonal antiserum argue for the presence of at least the FstE protein in the fimbrial structure, with a preference for the tip of the fiber (DE REE et al. 1987; RIEGMAN et al. 1988; Jann and Hoschützky, this volume).

4.2 Type 1

At the terminal parts of the genetic determinants *fim* and *pil* specific for type 1 fimbrial adhesins, three cistrons, which code for minor fimbrial proteins, were determined (KLEMM and CHRISTIANSEN 1987; MAURER and ORNDORFF 1987). The largest protein encoded by this region (FimH with a size of 32 kd, which is equivalent to pilE with a size of 31 kd) is indicated to be the adhesin mediating a mannose-specific binding to eukaryotic structures. Very recent chemical and immunologic data (HANSON and BRINTON 1988; HANSON et al. 1988) provide evidence that this protein represents the actual adhesin which is located at the tip of fimbriae as well as being interspersed at intervals along the fimbriae (KROGFLD and KLEMM 1988). It is interesting to note that the FimH protein seems to be conserved not only among different *E. coli* strains but also among other members of the *Enterobacteriaceae* family (ABRAHAM et al. 1988b). Mutations in the *fimG* but not in the *fimF* gene result in a hemagglutination-negative phenotype. In addition, both proteins seem to be involved in length modulation of the fimbriae, a fact which was also demonstrated for the *pilF* gene, the equivalent locus to *fimG* in strain J96 (MAURER and ORNDORFF 1987).

The DNA sequence of the *fimE, fimF,* and *fimG* cistrons and the predicted amino acid sequence of the proteins show a high degree of homology to the *fimA* (*pilA*) structural gene (KLEMM and CHRISTIANSEN 1987). It is interesting to note that the *fimA* gene products must be present on the cells to confer an adherence-positive phenotype to the strains. This is in contrast to what has been found for P and S fimbriae. In these cases subcloned plasmids can be isolated which are devoid of any fimbrial production but still exhibit hemagglutination (see HACKER et al. 1985).

4.3 S Type

The S-specific adhesin has been isolated and characterized as a 12- to 14-kd protein (depending on the gel system used; MOCH et al. 1987; SCHMOLL et al. 1989a). The gene *sfaS* is located next to the 3′ end of the *sfa* determinant (see Fig. 3a). The DNA sequence of this cistron (SCHMOLL et al. 1987; SCHMOLL et al. 1989a) shows homology to the sequence of the S fimbrial gene. A comparison of the SfaS sequence to the sequences specific for other sialyl-binding proteins (see JACOBS et al. 1986) shows a putative sialyl-binding site which includes a stretch of positively charged amino acids. Electron microscopic data show the presence of the SfaS adhesin protein at the fimbria with a predominance at the tip of the fiber (MOCH et al. 1987).

In contrast to the regions coding for the adhesins of the P fimbrial and type 1 fimbrial determinants, *sfaS* is not located at the extreme 3′ end of the *sfa* determinant. Another cistron (*sfaH*) is located distal to the *sfaS* gene (Fig. 3a). Mutations in the *sfaH* gene result in adhesin-positive but low fimbriated phenotypes, as also observed for P and type 1 fimbrial determinants (*papF* and *fimF* genes, see above).

It can be concluded that the functions of the minor components of the *sfa, pap,* and *fim* determinants resemble each other; their corresponding genes, however, are located at different positions in the determinants.

4.4 A (Non)-fimbrial Adhesins

Two of the cloned A(Non)-fimbrial adhesin determinants (see Table 1) have been analyzed in detail. The determinant coding for the A fimbrial adhesin I (*afaI*, LABIGNE-ROUSSELL et al. 1984, 1985; LABIGNE-ROUSSELL and FALKOW 1988) consists of five genes (Fig. 3b). One gene (*afaE*) located at the 3′ end of the gene cluster is the adhesin structural gene coding for the 16-kd adhesin protein found in the culture supernatant of AfaI-positive strains (see WALZ et al. 1985). The deduced amino acid sequence of AfaE has no homology to the sequence of any of the fimbrial subunit proteins show in Fig. 1. Another two genes, *afaB* and *afaC*, coding for proteins of 30 kd and 100 kd, respectively, are necessary for the expression of AfaI-specific hemagglutination (LABIGNE-ROUSSELL et al. 1985).

The gene organization of the blood group M agglutinin (*bma* described by RHEN et al. 1986a, b) resembles that published for *afaI* (Fig. 3b). The gene specific for the agglutination property is also located at the 3′ terminal end of the determinant and the putative amino acid sequence does not show any homology to the sequences of *afaE* and of fimbrial subunits.

5 Genes Involved in Transport and Assembly

As demonstrated in Sects. 3 and 4, major and minor fimbrial components are involved in assembly of fimbrial structures. In addition mutants have been isolated that show a low but significant intracellular pool of fimbrial subunit proteins but no fiber-like structures (NORGREN et al. 1984; HACKER et al. 1985; KLEMM et al. 1985; VAN DIE et al. 1984b; ORNDORFF and FALKOW 1984a). In all such mutants two or three gene products were abolished, a large protein of about 75–90 kd and polypeptides 20–30 kd (see Fig. 3a). These proteins are involved in the transport of the fimbrillin subunits and minor proteins to the periplasm and across the outer membrane. In the case of type 1 determinants the genes *fimC* and *fimD* (equivalent to *pilB* and *pilC*; ORNDORFF and FALKOW 1984a; KLEMM et al. 1985) seem to be involved in these functions. The *sfa* determinant codes for three proteins of 20, 26, and 89 kd which are necessary for transport functions (Ludwig and J. Hacker, unpublished observations).

In the *pap* system, the *papD* gene product, which is found in the periplasmic space, the PapC protein, which is part of the outer membrane, and the PapH product are involved in transport and assembly functions (NORGREN et al. 1987). The PapD protein seems to transport the major and minor subunits into the periplasm. The PapD protein forms complexes with PapA, PapE, and PapG and presumably also with other fiber-like proteins (NORMARK et al. 1988). The PapC protein, however, functions as an anchor for the fimbriae in the outer membrane, as demonstrated recently for the FaeD protein encoded by the K88 adhesin gene cluster (MOOI et al. 1986). The *fsoC* gene product, a 75-kd protein encoded by the F_{71} determinant, seems to have the same effect on fimbrial assembly as PapC, while *fsoC* frame shift mutants do not completely block fimbrial biogenesis (RIEGMAN, personal communication). Another Pap fimbriae-like protein, termed PapH, has been analyzed recently (BAGA et al. 1987). Mutants of this gene show a high degree of intact fimbriae in the supernatant. It is therefore suggested that PapH acts as terminator for fimbrial development. The PapH protein itself seems to be located at the base of the fiber which is in contact with the PapC outer membrane protein.

In *trans*-complementation experiments transposon-induced mutations in the regions involved in the transport and assembly of S fimbriae were reexpressed by *foc*-specific plasmids (coding for F1C fimbriae, see OTT et al. 1988). Similar experiments are described for both P fimbrial determinants and type 1-specific gene clusters (RHEN et al. 1985b; CLEGG et al. 1985a,b). In addition, DNA hybrid molecules which consisted of DNA fragments derived from plasmids coding for P fimbriae of different serotypes or for Sfa and F1C fimbriae were isolated (VAN DIE et al. 1986b; OTT et al. 1988). These hybrid DNAs, which were ligated in the transport/assembly regions, were able to produce intact fimbriae. This result suggests a functional complementation of the corresponding proteins. The complementation data support the idea that the determinants coding for Sfa/F1C, type 1, and P fimbriae form related classes of gene clusters, as already suggested on the basis of the fimbrillin-specific DNA sequence data (see Sect. 3).

6 Regulation of Expression

6.1 MR Fimbrial Adhesins

DNA-DNA hybridization studies have shown that the proximal (5′) ends of the determinants coding for MR fimbrial adhesins of P- and S-type and for F1C fimbriae are well conserved (OTT et al. 1987). Transposon-induced mutations in this region, which is located upstream of the fimbrial structural genes, led to a completely negative phenotype of the corresponding clones. Therefore, it was suggested that regulatory proteins which may act as *trans*-activators are encoded by this region (see Fig. 3a; UHLIN et al. 1985a; HACKER et al. 1986a). To substantiate this observation, *pap-lacZ* and *sfa-lacZ* operon fusions were constructed to map active promotors. In both cases (*pap* and *sfa*) two promotors, which were arranged in opposite orientations, were identified (BAGA et al. 1985; HACKER et al. 1986a). In addition, a week promotor was identified in front of the *sfaA* structural gene (see also Fig 3a).

Two gene products are encoded by the regulatory regions of *pap* and *sfa*. As demonstrated in Fig. 4, these proteins show a very high degree of homology to each other, thereby suggesting a similar function in the regulation of transcription. Another gene encoded by the adhesin determinant *fso* of serotype $F7_1$ shows nearly the same structure as the *papI* and *sfaC* genes (Fig. 4; see RHEN and VÄISÄNEN-RHEN 1987; see also LOW et al. 1987). It was further shown that the gene products encoded by these regions are *trans*-complementable between P, S, and F1C determinants (RHEN et al. 1985b; GÖRANSSON et al. 1988; SCHMOLL 1989).

The fact that the two proteins encoded by the 5' region of the *sfa* determinant act as positive regulators was further confirmed by the construction of *sfaA-phoA* operon fusions (see Table 3). Using the promoter-probe vector plasmid pCB267, *sfaA* was fused to the gene coding for the enzyme alkaline phosphatase (SCHNEIDER and BECK 1986). The amount of PhoA activity was used as an indicator for the transcriptional activity of the *sfaA* cistron. It is obvious from the data presented that both proteins SfaB and SfaC are necessary for transcriptional activity of *sfaA*. A positive *trans*-complementation of $sfaA\text{-}phoA^+$, $sfaB^+$, $sfaC^-$ clones was possible following cotransformation of a $sfaC^+$-specific plasmid.

```
                  10         20        30          40         50
SfaB:   MAQHEVITRG GTAFLLKLRE SALSSGSMSE EQFFLLIGIS SIHSTRVILA
PapB:   MAHHEVISRS GNAFLLNIRE SVLLFGSMSE MHFFLLIGIS SIHSTRVILA

                  60         70        80          90        100
SfaB:   MKTYLVSGHS RKTVCEKYQM NNGYFSTTLG RLTRLNVLVA RLAPYYTDSV
PapB:   MKTYLVGGHS RKTVCEKYQM NNGQFSTTLG RLIRLNALAA RLAPYYTDES

              110
SfaB:   SAIAEAASL*
PapB:   SAFD*


                  10         20        30          40         50
SfaC:        MQNEIM GFLSRHNVGK TAEIAEALAV TDYQARYYLL LLEKEGMVQR
PapI:   MSEYMKNEIL EFLNRHNHGK TAEIAEALAV TDYQARYYLL LLEKEGMVQR
Ks71:   MSEYMKNEIL EFLNRHNGGK TAEIAEALAV TDYQARYYLL LLEKAGMVQR

                  60         70
SfaC:   SPLRRGMATY WFLKGEMQAG QSCSSTT*
PapI:   SPLRRGMATY WFLKGEMQAG QNCSSTT*
Ks71:   SPLRRGMATY WFLKGEKQAG QSCSSTT*
```

Fig. 4. Comparison of the amino acid sequences of proteins involved in the regulation of transcription of P fimbrial (PapB, PapI, KS71) and S fimbrial (SfaB, SfaC) determinants. The amino acid sequences are given in single letter code. Homologies are indicated by *boxes*

Table 3. Operon fusions between *sfa*-specific cistrons and the gene *phoA* coding for alkaline phosphatase

Resident plasmid	Genotype	Complementary plasmid	Genotype	PhoA units
pTTS267-38	*sfaA-phoA*, B^+, C^+	pACYC184		859
pTTS267-38ΔC	*sfaA-phoA*, B^+, C^-	pACYC184		306
pTTS267-38ΔC	*sfaA-phoA*, B^+, C^-	pANN801-15ΔN	*sfaC*$^+$	1017
pTTS267-69	*sfaA-phoA*, B^-, C^-	pACYC184		146
pTTS267-69	*sfaA-phoA*, B^-, C^-	pANN801-15	*sfaA*$^+$, B^+, C^+	121
pTTS267-84	*sfaB-phoA*, A^-, C^+	pACYC184		2215
pTTS267-1/5	*sfaB-phoA*, A^-, C^-	pACYC184		658
pTTS267-1/5	*sfaB-phoA*, A^-, C^-	pANN801-15ΔN	*sfaC*$^+$	1230
pTTS267-1/8	*sfaC-phoA*, B^-, A^-	pACYC184		579
pTTS267-1/8	*sfaC-phoA*, B^-, A^-	pANN801-ΔEN	*sfaA*$^+$, B^+	2890
pCB267[a]	*phoA*			5

[a] Vector pCB267 was used for the cloning experiments (see SCHNEIDER and BECK 1986)

A positive *trans*-regulatory effect of PapB and PapI (equivalent to SfaB and SfaC) to the expression of the structural gene *papA* was also observed by Uhlin and co-workers (BAGA et al. 1985; UHLIN et al. 1985a). The intergenic regions between the two regulatory genes and the structural gene are necessary in *cis*-orientation for the thermoregulation of fimbrial production and they also play a role in the repressive effect of trimethoprim on P fimbrial expression (VÄISÄNEN-RHEN et al. 1988). In addition a putative Crp binding site was detected in the intergenic region which should have an influence on catabolic repression of the transcription of the *pap* operon (GÖRANSSON and UHLIN 1984; BAGA et al. 1985). Recently, it was reported that the differences in the expression of the genes *papB*, *papA*, and *papH*, which are all part of the same transcriptional unit, may result from different half-lives of the corresponding mRNA molecules (BAGA et al. 1988). This possibility, together with the presence of a teriminator downstream of *papA*, may explain why the amount of protein produced by the structural gene *papA* is somewhat larger than that produced by *papB* or papH.

6.2 Type 1 Fimbrial Adhesins

The region which regulates the expression of transcription of type 1 fimbrial determinants is also located at the proximal (5′) end of the gene cluster, upstream of the fimbrial structural gene. Sequence studies have shown that the regions involved in regulation of MR and MS fimbrial adhesins are completely different (BAGA et al. 1985; KLEMM 1986; SCHMOLL 1989). In addition it was demonstrated that the regulatory region of the type 1 determinant directs the process of phase variation, i.e., the on-off expression of type 1 fimbriae (EISENSTEIN 1981; FREITAG et al. 1985; ABRAHAM et al. 1985). An invertible DNA element of 314 base pairs (bp) located immediately upstream of the *fimA* gene is responsible for type 1 expression. This element, which is flanked on each side by a 9-bp inverted repeat sequence, carries the *fimA*-specific promoter region. Depending on the direction of integration of this promoter element, the *fimA* gene, together with the whole determinant, will

either be transcribed or not (see Fig. 3a). Two genes, *fimE* and *fimB*, located upstream
of the invertible element are also necessary for phase variation and seem to be
transcribed by their own promoters (KLEMM 1986). The respective proteins of 23 kd
and 25 kd seem to bind to the invertible DNA region and they switch the promoter
element in "on" (FimB) or "off" (FimE) orientation. Both proteins are highly basic
and have a high degree of homology to each other (48% identical amino acids).

It was shown that the type 1 phase variation process is independent of the action
of site-specific recombinases which mediate DNA inversion in *Salmonella typhimurium* (FREITAG et al. 1985). These recombinases, termed Hin proteins, are
involved in phase variation of the *Salmonella* H-antigen (SILVERMAN and SIMON
1980). Another *trans*-acting protein, the integration host factor (IHF), which is
encoded by the *E. coli* genes *himA* (map position at 37) and *hip/himD* (map position
at 20) and which is essential for phage integration and excision, is required for the
phase variation of type 1 fimbriae. Mutants in the gene loci *himA* and *himD* reduce
the frequency of switch events from 10^{-2}–10^{-3} to 10^{-5} per cell per generation
(EISENSTEIN et al. 1987). In addition, a sevenfold reduction of the transcriptional
activity of the *fim* determinant was observed in IHF negative strains as compared
with strains carrying intact IHF coding genes (DORMAN and HIGGINS 1987). It was
shown by sequence analysis that two IHF consensus sequences are located in the
type 1 regulatory region. It is suggested that the IHF protein shows recombinase
activity during the fimbrial switch (KLEMM 1986; EISENSTEIN et al. 1987; DORMAN and
HIGGINS 1987). Recently another DNA region involved in regulation of expression
of the type 1 determinant was identified (ABRAHAM et al. 1988a). A deletion located
downstream of the *fim*-specific structural genes led to an overproduction of the
adhesin protein FimH and the generation of so-called fimbriosomes, thereby
indicating the presence of a repressor presumably encoded by this DNA region.

The observations reviewed here were made for avirulent *E. coli* K-12 strains.
ABRAHAM et al. (1986) have shown however, that phase variation depending on the
switching activity of the invertible element also occurs in clinical isolates. In
addition, a locus termed *hyp* which is similar to the *fimE* gene of *E. coli* K-12 was
analyzed in the *pil* determinant of the pyelonephritic strain J96 (ORNDORFF and
FALKOW 1984b). Furthermore a gene coding for the PilG protein was described for
J96 and may also act as another *trans*-acting factor in type 1 fimbrial expression
(SPEARS et al. 1986). Therefore, it seems that processes similar to that found in *E. coli*
K-12 also occur in wild-type strains, thereby helping the bacteria to bind to mucosal
surfaces (in an "on" status) but preventing their adherence on phagocytic cells (in an
"off" status) which also carry mannose-specific receptor molecules (see OFEK and
SILVERBLATT 1982).

7 Adhesin Determinants Within Virulence Gene Blocks

7.1 Analysis of Structural Virulence Gene Blocks

Using adhesin-specific DNA probes it has been shown that the coding sequences responsible for one particular fimbrial adhesin may be homologous in *E. coli* strains

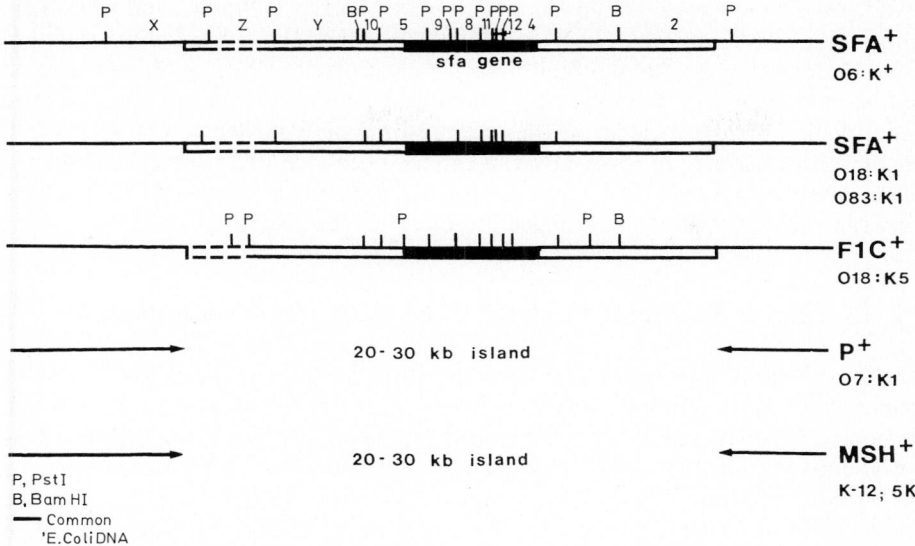

Fig. 5. Genetic maps of S fimbrial (*sfa*) and F1C fimbrial (*foc*) determinants and their flanking regions. The structural genes are indicated by *black boxes*. The *open boxes* represent homolog DNA belonging to the respective structural virulence gene blocks. The strains tested belong to serotypes O6:K[+], O18:K1, O83:K1, and O18:K5. As controls Sfa[-] F1C[-] strains exhibiting P fimbriae or type 1 (Msh) fimbriae were used

derived from different sources. Such an observation was made for DNA sequences which code for S fimbrial adhesins and F1C fimbriae of uropathogenic strains and meningitis isolates and for P adhesin determinants of uropathogenic *E. coli* (OTT et al. 1986, 1988; EKBÄCK et al. 1986). It is remarkable that the homology between the *sfa* and *foc* determinants is not restricted to their coding sequences. It is obvious that the regions flanking the *sfa* and *foc* determinants at both sides are also homologous to each other (see Fig. 5). These DNA regions, which consist of the structural genes for the fimbriae and the flanking sequences (regions of about 25 kb were tested), were termed "virulence gene blocks" because they seem to be specifically inserted into chromosomes of pathogenic *E. coli* isolates but are missing on the chromosomes of non-pathogenic strains (OTT et al. 1988; HACKER et al. 1988).

The flanking regions of the *sfa* and *foc* determinants consist of noncoding DNA sequences. Other adhesin determinants are, however, closely linked to gene clusters which also code for virulence factors. This has been described for sequences which code for G fimbriae and the M agglutinin, which are located very close to each other on the chromosome of a particular *E. coli* 02 strain (RHEN et al. 1986a). Such DNA regions are termed "structural virulence gene blocks" because the different virulence determinants are structurally linked on the chromosomes.

Very often P or P-related fimbrial genes are linked to gene clusters coding for the α-hemolysin, another virulence factor of *E. coli* (HULL et al. 1982; for review see HACKER and HUGHES 1985). Such a linkage of P and Hly determinants has been described for different *E. coli* serotypes, including O4, O6, and O18 strains (Low et al. 1984; HACKER et al. 1989b; HULL et al. 1988; M. OTT, L. BENDER and J. HACKER, unpublished observations). HIGH et al. (1988) described the linkage of five different adhesin determinants coding for P fimbriae of serotypes $F12_{rel}$, F13, F14, and an unknown F type, F1C fimbriae, and α-hemolysin (see also Table 2). The virulence gene block which is located on the chromosome of a particular O4:K12 strain encompasses a stretch of 50-kb DNA. It can be speculated that the physical linkage of six potential pathogenicity factors may have occurred by transpositional events between the chromosome and plasmid molecules (see Sect. 2).

7.2 Excision of Structural Virulence Gene Blocks from Chromosomes

The observation was made that structural virulence gene blocks are able to disappear from the chromosomes of different strains. It was shown that fimbriated, hemolytic *E. coli* strains are able to generate nonhemolytic, non-fimbriated mutants with relatively high frequencies (HACKER et al. 1983; KNAPP et al. 1984). As shown for the *E. coli* O6 strain 536 and for strains of serotypes O18:K1 and O4:K6, the structural genes coding for P (and P-related) fimbriae and for hemolysin were completely deleted from the chromosomes of the mutant strains (see Fig. 6).

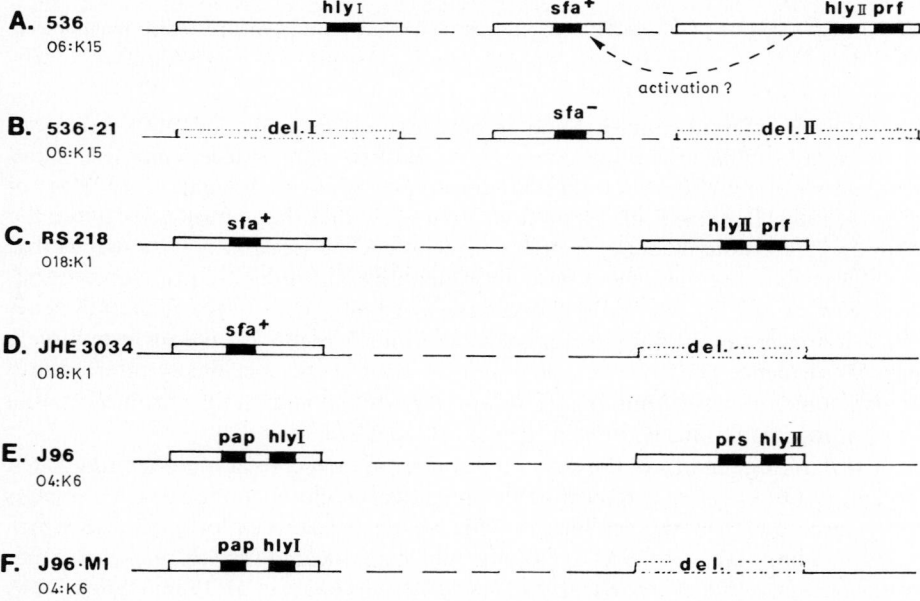

Fig. 6. Model of the chromosomes of wild-type strains and their deletion mutants. The coding regions for fimbriae or hemolysin are indicated by *black boxes*; the entire virulence gene blocks are indicated by *open boxes*. Strain 536 and the mutant 536-21 show structural and functional virulence gene blocks. In contrast the strains IHE 3034, RS218, J96, and J96-M1 only carry structural virulence gene blocks on their chromosomes. The *boxes with broken lines* represent chromosomal deletions (for references see text). *Abbreviations: hly,* hemolysin; *sfa,* S fimbrial adhesins; *prf,* P-related fimbriae; *del.,* deletion; *pap,* pili associated with pyelonephritis; *prs,* P-related sequence

Using DNA probes specific for the flanking sequences of the P-Hly coding regions it was demonstrated that at least in the case of the O6 strain 536 DNA regions of about 70–100 kb can be deleted from the chromosomes of the respective isolates (KNAPP et al. 1986). The deletion events are specific processes which include the left and right boundary sequences of the P-Hly virulence blocks. Such deletions of fimbrial determinants together with gene clusters coding for α-hemolysin from the chromosomes of wild-type strains may also occur in nature. It is generally accepted that the longer a patient suffer from a UTI, the greater the chances of isolating non-virulent strains from the urine. The non-virulent variants may be "deletion mutants" of former fully virulent strains which may have lost structural virulence gene blocks (see HACKER et al. 1989b).

7.3 Occurrence of Functional Virulence Gene Blocks in Strain 536

As already mentioned above, strain 536 carries gene clusters which code for two different α-hemolysins (hemolysin I and II) and P-related fimbriae (Prf). Mutants like strain 536-21 have lost these three determinants by deletional events. As indicated in Table 4, the production of S fimbrial adhesins and of type 1 fimbriae was also absent in these mutants.

Mutants which have lost the HlyII/Prf coding region but retained the HlyI coding region exhibited a negative phenotype for the two adhesins S fimbriae and type 1 fimbriae, whereas HlyI-negative, HlyII/Prf-positive strains were also Sfa and type I positive. Therefore, it can be concluded that the expression of the latter two fimbrial gene clusters is strongly associated with the presence of the HlyII/Prf coding DNA region. It was further demonstrated that the regulation, at least for the S fimbrial adhesin determinant, takes place at the transcriptional level (KNAPP et al. 1986). These data argue for the presence of a factor, encoded by the *hlyII/prf* virulence gene block, which functions as a positive *trans*-regulator, promoting the transcription of *sfa*. The fact that a region coding for two linked pathogenicity factors (Prf and Hly) directs the expression of another two nonlinked pathogenicity determinants (Sfa and type I) lends support to the existence of functional virulence gene blocks (see Fig. 6).

Similar phenomena were described for pathogenic strains of *Vibrio cholerae* and *Bordetella pertussis* (WEISS and HEWLETT 1986; MILLER et al. 1987). In such strains

Table 4. Properties of *E. coli* strain 536 and its mutants

Property	Strain			
	536	536-21	536-114	536-25
Hemolysin I	Active	Deleted	Deleted	Active
Hemolysin II	Active	Deleted	Active	Deleted
P-related	Active	Deleted	Active	Deleted
S fimbrial adhesin	Active	Reduced expression	Active	Reduced expression
Type 1	Active	Reduced expression	Active	Reduced expression

also *trans*-acting factors were necessary for the expression of different pathogenicity determinants, including gene clusters which code for adhesins. It therefore seems that the occurrence of functional virulence gene blocks is not restricted to *E. coli*.

8 Concluding Remarks

In the last decade, and especially the last 5 years, numerous determinants coding for fimbriae and adhesins have been cloned and genetically characterized (see Table 2). Despite this the following aspects have yet to be resolved.

1. Which sequences are necessary for the binding specificity of adhesins? It has been speculated that "consensus sequences" exist and these would explain the constant binding behavior of adhesins consisting of non-homologous proteins (e.g., different P fimbriae).

2. How do adhesin-specific gene products interact in the processes of transport and assembly? Different gene products involved in these processes have been determined but their complex interplay which leads to functional fimbrial adhesins has yet to be resolved.

3. How are adhesin genes regulated in wild-type strains? Studies on the regulation of transcription were performed with cloned determinants. The genetic "background" of the strains, however, seems to be important for regulatory events. In addition the copy effects of plasmids may influence the regulation of genes. Therefore the regulation of adhesin determinants should be studied in wild-type strains see (SCHMOLL et al. 1989b).

4. What kinds of molecular event are responsible for the phase variation of determinants other than type1? In the case of P and S fimbriae an on-off production of the fimbriae was described (RHEN et al. 1983b; NOWICKI et al. 1986; SAUKKONEN et al. 1988). The processes which lead to phase variation of these gene clusters seem to be different from those which are described for type 1 determinants.

5. How do *trans*-acting regulatory factors influence fimbrial expression? It has been shown that the expression of S fimbriae depends on factors which are encoded by genes not linked to the adhesin gene clusters. Factors like the IHF protein involved in type 1 expression also have to be determined and described for other adhesins.

6. What kinds of environmental signal play a role in fimbrial production? In order to describe the role of adhesins in infections more precisely the conditions of expression in wild-type strains and also in vivo have to be determined.

7. Does the genetic structure of *E. coli* adhesin determinants fit the "cassette model" of *Neisseria gonorrhoeae*? Some data, showing a conservation of special regions located next to sequences which are completely different among two or three adhesin determinants (e.g., regulatory genes in the case of P and S determinants or adhesins and minor proteins in the case of *pap* and *prs*), argue for an occurrence of processes similar to the generation of pili of *Neisseria gonorrhoeae* by a rearrangement of "cassettes" (see STERN and MEYER 1987). Do inactive copies of cassettes also exist on the *E. coli* chromosome?

8. How have adhesin-carrying strains evolved? Some data indicate a special development of extraintestinal *E. coli* pathogens which may include a "gene pick up" of fimbrial and hemolysin determinants (perhaps from plasmids or from different parts of the chromosomes, see HACKER et al. 1989b), the activation of genes during the acute phases of infection, and their subsequent loss (by deletion) in late phases of infectious processes. The dynamics of such processes have to be further elucidated.

In addition it is suggested that molecular studies with adhesin determinants should be yield new practical applications in order to improve diagnostic tools and to develop better drugs and vaccines for prevention.

Acknowledgments. I wish to thank many colleagues, too numerous to mention, who provided me with unpublished results, reprints, and helpful comments. In addition I thank Irma Van Die (Utrecht), Balakrishna Pillay (Würzburg), and Colin Hughes (Cambridge) for critical reading of the manuscript and Angelika Zirngibl (Würzburg) for editorial assistance. The work was supported by the Deutsche Forschungsgemeinschaft (Ha 1434/1-6).

References

Abe C, Schmitz S, Moser I, Boulnois G, High NJ, Orskov I, Orskov F, Jann B, Jann K (1987) Monoclonal antibodies with fimbrial F1C, F12, F13and F14 specificities with fimbriae from *E. coli* O4:K12:H⁻. Microb Pathogen 2: 71–77

Abraham JM, Freitag CS, Clements JR, Eisenstein BI (1985) An invertible element of DNA controls phase variation of type 1 fimbriae of *Escherichia coli*. Proc Natl Acad Sci USA 82: 5724–5727

Abraham JM, Freitag CS, Gander RM, Clements JR, Thomas VL, Eisenstein BI (1986) Fimbrial phase variation and DNA rearrangements in uropathogenic isolates of *Escherichia coli*. Mol Biol Med 3: 495–508

Abraham SN, Goguen JD, Beachey EH (1988a) Hyperadhesive mutant of type 1-fimbriated *Escherichia coli* associated with formation of FimH organelles (fimbriosomes). J Bacteriol 56: 1023–1029

Abraham SN, Sun D, Dale JB, Beachey EH (1988b) Conservation of the D-mannose-adhesion protein among type 1 fimbriated members of the family *Enterobacteriaceae*. Nature 336: 682–684

Achtman M, Pluschke G (1986) Clonal analysis of descent and virulence among selected *Escherichia coli*. Annu Rev Microbiol 4: 185–210

Bachmann BJ (1987) Linkage map of *Escherichia coli* K-12, Edition 7. In: Neidhardt FC, Ingraham JL, Low BK, Magasanik B, Schaechter M, Umbarger HE (eds) *Escherichia coli* and *Salmonella typhimurium*. Cellular and molecular biology, vol II. Washington, pp 807–876

Baga M, Normark S, Hardy J, O'Hanley P, Lark D, Olsson O, Schoolnik G, Falkow S (1984) Nucleotide sequence of the *papA* gene encoding the Pap pilus subunit of human uropathogenic *Escherichia coli*. J Bacteriol 157: 330–333

Baga M, Göransson M, Normark S, Uhlin BE (1985) Transcriptional activation of a Pap pilus virulence operon from uropathogenic *Escherichia coli*. EMBO J 4: 3887–3893

Baga M, Norgren M, Normak S (1988) Biogenesis of *E. coli* Pap Pili. PapH, a minor pilin subunit involved in cell anchoring and length modulation. Cell 49: 241–251

Baga M, Göransson M, Normark S (1988) Processed mRNA with differential stability in the regulation of *E. coli* pilin gene expression. Cell 52: 197–206

Berger H, Hacker J, Juarez A, Hughes C, Geobel W (1982) Cloning of the chromosomal determinants encoding hemolysin production and mannose-resistant hemagglutination in *Escherichia coli*. J Bacteriol 152: 1241–1247

Clegg S (1982) Cloning of genes determining the production of mannose-resistant fimbriae in a uropathogenic strain of *Escherichia coli* belonging to serogroup O6. Infect Immun 38: 739–744

Clegg S, Pierce JK (1983) Organization of genes responsible for the production of mannose-resistant fimbriae of a uropathogenic *Escherichia coli* isolate. Infect Immun 42: 900–906

Clegg S, Hull S, Hull R, Pruckler J (1985a) Construction and comparison of recombinant plasmids encoding type 1 fimbriae of members of the family *Enterobacteriaceae*. Infect Immun 48: 275–279

Clegg S, Pruckler J, Purcell BK (1985b) Complementation analysis of recombinant plasmids encoding type 1 fimbriae of members of the family *Enterobacteriaceae*. Infect Immun 50: 338–340

De Ree JM, Schwillens P, Van den Bosch JF (1985a) Molecular cloning of F11 fimbriae from a uropathogenic *Escherichia coli* and characterization of fimbriae with polyclonal and monoclonal antibodies. FEMS Microbiol Lett 29: 91–97

De Ree JM, Schwillens P, Van den Bosch JF (1985b) Monoclonal antibodies that recognize the P fimbriae $F7_1$, $F7_2$, F9 and F11 from uropathogenic *Escherichia coli*. Infect Immun 50: 900–904

De Ree JM, Schwillens P, Promes L, Van Die I, Bergmans H, Van den Bosch JF (1985c) Molecular cloning and characterization of F9 fimbriae from a uropathogenic *Escherichia coli*. FEMS Microbiol Lett 26: 163–169

De Ree JM, Schwillens P, Van den Bosch JF (1987) Monoclonal antibodies raised against Pap fimbriae recognize minor component(s) involved in receptor binding. Microb Pathogen 2: 113–121

Dorman CJ, Higgins CF (1987) Fimbrial phase variation in *Escherichia coli:* dependence on integration host factor and homologies with other site-specific recombinases. J Bacteriol 169: 3840–3843

Duguid JP, Smith IW, Dempster G, Edmunds PN (1955) Non-flagellar filamentous appendages ("fimbriae") and hemagglutination activity in *Bacterium coli*. J Pathol Bacteriol 70: 335–348

Eisenstein BI (1981) Phase variation of type 1 fimbriae in *Escherichia coli* is under transcriptional control. Science 214: 337–339

Eisenstein BI, Sweet DS, Vaughn V, Friedman DI (1987) Integration host factor is required for the DNA inversion that controls phase variation in *Escherichia coli*. Proc Natl Acad Sci USA 84: 6506–6510

Ekbäck G, Mörner S, Lund B, Normark S (1986) Correlation of genes in the *pap* cluster to expression of globoside-specific adhesin by uropathogenic *Escherichia coli*. FEMS Microbiol Lett 34: 355–360

Evans DJ, Evans DG (1983) Classification of pathogenic *Escherichia coli* according to serotype and the production of virulence factors, with special reference to colonization factor antigens. Rev Infect Dis 5 (S4) S682–S701

Freitag CS, Abraham JM, Clements JR, Eisenstein BI (1985) Genetic analysis of the phase variation control of expression of type 1 fimbriae in *Escherichia coli*. J Bacteriol 162: 668–675

Garcia E, Bergmans HEN, Van den Bosch JF, Orskov I, Van den Zeijst BAM, Gaastra W (1988a) Isolation and characterization of dog uropathogenic *Escherichia coli* strains and their fimbriae. Antonie van Leeuwenhoek 54: 149–163

Garcia E, Hamers A, Bergmans H, Bernard A, Van der Zeijst B, Jaastra W (1988b) Adhesion of canine and human uropathogenic *Escherichia coli* and *Proteus mirabilis* to canine and human epithelial cells. Curr Microbiol 17: 333–337

Goldhar J, Perry R, Golecki JR, Hoschützky H, Jann B, Jann K (1987) Nonfimbrial, mannose-resistant adhesins from uropathogenic *Escherichia coli* O83:K1:H4 and O14:K?:H11. Infect Immun 55: 1837–1842

Göransson M, Uhlin BE (1984) Environmental temperature regulates transcription of a virulence pili operon in *E. coli*. EMBO J 3: 2885–2888

Göransson M, Forsman K, Uhlin BE (1988) Functional and structural homology among regulatory cistrons of pili-adhesin determinants on an *Escherichia coli*. MGG 212: 412–417

Hacker J, Hughes C (1985) Genetics of *Escherichia coli* hemolysin. In: Current topics in microbiology and immunology, vol 118 Springer, Berlin Heidelberg New York, pp. 139–162

Hacker J, Knapp S, Goebel W (1983) Spontaneous deletions and flanking regions of the chromosomally inherited hemolysin determinant of an *Escherichia coli* O6 strain. J Bacteriol 154: 1146–1152

Hacker J, Schmidt G, Hughes C, Knapp S, Marget M, Goebel W (1985) Cloning and characterization of genes involved in production of mannose-resistant, neuraminidase-susceptible (X) fimbriae from a uropathogenic O6:K15:H31 *Escherichia coli* strain. Infect Immun 47: 434–440

Hacker J, Jarchau T, Knapp S, Marre R, Schmidt G, Schmoll T, Goebel W (1986a) Genetic and in vivo studies with S-fimbrial antigens and related virulence determinants of extraintestinal *Escherichia coli* strains. In: Lark D (ed) Protein–carbohydrate interactions in biological systems. Academic, London, pp 125–133

Hacker J, Ott M, Schmidt G, Hull R, Goebel W (1986b) Molecular cloning of the F8 fimbrial antigen from *Escherichia coli*. FEMS Microbiol Lett 36: 139–144

Hacker J, Ott M, Bender L, Schmittroth M, Tschäpe H, Achtmann M (1989a) Distribution of genetic determinants coding for hemolysin, aerobactin and adhesin production in *Escherichia coli* K1 isolates. Infect Immun (to be published)

Hacker J, Schmoll T, Ott M, Marre R, Hof H, Jarchau T, Knapp S, Then I, Goebel W (1989b) Genetic structure and expression of virulence determinants from uropathogenic *E. coli* strains. In: Kass E (ed.) Host–parasite interactions in urinary tract infections; Studies of infectious disease research, Univ of Chicago Press, pp. 144–156.

Hagberg L, Hull R, Falkow S, Freter R, Svanborg-Eden C (1983) Contribution of adhesion to bacterial persistence in the mouse urinary tract. Infect Immun 40: 265–272

Hales BA, Amyes SGB (1986a) Acquisition of a gene encoding mannose-resistant haemagglutionation fimbriae by a resistance plasmid during long-term urinary tract infection. J Med Microbiol 22: 297–301

Hales BA, Amyes SGB (1986b) The transfer of genes encoding production of mannose-resistant haemagglutinating fimbriae from uropathogenic *enterobacteria* J Gen Microbiol 132: 2243–2247

Hales BA, Beverley-Clarke H, High NJ, Jann K, Goldhar J, Boulnois GJ (1988) Molecular cloning and characterization of the genes for a non-fimbrial adhesin from *Escherichia coli*. Microb Pathogen 5: 9–17

Hanson MS, Brinton CC (1988) Identification and characterization of *E. coli* type I pilus tip adhesion protein. Nature 323: 265–268

Hanson MS, Hempel J, Brinton CC (1988) Purification of the *Escherichia coli* type I pilin and minor pilus proteins and partial characterization of the adhesin protein. J Bacteriol 170: 3350–3358

High NJ, Hales BA, Jann K, Boilnois GJ (1988) A block of urovirulence genes encoding multiple fimbriae and hemolysin in *Escherichia coli* O4:K12:H⁻. Infect Immun 56: 513–517

Hoekstra WPM, Felix HS, Theilemans M, Pere A, Korhonen TK (1986) Chromosomal location of P fimbrial gene clusters in uropathogenic *Escherichia coli* strains. Proceedings of the 14th International Congress of microbiology, IUMS, Manchester, England

Hopp TP, Woods KR (1981) Prediction of protein antigenic determinants from amino acid sequences. Proc Natl Acad Sci USA 78: 3824–3828

Hoschützky H, Lottspeich F, Jann K (1989) Isolation and characterization of the α-galactosyl-1, 4-β-galactosyl-specific adhesin (P adhesin) from fimbriated *Escherichia coli*. Infect Immun 57: 76–81

Hughes C, Hacker J, Roberts A, Goebel W (1983) Hemolysin production as a virulence marker in symptomatic and asymptomatic urinary tract infections caused by *Escherichia coli*. Infect Immun 39: 546–551

Hull RA, Gill RE, Hsu P, Minshew BH, Falkow S (1981) Construction and expression of recombinant plasmids encoding type 1 or D-mannose-resistant pili from a urinary tract infection *Escherichia coli* isolate. Infect Immun 33: 933–938

Hull SI, Hull RA, Minshew BH, Falkow S (1982) Genetics of hemolysin of *Escherichia coli*. J Bacteriol 151: 1006–1012

Hull S, Clegg S, Svanborg Eden C, Hull R (1985) Multiple forms of genes in pyelonephritogenic *Escherichia coli* encoding adhesins binding globoseries glycolipid receptors. Infect Immun 47:80–83

Hull R, Bieler S, Falkow S, Hull S (1986) Chromosomal map position of genes encoding P adhesins in uropathogenic *Escherichia coli*. Infect Immun 51: 693–695

Hull SI, Bieler S, Hull RA (1988) Restriction fragment length polymorphism and multiple copies of DNA sequences homologous with probes for p fimbriae and hemolysin genes among uropathogenic *Escherichia coli*. Can J Microbiol 34: 307–311

Jacobs AAC, van den Berg PA, Bak H, de Graaf FK (1986) Localization of lysine residues in the binding domain of the K99 fibrillar subunit of enterotoxigenic *Escherichia coli*. Biochim Biophys Acta 872: 92–97

Jann K, Jann B, Schmidt G (1981) SDS polyacrylamide gel electrophoresis and serological analysis of pili from *Escherichia coli* from different pathogenic origin. FEMS Microbiol Lett 11: 21–25

Jones GW, Rutter JM (1974) The association of K88 antigen with hemagglutination activity in porcine strains of *Escherichia coli*. J Gen Microbiol 84: 135–144

Källenius G, Möllby R, Svenson SB, Winberg J, Lundblad A, Svensson S (1980) The pᵏ antigen as receptor of pyelonephritic *E. coli*. FEMS Microbiol Lett 7: 297–300

Keith BR, Maurer L, Spears PA, Orndorff PE (1986) Receptor-binding function of type I pili effects bladder colonization by a clinical isolate of *Escherichia coli*. Infect Immun 53: 693–696

Klemm P (1984) The *fimA* gene encoding the type 1 fimbrial subunit of *Escherichia coli*. Eur J Biochem 143: 395–399

Klemm P (1985) Fimbrial adhesins. Rev Infect Dis 7: 321–340

Klemm P (1986) Two regulatory *fim* genes, *fimB* and *fimE*, control the phase variation of type 1 fimbriae in *Escherichia coli*. EMBO J 5: 1389–1393

Klemm P, Christiansen G (1987) Three *fim* genes required for the regulation of length and mediation of adhesion of *Escherichia coli* type 1 fimbriae. MGG 208: 439–445

Klemm P, Jorgensen BJ, Van Die I, De Ree H, Bergmans H (1985) The *fim* genes responsible for synthesis of type 1 fimbriae in *Escherichia coli*, cloning and genetic organization. MGG 199: 410–414

Knapp S, Hacker J, Then I, Müller D, Goebel W (1984) Multiple copies of hemolysin genes and associated sequences in the chromosome of uropathogenic *Escherichia coli* strains. J Bacteriol 159: 1027–1033

Knapp S, Hacker J, Jarchau T, Goebel W (1986) Large instable regions in the chromosome affect virulence properties of a uropathogenic *Escherichia coli* O6 strain. J Bacteriol 168: 22–30

Korhonen TK, Väisänen-Rhen V, Rhen M, Pere A, Parkkinen J, Finne J (1984) *Escherichia coli* fimbriae recognizing sialyl galactosides. J Bacteriol 159: 762–766

Korhonen TK, Valtonen MV, Paarkkinen J, Väisänen-Rhen V, Finne J, Orskove I, Orskov F, Svenson SB, Mäkelä PH (1985) Serotype, hemolysin production and receptor recognition of *Escherichia coli* strains associated with neonatal sepsis and meningitis. Infect Immun 48: 486–491

Krallmann-Wetzel U, Ott M, Hacker J, Schmidt G (1989) Chromosomal mapping of genes encoding mannose-sensitive (type I) and mannose-resistant F8 (P) fimbriae of *Escherichia coli* O18:K5:H5. FEMS Microbiol Lett 58: B15–B22

Krogfeld KA, Klemm P (1988) Investigation of minor components of *Escherichia coli* type I fimbriae: protein chemical and immunological aspects. Microb Pathogen 4: 231–238

Labigne-Roussell AF, Lark D, Schoolnik G, Falkow S (1984) Cloning and expression of an afimbrial adhesin (AfaI) responsible for blood group-independent, mannose-resistant hemagglutination from a pyelonephritic *Escherichia coli* strain. Infect Immun 46: 251–259

Labigne-Roussell A, Schmidt MA, Walz W, Falkow S (1985) Genetic organization of the afimbrial adhesin operon and nucleotide sequence from a uropathogenic *Escherichia coli* gene encoding an afimbrial adhesin. J Bacteriol 162: 1285–1292

Labigne-Roussell A, Falkow S (1988) Distribution and degree of heterogeneity of the afimbrial-adhesin-encoding operon (*afa*) among uropathogenic *Escherichia coli* isolates. Infect Immun 56: 640–648

Lindberg FP, Lund B, Normark S (1984) Genes of pyelonephritogenic *E. coli* required for digalactoside-specific agglutination of human cells. EMBO J 3: 1167–1173

Lindberg FP, Lund B, Normark S (1986) Gene products specifying adhesion of uropathogenic *Escherichia coli* are minor components of pili. Proc Natl Acad Sci USA 83: 1891–1895

Lindberg F, Lund B, Johansson L, Normark S (1987) Localization of the receptor-binding protein adhesin at the tip of the bacterial pilus. Nature 328: 84–87

Low D, David V, Lark D, Schoolnik G, Falkow S (1984) Gene clusters governing the production of hemolysin and mannose-resistant hemagglutination are closely linked in *Escherichia coli* serotype O4 and O6 isolates from urinary tract infections. Infect Immun 43: 353–358

Low D, Robinson EN, McGee ZA, Falkow S (1987) The frequency of expression of pyelonephritis-associated pili is under regulatory control. Mol Microbiol 1: 335–346

Lund B, Lindberg FP, Baga M, Normark S (1985) Globoside-specific adhesins of uropathogenic *Escherichia coli* are encoded by similar transcomplementable gene clusters. J Bacteriol 162: 1293–1301

Lund B, Lindberg F, Marklund BI, Normark S (1987) The PapG protein is the α-D-galactopyranosyl-(1-4)-β-D-galactopyranose-binding adhesin of uropathogenic *Escherichia coli*. Proc Natl Acad Sci USA 84: 5898–5902

Lund B, Lindberg F, Normark S (1988a) Structure and antigenic properties of the tip-located P-pilus proteins of uropathogenic *Escherichia coli*. J Bacteriol 170: 1887–1894

Lund B, Marklund BI, Strömberg N, Lindberg F, Karlsson KA, Normark S (1988b) Uropathogenic *Escherichia coli* can express serologically identical pili of different receptor binding specificities. Mol Microbiol 2: 255–263

Marre R, Hacker J (1987) Role of S- and common type I fimbriae of *Escherichia coli* in experimental upper and lower urinary tract infection. Microb Pathogen 2: 223–226

Maurer L, Orndorff PE (1985) A new locus, *pilE*, required for the binding of type 1 piliated *Escherichia coli* to erythrocytes. FEMS Microbiol Lett 30: 59–66

Maurer L, Orndorff PE (1987) Identification and characterization of genes determining receptor binding and pilus length of *Escherichia coli* type I pili. J Bacteriol 169: 640–645

Miller VL, Taylor RK, Mekalanos JJ (1987) Cholera toxin transcriptional activator ToxR is a transmembrane DNA binding protein. Cell 48: 271–279

Minion FC, Abraham SN, Beachey EH, Goguen JD (1986) The genetic determinant of adhesive function in type 1 fimbriae of *Escherichia coli* is distinct from the gene encoding the fimbrial subunit. J Bacteriol 165: 1033–1036

Mooi FR, Claasen I, Bakker D, Kuipters H, De Graaf FK (1986) Regulation and structure of an *Escherichia coli* gene coding for an outer membrane protein involved in export of K88ab fimbrial subunits. Nucleic Acids Res 14: 2443–2457

Moch T, Hoschützky H, Hacker J, Krönke KD, Jann K (1987) Isolation and characterization of the α-sialyl-β-2,3-galactosyl-specific adhesin from fimbriated *Escherichia coli*. Proc Natl Acad Sci USA 84: 3462–3466

Norgren M, Normark S, Lark D, O'Hanley P, Schoolnik G, Falkow S, Svanborg-Eden C, Baga M, Uhlin

BE (1984) Mutations in *E. coli* cistrons affecting adhesion to human cells do not abolish Pap pili fiber formation. EMBO J 3: 1159–1165

Norgren M, Baga M, Tennett JM, Normark S (1987) Nucleotide sequence, regulation and functional analysis of the *papC* gene required for cell surface localization of *pap* pili of uropathogenic *Escherichia coli*. Mol Microbiol 1: 169–178

Normark S, Lark D, Hull R, Norgren M, Baga M, O'Hanley P, Schoolnik G, Falkow S (1983) Genetics of digalactoside-binding adhesin from a uropathogenic *Escherichia coli* strain. Infect Immun 41: 942–949

Normark S, Baga M, Göransson M, Lindberg FP, Lund B, Norgren M, Uhlin BE (1986) Genetics and biogenesis of *Escherichia coli* adhesins. In: Mirelman D (ed) Microbial lectins and agglutinins: properties and biological activity. Wiley, New York, pp 113–143

Normark S, Hultgren S, Marklund BI, Strömberg N, Tennent J (1988) Biogenesis of pili adhesins associated with urinary tract infectious *Escherichia coli*. Antonie Van Leeuwenhoek 54: 405–409

Nowicki B, Vuopio-Varkila J, Viljanen P, Korhonen TK, Mäkelä PH (1986) Fimbrial phase variation and systemic *E. coli* infection studied in the mouse peritonitis model. Microb Pathogen 1: 335–347

Nowicki B, Barrish JP, Korhonen TK, Hull RA, Hull SI (1987) Molecular cloning of the *Escherichia coli* O75X adhesin. Infect Immun 55: 2268–2276

Nowicki B, Moulds J, Hull R, Hull S (1988) A hemagglutinin of uropathogenic *Escherichia coli* recognizes the Dr blood group antigen. Infect Immun 56: 1057–1060

Ofek I, Silverblatt F (1982) Bacterial surface structures involved in adhesion to phagocytic and epithelial cells. In: Schlessinger D (ed) Microbiology 1982. American Society for Microbiology, Washington, pp. 296–300

O'Hanley P, Lark D, Falkow S, Schoolnik G (1985) Molecular basis of *Escherichia coli* colonization of the upper urinary tract in BALB/c mice. J Clin Invest 75: 347–360

Orndorff PE, Falkow S (1984a) Organization and expression of genes responsible for type 1 piliation in *Escherichia coli*. J Bacteriol 159: 736–744

Orndorff PE, Falkow S (1984b) Identification and characterization of a gene product that regulates type 1 piliation in *Escherichia coli*. J Bacteriol 160: 61–66

Orndorff PE, Falkow S (1985) Nucleotide sequence of *pilA*, the gene encoding the structural component of type 1 pili in *Escherichia coli*. J Bacteriol 162: 454–457

Orskov I, Orskov F (1983) Serology of *Escherichia coli* fimbriae. Prog Allergy 33: 80–105

Orskov I, Orskov F (1985) *Escherichia coli* in extraintestinal infections. J Hyg Camb 95: 551–575

Orskov I, Orskov F, Jann B, Jann K (1977) Serology, chemistry and genetics of O and K antigens of *Escherichia coli* Microbiol Rev 41: 667–710

Ott M, Hacker J, Schmoll T, Jarchau T, Korhonen TK, Goebel W (1986) Analysis of the genetic determinants coding for the S-fimbrial adhesin (*sfa*) in different *Escherichia coli* strains causing meningitis or urinary tract infections. Infect Immun 54: 646–653

Ott M, Schmoll T, Goebel W, Van Die I, Hacker J (1987) Comparison of the genetic determinant coding for the S-fimbrial adhesin (*sfa*) of *Escherichia coli* to other chromosomally encoded fimbrial determinants. Infect Immun 55: 1940–1943

Ott M, Hoschützky H, Jann K, Van Die I, Hacker J (1988) Gene clusters for S fimbrial adhesin (*sfa*) and F1C fimbriae (*fod*) of *Escherichia coli*: comparative aspects of structure and function. J Bacteriol 170: 3983–3990

Parkkinen J, Rogers GN, Korhonen TK, Dahr W, Finne J (1986) Identification of the O-linked sialyloligosaccharides of glycophorin as the erythrocyte receptors for S-fimbriated *Escherichia coli*. Infect Immun 54: 37–42

Pawelzik B, Heesemann J, Hacker J, Opferkuch W (1988) Cloning and characterization of a new type of fimbriae (S/F1C related fimbria) expressed by an *Escherichia coli* O75:K1:H7 blood culture. Infect Immun 56: 2918–2924

Pere A, Leinonen M, Väisänen-Rhen V, Rhen M, Korhonen TK (1985) Occurrence of type 1C fimbriae on *Escherichia coli* strains isolated from human extraintestinal infections. J Gen Microbiol 131: 1705–1711

Rhen M (1985) Characterization of DNA fragments encoding fimbriae of the uropathogenic *Escherichia coli* strain KS71. J Gen Microbiol 131: 571–580

Rhen M, Väisänen-Rhen V (1987) Nucleotide sequence analysis of a P fimbrial regulatory element of the uropathogenic *Escherichia coli* strain KS71 (O4:K12). Microb Pathogen 3: 387–391

Rhen M, Knowles J, Penttila ME, Sarvas M, Korhonen TK (1983a) P fimbriae of *Escherichia coli*: molecular cloning of DNA fragments containing the structural genes. FEMS Microbiol Lett 19: 119–123

Rhen M, Mäkelä PH, Korhonen TK (1983b) P-fimbriae of *Escherichia coli* are subject to phase variation. FEMS Microbiol Lett 19: 267–271

Rhen M, Van Die I, Rhen V, Bergmans H (1985a) Comparison of the nucleotide sequence of the genes encoding the KS71A and F7 fimbrial antigens of uropathogenic *Escherichia coli*. Eur J Biochem 151: 573–577

Rhen M, Väisänen-Rhen V, Korhonen TK (1985b) Complementation and regulatory interaction between two cloned fimbrial gene clusters of *Escherichia coli* strain KS71 MGG 200: 60–64

Rhen M, Klemm P, Korhonen TK (1986a) Identification of two hemagglutinins of *Escherichia coli*, N-acetyl-D-glucosamine-specific fimbriae and a blood group M-specific agglutinin, by cloning the corresponding genes in *Escherichia coli* K-12. J Bacteriol 168: 1234–1242

Rhen M, Väisänen-Rhen V, Saraste M, Korhonen TK (1986b) Organization of genes expressing the blood-group-M-specific hemagglutinin of *Escherichia coli*: identification and nucleotide sequence of the M-agglutinin subunit gene. Gene 49: 351–360

Riegman N, Van Die I, Leunissen J, Hoekstra W, Bergmans H (1988) Biogenesis of $F7_1$ and $F7_2$ fimbriae of uropathogenic *Escherichia coli*: influence of the FsoF and FsoG proteins and localization of the Fso/FstE protein. Mol Microbiol 2: 73–80

Saukkonen KMJ, Nowicki B, Leinonen M (1988) Role of type I and S fimbriae in the pathogenesis of *Escherichia coli* O18:K1 bacteremia and meningitis in the infant rat. Infect Immun 56: 892–897

Schmidt MA, O'Hanley P, Lark D, Schoolnik GK (1988) Synthetic peptides corresponding to protective epitopes of *Escherichia coli* digalactoside-binding piling prevent infection in a murine pyelonephritis model. Proc Natl Acad Sci USA 85: 1247–1251

Schmoll T, Hacker J, Goebel W (1987) Nucleotide sequence of the *sfaA* gene coding for the S fimbrial protein subunit of *Escherichia coli*. FEMS Microbiol Lett 41: 229–235

Schmoll T (1989) Molekulargenetische Analyse einer Adhäsin–Determinante von *Escherichia coli*. Phil Diss Univ Würzburg

Schmoll T, Hoschützky H, Morschhäuser J, Lottspeich F, Jann K, Hacker J (1989a) Analysis of genes coding for the Sialic acid–binding adhesin and two other minor fimbrial subunits of the S fimbrial adhesin determinant of *Escherichia coli*. Mol Microbiol in press.

Schmoll T, Ott M, Hacker J (1989b) Environmental signals controlling the expression of the S fimbrial adhesin of an *Escherichia coli* pathogen: construction of a site specific wild-type fusion. J Bacteriol (to be published)

Schneider K, Beck CF (1986) Promoter probe vectors for the analysis of divergently promoters. Gene 42: 37–48

Silvermann M, Simon M (1980) Phase variation genetic analysis of switching mutants. Cell 19: 845–854

Spears PA, Schauer D, Orndorff PE (1986) Metastable regulation of type 1 piliation in *Escherichia coli* and isolation and characterization of a phenotypically stable mutant. J Bacteriol 168: 179–185

Stern A, Meyer TF (1987) Common mechanism controlling phase and antigenic variation in pathogenic *Neisseriae*. Mol Microbiol 1: 5–12

Väisänen-Rhen V, Saarela S, Rhen M (1988) Mutations in cloned *Escherichia coli* P fimbriae genes that makes fimbriae production resistant to suppression by trimethoprim. Microbiol Pathogen 4: 369–377

Svanborg-Eden C, Eriksson B, Hanson LA (1977) Adhesion of *Escherichia coli* to human uroepithelial cells in vitro. Infect Immun 18: 767–774

Svanborg Eden C, de Man P (1988) Bacterial virulence in urinary tract infection. Infect Dis Clin North Am 1: 731–750

Swaney LM, Liu YP, To CM, To CC, Ippen-Ihler K, Brinton CC (1977) Isolation and characterization of *Escherichia coli* phase variants and mutants deficient in type 1 pilus production. J Bacteriol 130: 495–505

Uhlin BE, Baga M, Göransson M, Lindberg FP, Lund B, Norgren M, Normark S (1985a) Genes determining adhesin formation in uropathogenic *Escherichia coli*. In: Current topics in microbiology and Immunology, vol 118 Springer, Berlin Heidelberg New York, pp. 163–178

Uhlin BE, Norgren M, Baga M, Normark S (1985b) Adhesion to human cells by *Escherichia coli* lacking the major subunit of a digalactoside-specific pilus-adhesin. Proc Natl Acad Sci USA 82: 1800–1804

Väisänen-Rhen V, Elo J, Väisänen E, Siitonen A, Orskov I, Orskov F, Svenson SB, Mäkelä PH, Korhonen TK (1984) P-fimbriated clones among uropathogenic *Escherichia coli* strains. Infect Immun 43: 149–155

Van Die I, Bergmans H (1984) Nucleotide sequence of the gene encoding the $F7_2$ fimbrial subunit of a uropathogenic *Escherichia coli* strain. Gene 32: 83–90

Van Die I, Van den Hondel C, Hamstra HJ, Hoekstra W, Bergmans H (1983) Studies on the fimbriae of an *Escherichia coli* O6:K2:H1:F7 strain: molecular cloning of a DNA fragment encoding a fimbrial antigen responsible for mannose-resistant hemagglutination of human erythrocytes. FEMS Microbiol Lett 19: 77–82

Van Die I, Van Geffen R, Hoekstra W, Bergmans H (1984a) Type 1C fimbriae of a uropathogenic

Escherichia coli strain: cloning and characterization of the genes involved in the expression of the 1C antigen and nucleotide sequence of the subunit gene. Gene 34: 187–196

Van Die I, Van Megen I, Hoekstra W, Bergmans H (1984b) Molecular organization of the genes involved in the production of F7$_2$ fimbriae, causing mannose resistant hemagglutination, of a uropathogenic *Escherichia coli* O6:K2:H1:F7 strain. MGG 194: 528–533

Van Die I, Spierings G, Van Megen I, Zuidweg E, Hoekstra W, Bergmans H (1985) Cloning and genetic organization of the gene cluster encoding F7$_1$ fimbriae of a uropathogenic *Escherichia coli* and comparison with the F7$_2$ gene cluster. FEMS Microbiol Lett 28: 329–334

Van Die I, Dijksterhuis M, de Cock H, Hoekstra W, Bergmans H (1986a) Structural variation of P fimbriae from uropathogenic *Escherichia coli*. In: Lark D (ed) Protein-carbohydrate interactions in biological systems. Academic, London, pp 39–46

Van Die I, Van Megen I, Zuidweg E, Hoekstra W, de Ree H, Van den Bosch H, Bergmans H (1986b) Functional relationship among gene clusters encoding F7$_1$, F7$_2$, F9 and F11 fimbriae of human uropathogenic *Escherichia coli*. J Bacteriol 167: 407–410

Van Die I, Zuidweg E, Hoekstra W, Bergmans H (1986c) The role of fimbriae of uropathogenic *Escherichia coli* as carriers of the adhesin involved in mannose-resistant hemagglutination. Microb Pathogen 1: 51–56.

Van Die I, Hoekstra W, Bergmans H (1987) Analysis of the primary structure of P fimbrillins of uropathogenic *Escherichia coli*. Microb Pathogen 3: 149–154

Van Die I, Riegman N, Gaykema O, Van Megen I, Hoekstra W, Bergmans H, De Ree H, Van den Bosch H (1988a) Localization of antigenic determinants of P fimbriae of uropathogenic *Escherichia coli*. FEMS Microbiol. Lett 49: 95–100

Van Die I, Wauben M, Van Megen I, Bergmans H, Riegman N, Hoekstra W, Pouwels P, Enger-Valk B (1988b) Genetic manipulation of major P-fimbrial subunits and consequences for formation of fimbriae. J Bacteriol 170: 5870–5876

Virkola R, Westerlund B, Holthöfer H, Parkkinen J, Kekomäki M, Korhonen TK (1988) Binding characteristics of *Escherichia coli* adhesins in human urinary bladder. Infect Immun 56: 2615–2622

Walz W, Schmidt A, Labigne-Roussell A, Falkow S, Schoolnik G (1985) AFA-I: a cloned afimbrial X-type adhesin from a human pyelonephritic *Escherichia coli* strain. Eur J Biochem 152: 315–321

Weiss AA, Hewlett EL (1986) Virulence factors of *Bordetella pertussis*. Annu Rev Microbiol 40: 661–686

Genetics of Adhesive Fimbriae of Intestinal
Escherichia coli

F. K. DE GRAAF

1 Introduction

Escherichia coli is a normal inhabitant of the intestinal tract of man and animals. Colonization by *E. coli* takes place soon after birth and once established, *E. coli* remains part of the fecal flora. In addition, *E. coli* has been recognized as an important pathogen involved in a variety of intestinal and extraintestinal diseases.

Relatively little is known about the factors that promote intestinal colonization in the newborn. It is likely that type 1 fimbriae, found on the majority of *E. coli* strains, play a role in the colonization process, but other less "common" adhesins, including M, S, P, and X fimbriae, are also frequently detected in the fecal flora of healthy individuals.

A wide variety of adherence factors have been identified in relation to pathogenicity and it has become evident that adhesins play a significant role in

Department of Molecular Microbiology, Biological Laboratory, Vrije Universiteit, de Boelelaan 1087, 1081 HV Amsterdam, The Netherlands

the colonization of habitats where commensal strains usually do not survive (DE GRAAF and MOOI 1986).

This review describes the genetics of fimbriae detected on strains causing intestinal infections with emphasis on the structural organization of the respective genetic determinants, regulation of expression, and the role of the various gene products in the biosynthesis of fimbriae. A summary of these adhesins is given in Table 1. Apart from type 1 fimbriae, the occurrence of adhesins appears to be host-specific in association with a limited number of serotypes capable of provoking a particular pathologic condition in man or domestic animals. The genetic determinants for fimbrial adhesins reside either on the bacterial chromosome or on large, sometimes conjugative plasmids.

2 Enterotoxigenic *E. coli* (ETEC) Strains

2.1 K88 and K99 Plasmids

The structural genes for the synthesis of K88 fimbriae were originally found to be located on transmissible plasmids that frequently also code for the ability to utilize raffinose (ORSKOV and ORSKOV 1966; SMITH and PARSELL 1975). However, conjugation experiments and electron microscopic studies by BAK et al. (1972) suggested that the transmissible nature of K88 plasmids was questionable and probably due to composite K88 plasmids that might dissociate into a fransfer factor and a K88 plasmid. A definite assignment of K88 plasmids was given by SHIPLEY et al. (1978), who reported that in most K88-producing strains the structural genes for

Table 1. Characteristics of fimbrial adhesins of intestinal *E. coli*

Fimbriae	Morphology	Mol. wt. subunit	Location of genes	Origin
Type 1	Rigid	15,700	Chromosome	—
K88	Flexible	27,600	Plasmid	Porcine
CS31A	Flexible	29,500	Plasmid	Bovine,
F41	Flexible	29,500	Chromosome	Porcine, bovine, ovine
CS1541	Flexible	18 000, 19 000		Porcine
K99	Flexible	16,500	Plasmid	Porcine, bovine, ovine
987P	Rigid	17,100	Chromosome	Porcine
F17	Flexible	19,500	Chromosome	Bovine
CFA/I	Rigid	15,000	Plasmid	Human
CFA/II				
CS1	Rigid	16,800	Plasmid	Human
CS2	Rigid	15,300	Plasmid	Human
CS3	Flexible	15,000	Plasmid	Human
CFA/III	Rigid	18,000		Human
CFA/IV				
CS4	Rigid	17,000	Plasmid	Human
CS5	Rigid	21,000	Plasmid	Human
CS6	Flexible?	14,500	Plasmid	Human

K88 production and raffinose utilization are located on a single nonconjugative plasmid of approximately 50 Md. Some strains contain larger conjugative plasmids, apparently resulting from a recombination between the Raf/K88 plasmid and a transfer factor. The K88 and Raf determinant are located on a DNA element flanked by direct repeats of the *E. coli* insertion sequence IS1 (SCHMITT et al. 1979). The K88 plasmids isolated from various strains had a very high degree of DNA homology (SHIPLEY et al. 1978). Recently, GIRARDEAU et al. (1988) described a new K88-related fimbrial antigen on bovine enterotoxigenic and septicemic *E. coli* strains. The genetic determinant of this antigen, designated CS31A, appears to be located on a 105 Md plasmid.

The synthesis of K99 fimbriae is controlled by a conjugative plasmid (SMITH and LINGGOOD 1972) of approximately 52 Md (So et al. 1974).

In contrast to the K88 and K99 determinants, the genetic determinants for the synthesis of F41 and, possibly, 987P fimbriae appeared to be located on the bacterial chromosome (MORRISSEY and DOUGAN 1986; MOSELEY et al. 1986; DE GRAAF and KLAASEN 1986).

Another fimbrial adhesin, F17, previously designated F(y) and identical to an adhesive antigen referred to as Att225, was observed on *E. coli* isolated from scouring calves (GIRARDEAU et al. 1980; POHL et al. 1982). Production of F(y) was not found to be associated with the synthesis of other adhesins, enterotoxins, or verotoxins (MORRIS et al. 1987). Most likely, the genetic determinant for F17 fimbriae is located on the bacterial chromosome (LINTERMANS et al. 1988).

2.2 CFA/I-ST Plasmids

In contrast to the K88- and K99-encoding plasmids occurring in porcine and bovine enterotoxigenic *E. coli* strains, the plasmids encoding fimbrial adhesins in ETEC strains of human origin usually also determine the synthesis of heat-stable (ST) and/or heat-labile (LT) enterotoxins.

EVANS et al. (1975) first demonstrated that the production of CFA/I by a strain of serotype O78:H11 correlated with the presence of a 60-Md plasmid. It appeared that several strains belonging to serogroup O78 contain a single plasmid which codes for both the production of CFA/I and heat-stable enterotoxin (SMITH et al. 1979). These plasmids are nonconjugative but can be mobilized into *E. coli* K-12. Similar CFA/I- ST plasmids ranging from 54 to 72 Md were observed in ETEC strains belonging to other serogroups which were isolated from various geographic areas (McCONNELL et al. 1981; MURRAY et al. 1983; REIS et al. 1980; WILLSHAW et al. 1982). The plasmids are fertility inhibition negative and belong to the same Inc group. A 65-Md plasmid coding for CFA/I, ST, and also LT was found in a strain of serogroup O63.

2.3 CFA/II-ST-LT Plasmids

Analysis of CFA/II-positive ETEC from different geographic locations and belonging to several O serogroups has indicated that the genetic determinants for the synthesis of CFA/II, ST, and LT all reside on a single plasmid which varies from 54 to 90 Md (PENARANDA et al. 1980, 1983). These plasmids are nonconjugative but can be mobilized into *E. coli* K-12 by co-conjugation with a drug resistance plasmid.

Studies on the expression of CFA/II-positive ETEC strains belonging to serotype O6:H16 using CFA/II antisera have shown that within this serotype three different antigenic components are detectable, designated as coli surface-associated antigens CS1, CS2, and CS3 (CRAVIOTO et al. 1982; SMYTH 1982). CS1 and CS2 are morphologically indistinguishable but are serologically distinct types of fimbriae (SMYTH 1984). In contrast to the rigid, 6–nm-wide CS1 and CS2 fimbriae, the CS3 antigen is composed of thinner and flexible fibrillae (LEVINE et al. 1984). Strains of serotype O6:H16, biotype A (rhamnose-negative) produce CS1 and CS3 whereas strains of biotypes B, C, and F (rhamnose-positive) possess the antigens CS2 and CS3. CFA/II-positive strains belonging to other serogroups usually produce CS3 only (CRAVIOTO et al. 1982; SMYTH 1982).

When CFA/II-ST-LT plasmids from O6:H16 strains of biotypes A, B, or C were transferred to *E. coli* K-12, all exconjugants produced CS3 only; when plasmids contained in *E. coli* K-12 were transferred back into wild-type *E. coli* O6:H16 strains of different biotype these strains again produced CS1 and CS3 (biotype A) or CS2 and CS3 (biotype B or C) (MULLANY et al. 1983; SMITH et al. 1983). Apparently, plasmids coding for the production of CFA/II, ST, and LT are very similar, if not identical, but the ability to synthesize the different CFA/II components depends on the serotype and biotype of the host. All CFA/II-positive strains, regardless of their serotype or biotype, produce CS3 but CS1 and CS2 are expressed exclusively by strains of serotype O6:H16 belonging to biotype A or to biotype B, C, or F, respectively. The rare CS2- only phenotype in serogroup O6 strains of biotypes B and C reported previously (SMYTH 1982) is likely the result of a defective CFA/II-ST-LT plasmid since introduction of other CFA/II-ST-LT plasmids into these strains results in the expression of both CS2 and CS3 (BOYLAN and SMYTH 1985). Furthermore, DNA isolated from strains producing only CS2 does not hybridize with a CS3-specific DNA probe (BOYLAN et al. 1987). BOYLAN and SMYTH (1985) suggest that possibly the expression of either CS1 or CS2 does not correlate with the ability to ferment rhamnose *per se,* since O6 serotypes other than O6:K15:H16 or H-, but capable of rhamnose fermentation, express only CS3. One exception described so far is a strain of serotype O139:H28 (SCOTLAND et al. 1985).

At present it is not known which chromosomal genes provoke the ability of O6:H16 strains to produce either CS1 or CS2.

2.4 CFA/IV-ST-LT Plasmids

The colonization factor CFA/IV of human ETEC strains, previously designated PCF8775, was also shown to consist of three antigenic components: CS4, CS5, and CS6. Strains of serogroup O25 produce CS4 and CS6, whereas strains of serogroups O6, O92, O115, and O167 produce the antigenic components CS5 and CS6. Antigens CS4 and CS5 were identified as fimbriae, whereas CS6 was reported to be nonfimbrial in nature (THOMAS et al. 1982, 1985; MCCONNELL et al. 1985). Strains producing only CS6 have been detected (MCCONNELL et al. 1985, 1986).

A plasmid with a molecular mass of 88 Md and coding for CS5, CS6, and ST has been demonstrated in strains of serogroup O167 (THOMAS et al. 1987). After mobilization of this plasmid into *E. coli* K-12, none of the transconjugants appear to cause mannose-resistant hemagglutination (MRHA), indicating that CS5 is not expressed in *E. coli* K-12. Expression of the MRHA-negative CS6 antigen was not tested. Mobilization of the 88-Md plasmid into strains of various serotypes demonstrated that expression of CS5 in wild-type *E. coli* strains is not affected by the serotype of the host. Recent analysis of a larger number of CFA/IV-positive strains indicated that the surface antigens CS4, CS5, and CS6 are usually carried on the same plasmid as the genes coding for ST or LT (MCCONNELL et al. 1988).

KNUTTON et al. (1987b) identified a new fimbrial structure consisting of curly fibrils in human ETEC strains of serotype O148:H28. They suggest that these fibrils may represent the CS6 antigen. Another new plasmid-encoded fimbrial adhesin was found on ETEC strains belonging to serotype O159:H4 (TACKET et al. 1987).

3 Enteropathogenic *E. coli* (EPEC) Strains

Adherence of EPEC strains to the mucosa of the small bowel is an important step in the infection process. Plasmid-encoded adhesins have been shown to be essential for the adherence of EPEC to cultured HEp-2 cells. BALDINI et al. (1983) demonstrated that HEp-2 adherence was encoded on a 60-Md plasmid (pMAR2) in an EPEC strain of serotype O127:H6. Using a DNA probe derived from this plasmid, the genes for HEp-2 adhesion were detectable in EPEC strains of O serogroups O55, O111, O119, and O127 (BALDINI et al. 1986).

Two patterns of adherence were distinguished: localized adherence, characterized by the formation of microcolonies on HEp-2 and HeLa cells, and diffuse adherence, in which bacteria are found in a uniform distribution all over the cells (NATARO et al. 1985). The two patterns are associated with distinct adhesins each encoded on plasmids varying in size from 55 to 70 Md. In strains of serogroup O55, the structural genes for the localized adherence phenotype are located on a plasmid also carrying drug resistance markers (LAPORTA et al. 1986). *E. coli* K-12 acquires adhesiveness after receiving this resistance factor or the previously identified plasmid pMAR2. Cells cured of these plasmids lose their ability to adhere (LAPORTA et al. 1986; KNUTTON et al. 1987c). Plasmid-containing *E. coli* K-12 cells, however, do not show the intimate attachment to HEp-2 cells which is characteristic for all wild-type EPEC cells (KNUTTON et al. 1987a). Probably, the pMAR2 plasmid codes for a fimbrial adhesin that promotes HEp-2 adherence but other chromosomally encoded factors are required for the characteristic intimate attachment of EPEC cells. The authors suggest that two steps may be involved in the adherence of EPEC strains to enterocytes: an initial attachment mediated by plasmid-encoded adhesins,

followed by effacement of brush border microvilli and intimate attachment
(KNUTTON et al. 1987a). Recently, SCALETSKY et al. (1988) identified the binding
factor responsible for localized adherence as a protein doublet with molecular
weights of 29 000 and 32 000. Both proteins appear to be components of the bacterial
outer membrane.

4 Enteroinvasive *E. coli* (EIEC) Strains

EIEC strains associated with dysentery-like diarrheal disease also adhere to the
surface of cultured HEp-2 cells. The genetic determinant(s) for the adhesive
phenotype is probably encoded by the bacterial chromosome (NANDADASA et al.
1981).

The mannose-resistant hemagglutinins of two EIEC strains were initially
described as nonfimbrial agglutinins essential for adherence to erythrocytes and
HEp-2 cells (WILLIAMS et al. 1984). MRHA-deficient mutants have lost their ability
to adhere to or to penetrate cultured cells (KNUTTON et al. 1984). Recently, it was
demonstrated that the ability to adhere to human colonocytes, originally attributed
to nonfimbrial adhesins, was promoted by fine fibrillae (HINSON et al. 1987).

An invasive *E. coli* strain of serotype O15:K + :H21 isolated from a septicemic
lamb was shown to possess a virulence plasmid, denoted Vir, which is involved in
attachment of Vir-positive strains to intestinal brush borders (MORRIS et al. 1982).
The presence of the Vir plasmid could be associated with the ability to produce a
fimbrial adhesin (F.K. DE GRAAF, unpublished results).

Another fimbrial adhesin, designated F165, was found on *E. coli* strains of
serogroup O115, associated with hemorrhagic enteritis and septicemia in calves and
piglets (FAIRBROTHER et al. 1986).

5 Structural Organization of Gene Clusters Involved in the Biosynthesis of Fimbrial Adhesins

5.1 The K88, F41, and 987P Determinants

Analysis of genetic determinants coding for the biosynthesis of fimbriae has evolved
rapidly and in particular the gene clusters involved in the production of type 1, P,
K88, and K99 fimbriae have been studied in detail. The organization of these gene
clusters was analyzed after cloning the respective genetic determinants in a suitable
plasmid vector, construction of deletion and transposon-insertion mutants, analysis
of gene expression in *E. coli* minicells, and nucleotide sequence analysis.

The genes involved in the biosynthesis of the two K88 variants K88ab and
K88ac appeared to be located on an 8-Md DNA fragment, and derived from wild-
type K88/Raf plasmids (MOOI et al. 1979; SHIPLEY et al. 1981). Comparison of the

restriction endonuclease cleavage patterns of the cloned fragments and of the parental plasmids suggests that both fragments are essentially the same. Most probably, the K88ab and K88ac encoding plasmids are derived from a common ancestral plasmid. The smallest fragment of cloned K88-DNA that contains sufficient information for the biosynthesis of K88 fimbriae is about 6.5 kb pair in size. *E. coli* K-12 cells harboring either pFM205 (K88ab) or pMK005 (K88ac) agglutinate with anti-K88 antisera, cause MRHA of guinea pig erythrocytes, and adhere to the brush borders of isolated porcine intestinal epithelial cells.

Expression of the recombinant plasmids *E. coli* minicells reveals the synthesis of several proteins ranging from 16 to 82 kd. The fimbrial subunits with a molecular weight of 27 000 have been identified by immunoprecipitation with anti-K88 antiserum (MOOI et al. 1981; SHIPLEY et al. 1981).

Subsequently, the identification and the genetic organization of the K88 gene cluster was analyzed by studying the expression of various deletion and transposon-insertion mutants in *E. coli* minicells (KEHOE et al. 1981, 1983; MOOI et al. 1981, 1982, 1984). Six structural genes (*faeC–H*) were identified on the cloned DNA contained in pFM205 (K88ab) encoding polypeptides with apparent molecular weights of 16 900, 82 100, 26 300, 15 400, 27 600, and 27 500 respectively (Fig. 1). Mutational inactivation of *faeC, faeD,* or *faeE* results in a K88⁻, MRHA⁻, Adh⁻ phenotype, although some of these mutants are capable of expressing a low level of K88 subunits. Gene *faeG* encodes the K88ab fimbrial subunit. Inactivation of gene *faeH* results in a strong reduction in the biosynthesis of K88ab fimbriae and in the ability of the cells to agglutinate erythrocytes and to adhere to intestinal epithelial cells (MOOI et al. 1981, 1982, 1984). Complementation with FaeH restores normal fimbrial production, indicating that this protein is involved in K88 biosynthesis (D. BAKKER, unpublished results). Nucleotide sequence analysis of a K88ab clone which contains the complete genetic determinant for the synthesis of K88 fimbriae revealed the presence of at least two more K88 genes, designated *faeB* and *faeI*, located upstream of *faeC* and downstream of *faeH*, respectively (D. BAKKER, unpublished results).

Five structural gene (*adhA-E*) were located on the cloned DNA contained in pMK005 (K88ac). The K88ac fimbrial subunits are encoded by *adhD*. The genes *adhE, adhA, adhB,* and *adhC* correspond, respectively, to *faeC, faeD, faeE,* and *faeF* in the K88ab gene cluster (Fig. 1). The K88ac gene corresponding to *faeH* was not detected. Expression of all *adh* genes is essential for high-level expression of K88ac fimbriae (KEHOE et al. 1981, 1983).

All the structural genes identified code for polypeptides which contain an amino-terminal leader sequence essential for initiation of their translocation across the cytoplasmic membrane.

The genetic determinant for production of F41 fimbriae was isolated by cosmid cloning (MOSELEY et al. 1986). DNA probes derived from the K88ac-encoded plasmid pMK005 hybridize with the cloned F41-DNA and transposon insertions which inactivate F41 production were mapped to the region of homology, indicating a genetic relatedness between the F41 and K88 determinants. Recently, the genetic organization of the polypeptides involved in F41 biosynthesis was investigated (ANDERSEN and MOSELY 1988). Genes encoding polypeptides with apparent molecular weights of 29 000, 30 000, 32 000, and 86 000 were indicated. The 29-kd protein was identified as the F41 fimbrial subunit. Extensive nucleotide sequence homology

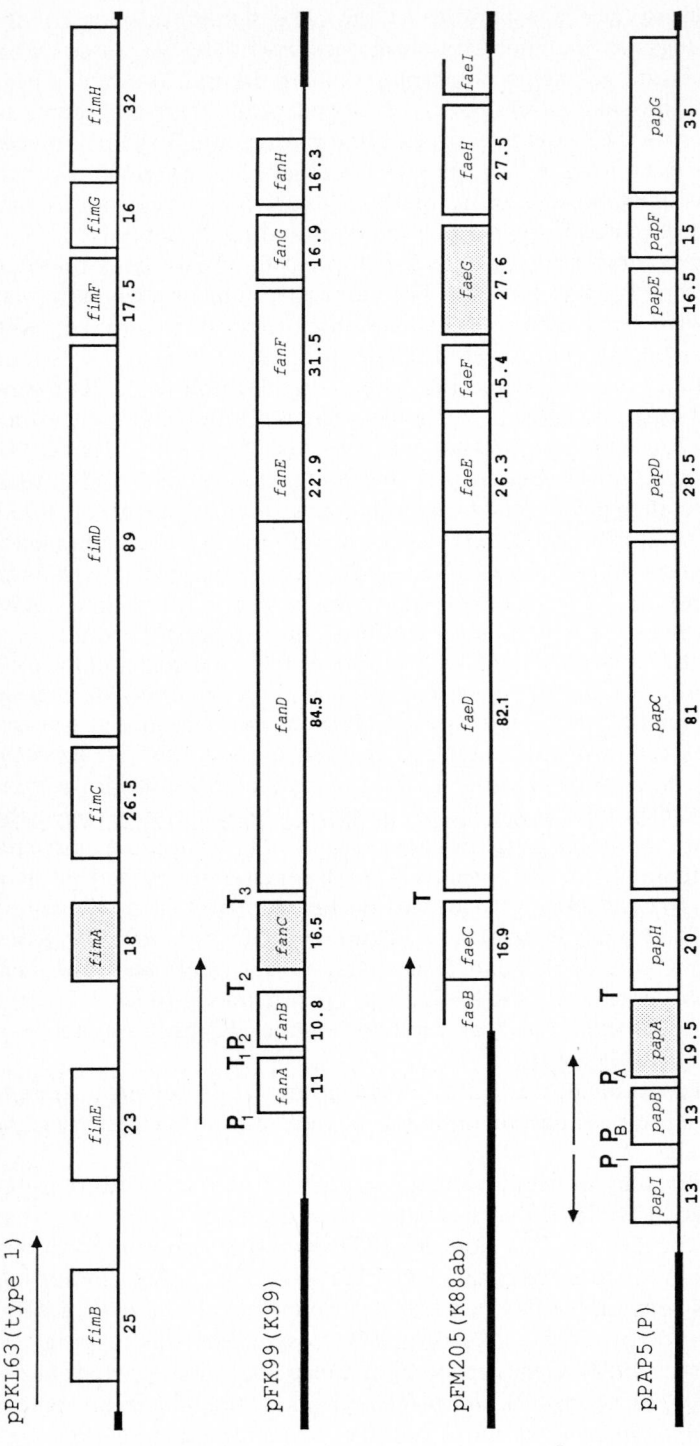

Fig. 1. Structural organization of genetic determinants encoding the biogenesis of fimbrial adhesins. The location of the genes is indicated by *boxes*; the molecular mass of the gene products is printed in kd below the boxes. As far as the nucleotide sequence of the respective genes has been determined, the molecular mass of the proteins has been given on the basis of these data. Promoter (*P*) and terminator (*T*) sequences are indicated. *Arrows* represent the direction of transcription. The genes encoding the major fimbrial subunits are *dotted*

was observed between the region encoding the putative signal sequences of the F41 and K88 subunits and the region immediately upstream of the subunit genes. A K88 probe hybridized at high stringency to all DNA fragments shown to be essential for F41 production. Furthermore, hybridization experiments with the K88 gene probe and plasmid or total DNA prepared from a variety of ETEC strains indicated that F41 is chromosomally encoded.

Also the genetic determinant for the biosynthesis of 987P fimbriae was isolated by cosmid cloning (DE GRAAF and KLAASEN and 1986; MORRISSEY and DOUGAN 1986). Cells of *E. coli* HB101 harboring the recombinant plasmid produced fimbriae which could not be distinguished from the 987P fimbriae produced by the parental strain. A 12.4-kb pair derivative, designated pPK180, was used to analyze the genetic organization of the 987P determinant (DE GRAAF and KLAASEN 1986). Five structural genes encoding polypeptides with apparent molecular weights of 16 500, 20 500, 28 500, 39 000, and 81 000 could be located on pPK180. The 20.5-kd polypeptide was identified as the fimbrial subunit (DE GRAAF and KLAASEN 1987). Recently, BROES et al. (1988) described a new fimbrial antigen (CS 1541), isolated from porcine enterotoxigenic *E. coli* and antigenically related to the 987P antigen. Like 987P fimbriae, the CS 1541 fimbriae possess no hemagglutinating properties. The CS 1541 fimbriae consist of two polypeptides of about 18 and 19 kd.

5.2 The K99 Determinant

The genetic determinant for the biosynthesis of K99 fimbriae is located on a 7.1-kb pair DNA fragment cloned into pBR325 (VAN EMBDEN et al. 1980). *E. coli* K-12 harboring this recombinant DNA synthesizes K99 fimbriae, causes MRHA of horse erythrocytes, and adheres to the brush border of porcine intestinal epithelial cells.

Analysis of the genetic organization of the K99 determinant in minicells revealed the presence of six structural genes, designated *fanC-H* (Fig. 1), and encoding polypeptides ranging in molecular weight from 16 300 to 84 500 (DE GRAAF et al. 1984; ROOSENDAAL 1987). The *fanC* gene product was identified as the K99 fimbrial subunit. Mutational inactivation of *fanC*, *fanD*, or *fanE* results in a K99$^-$, MRHA$^-$, Adh$^-$, phenotype. Inactivation of *fanF*, *fanG*, or *fanH* results in a simultaneous decrease in antigen production and adhesive capacity (JACOBS 1987; ROOSENDAAL et al. 1987b). All six polypeptides are synthesized with a leader sequence. Analysis of the nucleotide sequence of the region at the proximal end of the fimbrial subunit gene *fanC* revealed the presence of two open reading frames, designated *fanA* and *fanB* (ROOSENDAAL et al. 1987a). Frame shift mutations in both genes and complementation studies have indicated that *fanA* and *fanB* code for regulatory *trans*-acting polypeptides involved in the biogenesis of K99 fimbriae.

5.3 The CFA Determinants

SMITH et al. (1982) investigated the CFA/I determinant located on the 58-Md, CFA/I-ST plasmid NTP113. They were able to show that two regions of this plasmid are required for expression of CFA/I. Strains harboring plasmids with mutations in either region do not produce fimbriae. The two regions involved are separated by a DNA segment of about 25 Md. Both regions were cloned separately on compatible plasmid vectors (WILLSHAW et al. 1983). *E. coli* K-12 carrying both recombinant plasmids produces similar levels of CFA/I as cells harboring the wild-type plasmid NTP113. Analysis of the cloned DNA regions using *E. coli* minicells showed that region 1 encodes at least six independent proteins, including the fimbrial subunit, whereas region 2 codes for three polypeptides. At least one of the polypeptides encoded by region 2 is essential for fimbrial assembly (WILLSHAW et al. 1985).

The genes encoding the biosynthesis of CS3 fimbriae by CFA/II-positive ETEC strains have been cloned in pBR322 (MANNING et al. 1985). Recombinant plasmids expressing CS3 have a 4.6 kb pair DNA fragment in common. Expression of the cloned determinant in minicells showed that at least two polypeptides with molecular weights of 94 000 and 16 000 are essential for CS3 production. A more detailed analysis of the CS3 determinant was recently described by BOYLAN et al. (1987), who detected polypeptides of 94, 26, 24, 17, and 15 kd in *E. coli* minicells harboring the recombinant plasmid pCS100. The 15-kd protein and its 17-kd precursor were identified as the CS3 fimbrial subunit (BOYLAN et al. 1988). Both groups describe indications for the existence of overlapping genes in the CS3 gene cluster, but control of initiation and termination of transcription in minicells seems to be less accurate, and both internal initiations and premature terminations have been observed.

Colonization factor CFA/III has been identified by HONDA et al. (1984). The CFA/III fimbriae are composed of subunits with a molecular weight of 18 000.

The *E. coli* surface antigens CS4, CS5, and CS6 represent the colonization factor CFA/IV. CS4 and CS5 are fimbrial antigens with subunit molecular weights of 17 000 and 21 000, respectively (McCONNELL et al. 1988). The molecular weight of the CS6 antigen appeared to be 14 500 in strains of serogroups O25 and O27. CS6-positive strains of serogroups O148, O159, and O167 produce CS6 antigen consisting of two polypeptides with molecular weights of 14 500 and 16 000.

5.4 The Type 1 Determinant

Type 1-fimbriated *E. coli* strains bind to a large variety of eukaryotic cells, including erythrocytes, phagocytes, and epithelial cells. The genes involved in the production of type 1 fimbriae reside on the bacterial chromosome at a position corresponding to about 98 minutes on the *E. coli* linkage map (BRINTON et al. 1961) in between the markers *valS* and *uxuA* (FREITAG and EISENSTEIN 1983). Complementation studies with mutants defective in type 1 fimbriation indicated the existence of at least three genes involved in the biosynthesis of the fimbriae (SWANEY et al. 1977).

ORNDORFF and FALKOW (1984a) cloned a 11.2-kb pair DNA fragment coding for the biosynthesis of type 1 fimbriae and described the localization of five genes associated with fimbrial expression (Fig. 1). Three polypeptides with apparent molecular masses 17 000, 30 000, and 86 000 are involved in fimbrial assembly. The

17-kd polypeptide was identified as the type 1 fimbrial subunit on the basis of its reaction with anti-type 1 antibodies. Similar results were reported by KLEMM et al. (1985), who clones the type 1 determinant on an 8-kb pair DNA segment which expresses polypeptides with molecular masses of 16 500, 26 000, and 89 000. A fourth gene, designated *hyp*, encoding a 23-kd polypeptide, was found proximal to the fimbrial subunit gene (ORNDORFF and FALKOW 1984b; KLEMN et al. 1985). Insertional inactivation of this gene resulted in hyperfimbriation due, at least in part, to an increased number of fimbriae per cell (ORNDORFF and FALKOW 1984b). The protein appears to act *in trans* and to regulate the level of fimbriation.

Expression of type 1 fimbriae is subject to phase variation control which causes an oscillation between an "on" and an "off" mode. Recently, KLEMM (1986) reported that the phase-dependent expression of type 1 fimbriae actually involves two regulatory genes, *fimB* and *fimE*, located upstream of the fimbrial subunit gene and directing the phase switch into the "on" and "off" positions, respectively.

Besides the genes involved in the control and assembly of the fimbriae, the genetic determinant for type 1 fimbriae production appears to include genes which are responsible for determining receptor binding and length. The existence of these genes was indicated by MAURER and ORNDORFF (1985), who isolated fully fimbriated mutants that fail to agglutinate guinea pig erythrocytes. A comparable mutant was reported by MINION et al. (1986). Fimbriae produced by these mutants were morphologically and antigenetically indistinguishable from fimbriae produced by the parental strain. The genetic locus associated with the adhesion-negative phenotype was identified and found to be located very near the locus which encodes the 14-kd protein originally described as a minor fimbrial component (ORNDORFF and FALKOW 1984a; MINION et al. 1986). Recently, the coding region for receptor binding and fimbrial length control were fully analyzed and three open reading frames designated *fimF*, *fimG*, and *fimH* (KLEMM and CHRISTANSEN 1987) were identified. The molecular masses of the corresponding proteins were determined as 16 700, 14 900, and 28 900, respectively. Similar gene products designated PilD, PilF, and PilE were detected by MAURER and ORNDORFF (1987).

6 Regulation of Expression

6.1 Effects of Growth Conditions

The biosynthesis of fimbrial adhesins is regulated by growth conditions (DE GRAAF and MOOI 1986). A common feature of all fimbrial adhesins present on pathogenic *E. coli* strains is the repression of their production at low temperatures. In part, this effect may be explained by temperature-dependent promoter activity. For instance, the activity of the P_1 promoter in K99 operon (Fig. 1) is regulated by temperature (ROOSENDAAL 1987). On the other hand, the biosynthesis of fimbriae is regulated by the specific growth rate, as demonstrated for the biosynthesis of K88, K99, and F41 fimbriae. The highest production is found at the maximal growth rate; lower specific growth rates result in a gradual decrease in fimbrial production (VAN VERSEVELD et al. 1985; JACOBS and DE GRAAF 1985). Also the variations in the level of fimbrial production observed in different growth media are probably in part due to the regulatory effects of growth rate on the biosynthesis of fimbriae. At present, little information is available about the genetic background of the regulatory influence of growth rate on the biosynthesis of fimbriae.

Glucose and other fermentable sugars suppress the synthesis of type 1 (EISENSTEIN and DODD 1982) and K99 fimbriae (ISAACSON 1980; OLLIER and GIRARDEAU 1983). Indications have been obtained that this phenomenon is probably not due to the classic mechanism of catabolite repression but to a selective outgrowth of nonfimbriated versus fimbriated cells or a decrease in growth rate at lower pH due to an accumulation of fermentation products (VAN VERSEVELD et al. 1985).

6.2 Phase Variation

Expression of the chromosomally encoded type 1 and 987P fimbriae exhibits phase variation whereby individual *E. coli* cells can switch from a fimbriated to a nonfimbriated state and vice versa (BRINTON 1959; NAGY et al. 1977). A possibly related phenomenon was described for the expression of P fimbriae present on uropathogenic *E. coli* strains (RHEN et al. 1983).

Using a fusion of the *lacZ* gene to the DNA region encoding type 1 fimbriae, EISENSTEIN (1981) was able to demonstrate that phase variation between the fimbriate and non-fimbriate state is under transcriptional control. The transition rate from the fimbriate to the nonfimbriate state is 1.05×10^{-3} per bacterium per generation and 3.12×10^{-3} in the reverse direction. Apparently, the oscillation between both states occurs randomly at a frequency of approximately one switch per 1000 cells per generation. Subsequently, it was demonstrated that the region controlling the phase variation is directly adjacent to the gene encoding the fimbrial subunit and codes for a *cis*-acting element (FREITAG et al. 1985). Furthermore, these authors reported that besides the *cis*- active element, a *trans*-acting permissive factor supplied by the host strain is also required. F factors carrying the *fim* region were able to complement phase-locked mutants, indicating that the *trans*-acting element must map near the *fim* operon at 98 min.

Using Southern blot and DNA sequence analysis ABRAHAM et al. (1985) detected a genomic rearrangement in the switch region immediately upstream of the fimbrial structural gene. The switch consists of a small invertible segment of about 300 base pairs carrying the *fim* promoter flanked by a nine-base pair inverted repeat. The invertible element does not encode its own recombinase and is therefore different from other DNA inversion systems characterized by the site-specific recombinases *hin*, *gin*, *cin* or *pin*, which act as invertases (CRAIG 1985; PLASTERK et al. 1983; ZIEG and SIMON 1980). Similar results were recently reported by KLEMM (1986), who detected that the phase switch is controlled by the products of two regulatory genes, *fimB* and *fimE*, located upstream of the fimbrial subunit gene (Fig. 1). The FimB and FimE proteins direct the phase switch into the "on" and "off" positions, respectively. Both proteins have a molecular weight of about 23 000 and contain a high number of positively charged amino acid residues. Furthermore, both protein show a striking degree of sequence homology. The proteins contain an unequal number of cysteine residues and it was assumed that they may exist as dimers or multimers in their natural state (KLEMM 1986).

Significant homologies were found between a part of the FimB and FimE proteins and the active sites of site-specific recombinases known as the integrases of various phages (DORMAN and HIGGINS 1987). Therefore, it seems likely that FimB- and FimE- directed phase variation is based on a mechanism that is comparable to the Int-Xis system of phage λ. Further support for such a mechanism was provided by the observation that expression of type 1 fimbrial subunits is dependent on the presence of the integration host factor (IHF), a protein that is also required for the functioning of the λ integrase (EISENSTEIN et al. 1987; DORMAN and HIGGINS 1987). Two nucleotide sequences which match the consensus sequence for the IHF binding site are detected in the region containing the invertible DNA element. IHF is composed of two subunits α and β encoded by the *himA* and *himD* genes, respectively. Insertional inactivation of either of these two genes prevents phase variation and locks expression of *fimA* in either the "on" or the "off" state. In addition to the role of IHF in the inversion of the phase variation switch, DORMAN and HIGGINS (1987) observed that IHF is also required for efficient expression controlled by the *fimA* promoter. It is suggested that IHF facilitates the adoption of a DNA conformation favorable for FimB- and FimE-directed recombination. The conformational change induced by IHF probably involves bending DNA.

6.3 Transcriptional Control of K88 and K99 Expression

Analysis of the nucleotide sequence of the DNA region located upstream of the K99 fimbrial subunit gene (*fanC*) made possible the detection of two open reading frames and several transcriptional signals (Fig. 1). The open reading frames, designated *fanA* and *fanB*, code for positively charged proteins with a molecular mass of about 11 000 (ROOSENDAAL et al. 1987a). Both proteins exhibit a remarkable sequence homology. Furthermore, they show significant homology with another regulatory protein (PapB) encoded by the Pap operon (BAGA et al. 1985). No indications for a leader sequence were detected, suggesting that both proteins reside in the bacterial cytoplasm. Expression of FanA and FanB is very low since neither protein is detectable in minicells harboring the complete K99 operon. Both proteins, however, are clearly detectable in an in vitro DNA-directed translation system (ROOSENDAAL et al. 1988).

The possible role of FanA and FanB was investigated by constructing a frame-shift mutation in each of the respective genes. The mutational inactivation resulted in 8- and 16-fold reductions in K99 expression, respectively. Complementation of the mutations *in trans* restored K99 production to about 75% of the wild-type level. Apparently, FanA and FanB are *trans*-acting regulatory proteins affecting the biosynthesis of K99 fimbriae (ROOSENDAAL et al. 1987).

Putative promoters were observed preceding *fanA*, *fanB*, and *fanC*, respectively. No other promoter consensus sequences were detectable downstream of *fanC*, and

apparently the whole operon is controlled by the regulatory region upstream of
fanC. The activity of the putative promoters was tested after cloning of the
appropriate DNA fragments in a promoterless transcription vector which contains
the *E. coli* galactokinase gene as assayable marker (ROOSENDAAL et al. 1988). A very
low activity was detected for the putative promoter sequence preceding the fimbrial
subunit gene *fanC*. Expression of the regulatory proteins FanA and FanB did not
stimulate this putative transcription signal, indicating that the *fanC* promoter is of
minor importance in the transcriptional control of the K99 gene cluster (Fig. 1). In
this context it should be mentioned that also the putative promoter sequences
preceding the gene for the fimbrial subunit-like protein encoded by the K88 operon
(*faeC*) or the 987P fimbrial subunit gene exhibit low activity (MOOI et al. 1984; F. K.
DE GRAAF, unpublished results). Obviously, the control of expression of gene
clusters coding for the biosynthesis of fimbriae is more complicated and requires
expression of regulatory genes at the 5′ end of the respective clusters.

The recombinant plasmids pFM205 and pMK005 contain all necessary
information for the biosynthesis of K88ab and K88ac fimbriae, respectively, but
only some of the regulatory elements were cloned from the original parental
plasmids (KEHOE et al. 1983; MOOI et al. 1982). Therefore, expression of K88ab or
K88ac fimbriae by both recombinant strains is controlled by the P1 promoter of
vector pBR322. An open reading frame strongly homologous to the *fanB* gene of
pFK99 was observed in pFM205 upstream of *faeC*. This open reading frame (*faeB*),
however, is not complete because the nucleotide sequence encoding the amino-
terminal end of the corresponding polypeptide, FaeB, is not contained in pFM205
(Fig. 1). Recently the complete K88ab determinant was cloned into pBR322. This
clone exhibits a ten fold higher production of K88 fimbriae, indicating that K88
regulatory proteins also have a positive effect on the synthesis of fimbriae (D.
Bakker, unpublished results).

Further analysis of the K99 regulatory region indicated that the promoters P_1
and P_2 possess a high and an intermediate activity, respectively (ROOSENDAAL et al.
1988). The activity of these promoters, cloned into the promoterless transcription
vector, is not affected by FanA or FanB, but both regulatory proteins appear to
affect the transcriptional activity of a DNA fragment containing P1, *fanA,* and P2,
indicating a possible interference with other control elements located within this
region.

Factor-dependent terminators of transcription were detected and are probably
located in the intercistronic regions between *fanA* and *fanB* (T_1) and between *fanB*
and *fanC* (T_2) (ROOSENDAAL et al. 1988). The identity of these terminators (Fig. 1) is
not known. The intercistronic regions between *fanA* and *fanB* and between *fanB*
and *fanC* do not contain sequences known to be involved in a transcriptional pause
(ROOSENDAAL et al. 1987a). Factor-dependent terminators, however, do not appear
to correspond to any specific sequence requirements but require additional
sequences considerably upstream of the termination endpoints to convey the signal
for termination (PLATT 1986). A region showing dyad symmetry that may function
as a potential binding site for protein factors that regulate transcription is located
within *fanA* (ROOSENDAAL et al. 1987a). This region shows similarities to the Box B
region of the lambda *nut*-site which is part of the lambda pN antitermination system

(FRIEDMAN and GOTTESMAN 1983, PALO and SAUER 1984). Most likely, the regulatory proteins FanA and FanB both function as antiterminators of transcription (ROOSENDAAL et al. 1988).

A third terminator of transcription (T_3) is located in the intercistronic region between *fanC* and *fanD* (Fig. 1) and has an efficiency of 90% (ROOSENDAAL et al. 1984; 1988). This terminator is composed of a GC-rich region of dyad symmetry, thus allowing the formation of an intramolecular hairpin structure in the RNA transcript but without the usual series of uridine residues at the 3' end. This might indicate that either T_3 must be considered as a factor (rho)-dependent terminator or the pausing of the RNA-polymerase at the hairpin is sufficient to allow 'translational attenuation.' The efficiency of the termination at T_3 is in agreement with the observation that the fimbrial subunit is expressed at a much higher level than the auxiliary proteins FanD–FanH.

Other indications that the region of dyad symmetry in the intercistronic region between *fanC* and *fanD* is involved in translational control have been derived from the K88 operon, where a comparable region contains the 3' end of *faeC*, the ribosome binding site of *faeD*, and the translation initiation codon of this gene (MOOI et al. 1986; ROOSENDAAL 1987). Translation of *faeC* is terminated within the ribosome-binding site preceding the start codon of *faeD*. Therefore, it is conceivable that a ribosome terminating translation of *faeC* could reinitiate translation at *faeD* without being released from the mRNA molecule (translational coupling). A study of this possible function for translational control of the *faeD* gene has shown that if translation of the 3' end of *faeC* is prevented, the synthesis of FaeD is descreased. Most likely, the ribosome binding site of *faeD* is masked by mRNA secondary structure if the 3' end of *faeC* is not translated (MOOI et al. 1986).

7 Biosynthesis of Fimbriae

The biosynthesis of fimbriae is rather complicated process which requires at least (a) the transport of fimbrial subunits across the cytoplasmic membrane, the periplasm, and the outer membrane; (b) the polymerization of fimbrial subunits at the cellular surface, and (c) the anchoring of fimbriae to the cell envelope. Furthermore, it seems likely that the bacteria are able to regulate the length of their fimbriae. Functionally, the capacity of fimbriae to adhere to their respective receptors on the intestinal mucosa or the underlying enterocytes may be encoded by the fimbrial subunits or specific adhesin molecules attached to the fimbrial structure.

Analysis of the gene clusters responsible for the biosynthesis of various fimbriae has shown that these clusters indeed contain much more information than is required for the production of fimbrial subunits, and in general at least five or six auxiliary or "helper" proteins with distinct functions are encoded by the various genetic determinants for fimbrial production. The subcellular location of the proteins encoded by the K88 and K99 operons has been determined (JACOBS 1987; VAN DOORN et al. 1982).

Comparison of the gene clusters (Fig. 1) shows that each cluster analyzed so far contains a gene encoding a large outer membrane protein (molecular mass of about 85000). This protein may have several functions in the biosynthesis of fimbriae: as a pore protein that facilitates the translocation of fimbrial subunits across the outer membrane, as an anchor for assembled fimbriae, or as a sort of catalyst for subunit polymerization. Mutants impaired in the synthesis of this protein do not possess fimbriae and accumulate fimbria precursor complexes in their periplasm (MOOI et al. 1982, 1983). The topology of the protein in the outer membrane is not known, but analysis of the secondary structure of the FaeD protein indicates a large central region in the protein with an alternation of turns and β sheets that may form loops in the membrane bilayer (MOOI et al. 1986). Comparable features have been found in the large outer membrane protein FanD encoded by the K99 operon (E. ROOSENDAAL, unpublished results). Both proteins exhibit considerable amino acid sequence homology.

A second gene present in all operons described encodes a smaller periplasmic protein (molecular weight about 26000, Fig. 1) which probably also has several functions in fimbrial biogenesis. Not only do mutations in this gene abolish the formation of fimbriae, but in the absence of this protein none of the fimbrial subunits are detectable in the periplasm (DE GRAAF and KLAASEN 1986; DE GRAAF et al. 1984; JACOBS 1987; KLEMM et al. 1985; MOOI et al. 1983; ORNDORFF and FALKOW 1984a). Apparently, this protein is involved in the stabilization of nonpolymerized fimbrial subunits. Complexes of FaeE and FaeG have been isolated from K88-producing cells (MOOI et al. 1983). Also the K99 fimbrial subunits (FanC) appear to form a complex with the periplasmic protein FanE encoded by the K99 operon. Immunoprecipitation of ultrasonic extracts of K99-producing cells with anti-K99 antiserum results in a coprecipitation of FanE with the fimbrial subunits (DE GRAAF et al. 1984). The interaction of the 26-kd protein with other proteins is not restricted to the major fimbrial subunit because in the absence of the 26-kd protein other minor components of the fimbriae are also no longer detectable (JACOBS 1987; NORGREN et al. 1984). Apparently, all proteins that are translocated through the cell envelope and assembled into the fimbrial structure are protected by the 26-kd protein. We assume that this periplasmic protein may have the following functions: (a) to prevent premature subunit polymerization in the periplasmic space, (b) to act as a shuttle that transports the major and minor fimbrial subunits from the inner to the outer membrane, and (c) to deliver the energy necessary for transport and/or polymerization of subunits at the outer membrane. The protein may be considered as an "unfoldase" which keeps the fimbrial subunits in an energetically unfavourable extended conformation necessary for export across the outer membrane and the delivery of the conformational energy during the transport and/or polymerization process.

In principle, the synthesis of fimbrial subunits together with the periplasmic carrier protein and the outer membrane pore protein might be sufficient for the biogenesis of fimbriae, as has been observed for type 1 and P fimbriae (KLEMM and CHRISTIANSEN 1987; MAURER and ORNDORFF 1985; NORGREN et al. 1984). Such fimbriae, however, are longer than the fimbriae produced by cells containing the complete genetic determinant (KLEMM and CHRISTIANSEN 1987; MAURER and ORNDORFF 1987), indicating that other proteins are required to regulate fimbrial length.

Another group of proteins which play a role in the biosynthesis of fimbriae are the so-called fimbrial subunit-like proteins or minor fimbrial subunits, which are supposed to act as initiators or terminators of subunit polymerization. These proteins show a substantial amino acid sequence homology with the major fimbrial subunit, they have a comparable molecular weight, and in some cases they were detected as minor components of fimbriae.

For type 1 fimbriae, the two minor fimbrial subunits FimF (PilD) and FimG (PilF), have been identified. The FimG (PilF) protein regulates the length of the fimbriae. Mutants with a lesion in this gene produce very long fimbriae (MAURER and ORNDORFF 1987; KLEMM and CHRISTIANSEN 1987). The easiest explanation for the regulatory function of this protein is that subunit polymerization stops upon incorporation of a FimG (PilF) molecule in the fimbrial structure. The length of the fimbriae would depend upon the ratio of FimA (PilA) and FimG (PilF) molecules available. In this model FimG (PilF) acts as a competitive inhibitor of fimbrial polymerization: inactivation of the protein results in longer fimbriae, whereas overproduction of the protein results in shorter fimbriae. A similar model has been proposed for the PapH protein, which modulates the length of P fimbriae (BAGA et al. 1987). Little information is available for the role of the second minor fimbrial subunit FimF (PilD). We assume that this minor component acts as an initiator of polymerization and will discuss this possibility in relation to the expression of another gene contained in the genetic determinant of the various fimbriae and encoding a 30-kd protein.

The K99 operon contains two similar genes, designated *fanG* and *fanH*, which encode fimbrial subunit-like proteins showing significant amino acid sequence homology with the major fimbrial subunit FanC (ROOSENDAAL et al. 1987b). Mutational inactivation of *fanG* or *fanH* resulted in a simultaneous decrease in the production of K99 fimbriae and in adhesive capability. FanG and FanH are associated with the bacterial outer membrane but have not been detected so far in purified fimbriae. Their role in K99 biosynthesis is not known but we assume that these proteins are functionally comparable to the minor subunit components of type 1 and P fimbriae.

Analysis of the K88ab determinant contained in pFM205 revealed the presence of *faeH* encoding a polypeptide showing significant sequence homology with the major fimbrial subunit FaeG (D. Bakker, unpublished results). Mutational inactivation of *faeH* results in a simultaneous decreased in the production of K88 fimbriae and adhesive capability (MOOI et al. 1982). A second fimbrial subunit-like protein (FaeI) is probably encoded by the K88 operon but the reading frame of the corresponding gene (*faeI*) is only partially contained in pFM205 (Fig. 1).

As depicted in Fig. 1, all gene clusters analyzed contain an open reading frame encoding a protein of about 30-kd. The *fimH* (*pilE*) gene contained in the genetic determinant for the expression of type 1 fimbriae appeared to be responsible for the mannose-specific adherence of these fimbriae (KLEMM and CHRISTIANSEN 1987; MAURER and ORNDORFF 1985, 1987). Overproduction of the mannose-specific adhesin FimH results in the formation of 10-nm-diameter rounded structures (fimbriosomes) composed of FimH protein and detectable in association with type 1 fimbriae or in the culture medium (ABRAHAM et al. 1988). Mutants with a lesion in *fimH* (*pilE*) are adhesion-negative and fail to agglutinate erythrocytes. Furthermore, mutational inactivation of FimH (type 1) reduces the amount of fimbriae produced (KLEMM and CHRISTIANSEN 1987), indicating that the protein is involved in the biogenesis of fimbriae. A similar role as described for FimH has been found for the PapG protein,which could be identified as the Gal α (1 → 4) Gal-specific adhesin of P fimbriae (LUND et al. 1987). Mutants defective in the synthesis of the 30-kd FanF (K99) protein are unable to agglutinate horse erythrocytes but are also strongly hampered in the synthesis of K99 fimbriae (JACOBS 1987). Similarly, mutation in the 33-kd protein encoded by the 987P gene cluster results in a strong reduction of 987P production (DE GRAAF and KLAASEN 1986). Unfortunately, the adhesive capability of this mutant was not tested.

The available data clearly indicate that at least some fimbriae, i.e., 1 and P, carry separate adhesin molecules, while in other cases, i.e., K88 and K99 fimbriae, such indications have not yet been obtained. Studies on the structure–function relationship of K88 and K99 fimbrial subunits have indicated that the adhesive capability of these fimbriae is an instrinsic property of the major fimbrial subunit. Three different tripeptides, Ile-Ala-Phe, Ser-Leu-Phe, and Ala-Ile-Phe, derived from the K88 fimbrial subunit, were found to inhibit the binding of K88 fimbriae to guinea pig erythrocytes or pig intestinal epithelial cells in vitro (JACOBS et al. 1987c). Subsequently, it has been shown that in vivo, phenylalanine residue 150 plays an essential role in the binding of K88 fimbrial subunits to guinea pig erythrocytes (JACOBS et al. 1987a). Recently, however, indications were obtained that the high affinity binding of K88ab fimbriae to chicken erythrocytes is not mediated by the fimbrial subunits but most likely by one of the minor components of the K88 fimbriae, either FaeC, or FaeF (D. BAKKER, unpublished results). In vitro mutagenesis of the K99 fimbrial subunit Fan C has demonstrated that lysine residue 132 and arginine residue 136 are essential for the adhesive capability of these fimbriae (JACOBS et al. 1987b).

In this context, the question arises whether (a) the adhesive capability of the 30-kd minor components of type 1 and P fimbriae represents their unique biologic activity or (b) the 30-kd proteins encoded by all gene clusters share another common function in the biosynthesis of fimbriae. Comparison of the primary structure of the 30-kd proteins encoded by the type 1, K99, and P determinants revealed that these proteins are about twice the size of fimbrial subunits and contain four cysteine residues. The distance between two of these residues in the C-terminal domain of the protein is similar to the distance found for the two cysteine residues present in the various fimbrial subunits. Furthermore, the C-terminal domain of the 30-kd proteins shows substantial sequence homology with major and minor fimbrial subunits. We suppose that each of the different fimbrial subunits encoded by a particular operon possesses a distinct but different affinity to each of the other fimbrial subunits and to the 30-kd protein of that operon. In addition it is likely that major and minor fimbrial subunits and the 30-kd protein possess affinity for the pore protein which is involved in their translocation to the cellular surface.

On the basis of these assumptions and the experimental data available, we propose the following model for the biosynthesis of fimbriae:

1. Major and minor fimbrial subunits as well as the 30-kd protein form a complex with the periplasmic 26-kd carrier protein upon their translocation through the cytoplasmic membrane. This interaction prevents preliminary polymerization of subunits and keeps the various polypeptides in a conformation favorable to their translocation across the outer membrane. Furthermore, the interaction with the carrier protein prevents the proteolytic degradation of "unfolded" fimbrial subunits.

2. The 30-kd protein possesses affinity for the outer membrane pore protein and is the first protein to be inserted in the fimbrial growth point. Once detached from the carrier protein and delivered to the pore protein 30-kd protein blocks this pore in the absence of a minor fimbrial subunit, probably because of the formation of disulfide bridges, which changes the conformation of the 30-kd protein to an export-unfavorable form.

   Indications for such a possibility can be derived from an experiment described by KLEMM and CHRISTIANSEN (1987), who observed that fimbriated cells harboring a plasmid pPK5 encoding *fimA, fimC,* and *fim D* became bald when another plasmid was introduced in these cells, coding for the 30-kd FimH protein only.

3. In the presence of a minor fimbrial subunit with high affinity for the C-terminal domain of the 30-kd protein, the transition to an export-unfavorable conformation is prevented and the pore remains open for transport of fimbrial subunits. This minor fimbrial subunit may be considered as an initiator protein, which together with the 30-kd protein forms an initiation complex.

4. Interaction between the 30-kd protein and one of the minor fimbrial subunits does not appear to be sufficient to allow efficient export of the major fimbrial subunits, possibly because the N-terminal domain of the 30-kd protein still interacts with the pore protein at the cell surface and/or because the affinity of the major fimbrial subunit to this initiation complex is relatively low. A second minor fimbrial subunit with a comparable affinity for the first minor as well as the major fimbrial subunit is required to detach the initiation complex from the pore protein and to keep the pore open for subsequent export of the major fimbrial subunits. In addition, this second minor fimbrial subunit may facilitate the translocation of major fimbrial subunits through the outer membrane. The initiation complex remains attached to the tip of the fimbrial structure. It seems likely that the presence of the minor fimbrial subunits in the fimbrial growth point allow not only efficient transport of major fimbrial subunits but also the transport and polymerization of the 30-kd protein depending on the relative concentration of 30-kd protein and major fimbrial subunits. This situation may result in the formation of fimbriosomes (ABRAHAM et al. 1988). Furthermore, the concentration of minor fimbrial subunits may be high enough to allow their integration into the fimbrial structure (ABRAHAM et al. 1987; LINDBERG et al. 1987).

5. As soon as another 30-kd protein or an initiation complex is delivered to a functional growth point we presume that it prevents further subunit export. The

C-terminal domain of the 30-kd protein or the attached initiator protein interacts with the last major fimbrial subunit inserted in the growth point. Further translocation of the 30-kd protein is arrested because it attaches to the growth point in the opposite orientation. In this model the 30-kd protein is a multifunctional protein. In the wild-type situation this protein is part of the initiation complex but it also acts as terminator of subunit polymerization.

The number and length of fimbriae produced depend entirely on the concentration of major subunits versus the other (auxiliary) proteins which are all equally essential for efficient synthesis of fimbriae. This distinction between major subunits on the one hand and other proteins on the other is reflected in the regulation of expression of the fimbrial gene clusters. Mutations is one of the minor fimbrial subunits or the 30-kd protein affect the regulation of the number and length of fimbriae produced. Inactivation of the 30-kd protein, for instance, may allow the synthesis of longer fimbriae at low efficiency because both initiation and termination of the polymerization are hampered. On the other hand overproduction of this protein should reduce the length of the fimbriae without having much effect on the number of fimbriae per cell.

Although our model presents a common pathway in fimbriae biosynthesis, it is likely that variations have developed in the course of the evolution. In some systems, e.g., as described for P fimbriae, an additional minor fimbrial subunit protein may act as terminator of polymerization, for instance the PapH protein encoded by the Pap operon (BAGA et al. 1987). In other systems the expression of major fimbrial subunits may be hampered and the complex of the 30-kd protein with the minor fimbrial subunits remains in contact with the outer membrane, where it serves as nonfimbrial adhesin. An interesting variation on this model is the biosynthesis of K88 fimbriae. In the K88 operon, the open reading frame *faeC* located at the position of the major fimbrial subunit gene in other operons, is expressed at a very low level, while *faeG,* corresponding to the position of a minor fimbrial subunit, is expressed at a high level (MOOI et al. 1984). Consequently, K88 fimbriae may be considered as composed of lectin molecules or of "originally" minor fimbrial subunits which have taken over the function of FaeC. The "originally" major fimbrial subunit (FaeC) is located at the tip of the K88 fimbriae (B. Oudega, unpublished results). Possibly FaeC is responsible for the high affinity of K88 fimbriae to chicken erythrocytes while the major fimbrial subunits (FaeG) confer the adhesive capability for guinea pig erythrocytes.

It seems likely that the variations on the "common" pathway described have developed by a natural selection of mutants with a more efficient and specific adherence for a particular host epithelium.

References

Abraham, JM, Freitag, CS, Clements, JR, Eisenstein BI (1985) An invertible element of DNA controls phase variation of type 1 fimbriae of *Escherichia coli.* Proc Natl Acad Sci USA 82: 5724–5727
Abraham SN, Goguen JD, Sun D, Klemm P, Beachey EH (1987) Identification of two ancillary subunits

of *Escherichia coli* type 1 fimbriae by using antibodies against synthetic oligopeptides of *fim* gene products. J Bacteriol 169: 5530–5536

Abraham SN, Goguen JD, Beackey EH (1988) Hyperadhesive mutants of type 1-fimbriated *Escherichia coli* associated with formation of FimH organelles (fimbriosomes). Infect Immun 56: 1023–1029

Anderson DG, Mosely SL (1988) *Escherichia coli* F41 adhesin: genetic organization, nucleotide sequence, and homology with the K88 determinant. J Bacteriol 170: 4890–4896

Baga M, Göransson M, Normark S, Uhlin BE (1985) Transcriptional activation of a Pap pilus virulence operon from uropathogenic *Escherichia coli* EMBO J 4: 3887–3893

Baga M, Norgren M, Normark S (1987) Biogenesis of E. coli Pap pili: Pap H, a minor pilin subunit involved in cell anchoring and length modulation. Cell 49: 241–251

Bak AL, Christiansen G, Christiansen C, Stenderup A, Orskov I, Orskov F (1972) Circular DNA molecules controlling synthesis and transfer of the surface antigen (K88) in *Escherichia coli.* J Gen Microbiol 73: 373–385

Baldini MM, Kap JB, Levine MM, Candy DCA, Moon HW (1983) Plasmid-mediated adhesion in enteropathogenic *Escherichia coli.* J Pediatr Gastroenterol Nutr 2: 534–538

Baldini MM, Nataro JP, Kaper JB (1986) Localization of a determinant for HEp-2 adherence by enteropathogenic *Escherichia coli.* Infect Immun 52: 334–336

Boylan M, Smyth CJ (1985) Mobilization of CS fimbriae-associated plasmids of enterotoxigenic *Escherichia coli* of serotype O6:K15:H16 or H- into various wild-type hosts. FEMS Microbiol Lett 29: 83–89

Boylan M, Coleman DC, Smyth CJ (1987) Molecular cloning and characterization of the genetic determinant encoding CS3 fimbriae of enterotoxigenic *Escherichia coli.* Microb Pathogen 2: 195–209

Boylan M, Smyth CJ, Scott JR (1988) Nucleotide sequence of the gene encoding the major subunit of CS3 fimbriae of enterotoxigenic *Escherichia coli.* Infect Immun 56: 3297–3300

Brinton CC (1959) Non-flagellar appendages of bacteria. Nature 183: 782–786

Brinton CC, Gemski P, Falkow S, Baron LS (1961) Location of the piliation factor on the chromosome of *Escherichia coli.* Biochem Biophys Res Commun 5: 293–298

Bores A, Fairbrother JM, Jacques M, Larivière S (988) Isolation and characterization of a new fimbrial antigen (CS1541) from a porcine enterotoxigenic *Escherichia coli* 08:KX105 strain. FEMS Microbiol Lett 55: 341–348

Craig NL (1985) Site-specific inversion: enhancers, recombination proteins and mechanism. Cell 41: 649–650

Cravioto A, Scotland SM, Rowe B (1982) Hemagglutination acitivity and colonization factor antigens I and II in enterotoxigenic and non-enterotoxigenic strains of *Escherichia coli* isolated from humans. Infect Immun 36: 189–197

De Graaf FK, Klaasen P (1986) Organization and expression of genes involved in the biosynthesis of 987P fimbriae, MGG 204: 75–81.

De Graaf FK, Klaasen P (1987) Nucleotide sequence of the gene encoding the 987P fimbrial subunit of *Escherichia coli.* FEMS Microbiol Lett 42: 253–258

De Graaf FK, Mooi FR (1986) The fimbrial adhesins of *Escherichia coli.* Adv Microbiol Physiol 28: 65–143.

De Graaf FK, Krenn BE, Klaasen P (1984) Organization and expression of genes involved in the biosynthesis of K99 fimbriae. Infect Immun 43: 508–514

Dorman CJ, Higgins CF (1987) Fimbrial phase variation in *Escheriachia coli*: dependence on integration host factor and homologies with other site-specific recombinases. J Bacteriol 169: 3840–3843

Eisenstein BI (1981) Phase variation of type 1 fimbriae in *Escherichia coli* is under transcriptional control. Science 214: 337–339

Eisenstein BI, Dodd DC (1982) Pseudocatabolite repression of type 1 fimbriae of *Escherichia coli.* J Bacteriol 151: 1560–1567

Eisenstein BI, Sweet DS, Vaughn V, Friedman D (1987) Integration host factor is required for the DNA inversion that controls phase variation in *Escherichia coli.* Proc Natl Acad Sci USA 84: 6506–6510

Evans DG, Silver RP, Evans DJ, Chase DG, Gorbach SL (1975) Plasmid-controlled colonization factor associated with virulence in *Escherichia coli* enterotoxigenic for humans. Infect Immun 12: 656–667

Fairbrother JM, Larivière S, Lallier R (1986) New fimbrial antigen F165 from *Escherichia coli* serogroup O115 strains isolated from piglets with diarrhea. Infect Immun 51: 10–15

Freitag CS, Eisenstein BI (1983) Genetic mapping and transcriptional orientation of the *fimD* gene. J Bacteriol 156: 1052–1058

Freitag CS, Abraham JM, Clements JR, Eisenstein BI (1985) Genetic analysis of the phase variation control of expression of type 1 fimbriae in *Escherichia coli.* J Bacteriol 162: 668–675

Friedman DI, Gottesman M (1983) Lytic mode of lambda development. In: Hendrix RW, Roberts JW, Stahl FW, Weisberg RA (eds) Lambda II. Cold Spring Harbor Lab, New York, pp 21–51

Girardeau JP, Dubourguier HC, Contrepois M (1980) Attachement des *E. coli* entéropathogènes à la muqueuse intestinale. Bull Group Technol Vet 80-4-B-190: 49–59

Girardeau JP, Der Vartanian M, Ollier JL, Contrepois M (1988) CS31A, a new K88-related fimbrial antigen on bovine enterotoxigenic and septicemic *Escherichia coli* strains. Infect Immun 56: 2180–2188

Hinson G, Knutton S, Lam-Po-Tang KL, McNeish HS, Williams PH (1987) Adherence to human colonocytes of an *Escherichia coli* strain isolated from severe infantile enteritis: molecular and ultrastructural studies of a fibrillar adhesin. Infect Immun 55: 393–402

Honda T, Arita M, Miwatani T (1984) Characterization of new hydrophobic pili of human enterotoxigenic *Escherichia coli*: a possible new colonization factor. Infect Immun 43: 959–965

Isaacson RE (1980) Factors affecting expression of the *Escherichia coli* pilus K99. Infect Immun 28: 190–194

Jacobs AAC (1987) Structure-function relationships of the K88 and K99 fibrillar adhesins of enterotoxigenic *Escherichia coli*. Thesis, Krips Repro, Meppel, The Netherlands

Jacobs AAC, de Graaf FK (1985) Production of K88, and F41 fibrillae in relation to growth phase and a rapid procedure for adhesin purification. FEMS Microbiol Lett 26: 15–19

Jacobs AAC, Roosendaal B, van Breemen JFL, de Graaf FK (1987a) Role of phenylalanine 150 in the receptor-binding domain of the K88 fibrillar subunit. J Bacteriol 169: 4907–4911

Jacobs AAC, Simons BH, de Graaf FK (1987b) The role of lysine-132 and arginine-136 in the receptor-binding domain of the K99 fibrillar subunit. EMBO J 6: 1805–1808

Jacobs AAC, Venema J, Leeven R, van Pelt-Heerschap H, de Graaf FK (1987c) Inhibition of adhesive activity of K88 fibrillae by peptide derived from the K88 adhesin. J Bacteriol 169: 735–741

Kehoe M, Sellwood R, Shipley P, Dougan G (1981) Genetic analysis of K88-mediated adhesion of enterotoxigenic *Escherichia coli*. Nature 291: 122–126

Kehoe M, Winther M, Dougan G (1983) Expression of a cloned K88ac adhesion antigen determinant: identification of a new adhesion cistron and role of a vector-encoded promoter. J Bacteriol 155: 1071–1077

Klemm P (1986) Two regulatory *fim* genes, *fimB* and *fimE*, control the phase variation of type 1 fimbriae in *Escherichia coli*. EMBO J 5: 1389–1393

Klemm P, Christiansen G (1987) Three *fim* genes required for the regulation of length and mediation of adhesion of *Escherichia coli* type 1 fimbriae. MGG 208: 439–445

Klemm P, Jorgensen BJ, van Die I, de Ree H, Bergmans H (1985) The *fim* genes responsible for synthesis of type 1 fimbriae in *Escherichia coli*, cloning and genetic organization. MGG 199: 410–414

Knutton S, Baldini MM, Kaper JB, McNeish AS (1987c) Role of plasmid-encoded adherence factors in adhesion of enteropathogenic *Escherichia coli* to HEp-2 cells. Infect Immun 55: 78–85

Knutton S, Lloyd DR, McNeish AS (1987a) Adhesion of enteropathogenic *Escherichia coli* to human intestinal enterocytes and cultured human intestinal mucosa. Infect Immun 55: 69–77

Knutton S, Lloyd DR, McNeish AS (1987b) Identification of a new fimbrial structure in enterotoxigenic *Escherichia coli* (ETEC) serotype O148:H28 which adheres to human intestinal mucosa: a potentially new human ETEC colonization factor. Infect Immun 55: 86–92

Knutton S, Williams PH, Lloyd DR, Candy DCA, McNeish AS (1984) Ultrastructural study of adherence to and penetration of cultured cells by two invasive *Escherichia coli* strains isolated from infants with enteritis. Infect Immun 44: 599–608

Laporta MZ, Silva MLM, Scaletsky ICA, Trabulski LR (1986) Plasmids coding for drug resistance and localized adherence to Hela cells in enteropathogenic *Escherichia coli* O55:H- and O55:H6. Infect Immun 51: 715–717

Levine MM, Ristaino P, Marley G, Smyth C, Knutton E, Boedeker R, Black R, Young C, Clements ML, Cheney C, Patnaik R (1984) Coli surface antigens 1 and 3 of colonization factor antigen II-positive enterotoxigenic *Escherichia coli*: morphology, purification, and immune responses in humans. Infect Immun 44: 409–420

Lindberg F, Lund B, Johansson L, Normark S (1987) Localization of the receptor-binding protein adhesin at the tip of the bacterial pilus. Nature 328: 84–87

Lintermans P, Pohl P, Deboeck F, Bertels A, Schlicker C, VandeKerckhove J, van Damme J, van Montagu M, de Greve H (1988) Isolation and nucleotide sequence of the F17-A gene encoding the structural protein of the F17 fimbriae in bovine enterotoxigenic *Escherichia coli*. Infect Immun 56: 1475–1484

Lund B, Lindberg F, Marklund BI, Normark S (1987) The Pap G protein is the α-*D*-galactopyranosy-(1 → 4)β-D-galactopyranose-binding adhesin of uropathogenic *Escherichia coli*. Proc Natl Acad Sci USA 84: 5898–5902

Manning PA, Timmis KN, Stevenson G (1985) Colonization factor antigen II (CFA/II) of enterotoxigenic *Escherichia coli*: molecular cloning of the CS3 determinant. MGG 200: 322–327

Maurer L, Orndorff PE (1985) A new locus, *pilE*, required for the binding of type 1 piliated *Escherichia coli* to erythrocytes. FEMS Microbiol Lett 30: 59–66

Maurer L, Orndorff PE (1987) Identification and characterization of genes determining receptor binding and pilus length of *Escherichia coli* type 1 pili. J Bacteriol 169: 640–645

McConnell MM, Smith HR, Willshaw GA, Field AM, Rowe B (1981) Plasmids coding for colonization factor antigen I and heat-stable enterotoxin production isolated from enterotoxigenic *Escherichia coli*: comparison of their properties. Infect Immun 32: 927–936

McConnel MM, Thomas LV, Day NP, Rowe B (1985) Enzyme-linked immunosorbent assays for the detection of adhesion factor antigens of enterotoxigenic *Escherichia coli*. J Infect Dis 152: 1120–1127

McConell MM, Thomas LV, Scotland SM, Rowe B (1986) The possession of coli surface antigen CS6 by enterotoxigenic *Escherichia coli* of serogroups O25, O27, O148 and O159: a possible colonisation factor? Curr Microbiol 14: 51–54

McConnell MM, Thomas LV, Willshaw GA, Smith HR, Rowe B (1988) Genetic control and properties of coli surface antigens of colonization factor antigen IV (PCF 8775) of enterotoxigenic *Escherichia coli*. Infect Immun 56: 1974–1980

Minion FC, Abraham SN, Beachey EH, Goguen JD (1986) The genetic determinant of adhesive function in type 1 fimbriae of *Escherichia coli* is distinct from the gene encoding the fimbrial subunit. J Bacteriol 165: 1033–1036

Mooi FR, De Graaf FK, van Embden JDA (1979) Cloning, mapping and expression of the genetic determinant that encodes for the K88ab antigen. Nucleic Acids Res. 6: 849–865

Mooi FR, Harms N, Bakker D, De Graaf FK (1981) Organization and expression of genes involved in the production of the K88ab antigen. Infect Immun 32: 1155–1163

Mooi FR, Wouters C, Wijfjes A, De Graaf FK (1982) Construction and characterization of mutants impaired in the biosynthesis of the K88ab antigen. J Bacteriol 150: 512–521

Mooi FR, Wijfjes A, De Graaf FK (1983) Identification and characterization of precursors in the biosynthesis of the K88ab fimbria of *Escherichia coli*. J Bacteriol 154: 41–49

Mooi FR, van Buuren M, Koopman G, Roosendaal B, De Graaf FK (1984) A K88ab gene of *Escherichia coli* encodes a fimbria-like protein distinct from the K88ab fimbrial adhesin. J Bacteriol 159: 482–487

Mooi FR, Claasen I, Bakker D, Kuipers H, De Graaf FK (1986) Regulation and structure of an *Escherichia coli* gene coding for an outer membrane protein involved in export of the K88ab fimbrial subunits. Nucleic Acids Res 14: 2443–2457

Morris JA, Thorns CJ, Scott AC, Sojka WJ (1982) Adhesive properties associated with the Vir plasmid: a transmissible pathogenic characteristic associated with strains of invasive *Escherichia coli*. J Gen Microbiol 128: 2097–2103

Morris JA, Chanter N, Sherwood D (1987) Occurrence and properties of FY (Att25)[+] *Escherichia coli* associated with diarrhoea in calves. Vet Rec 121: 189–191

Morrissey PM, Dougan G (1986) Expression of a cloned 987P adhesion-antigen fimbrial determinant in *Escherichia coli* K-12 strain HB101. Gene 43: 79–84

Moseley SL, Dougan G, Schneider RA, Moon HW (1986) Cloning of chromosomal DNA encoding the F41 adhesin of enterotoxigenic *Escherichia coli* and genetic homology between adhesins F41 and K88. J Bacteriol 167: 799–804

Mullany P, Field AM, McConnell MM, Scotland SM, Smith HR, Rowe B (1983) Expression of plasmids coding for colonization factor antigen II (CFA/II) and enterotoxin production in *Escherichia coli*. J Gen Microbiol 129: 3591–3601

Murray BE, Evans DJ, Penaranda ME, Evand DG (1983) CFA/I-ST plasmids: comparison of enterotoxigenic *Escherichia coli* (ETEC) of serogroups O25, O63, O78 and O128 and mobilisation from an R-factor-containing epidemic ETEC isolate. J Bacteriol 153: 566–570

Nagy B, Moon HW, Isaacson RE (1977) Colonization of porcine intestine by enterotoxigenic *Escherichia coli*: selection of piliated forms in vivo, adhesion of piliated forms to epithelial cells in vitro, and incidence of pilus antigen among porcine enteropathogenic *E. coli*. Infect Immun 16: 344–352

Nandadasa HG, Sargent GF, Brown MGM, McNeish AG, Williams PH (1981) The role of plasmids in adherence of invasive *Escherichia coli* to mammalian cells. J Infect Dis 143: 286–290

Nataro JP, Scaletsky ICA, Kaper JB, Levine MM, Trabulski LR (1985) Plasmid-mediated factors conferring diffuse and localized adherence of enteropathogenic *Escherichia coli*. Infect Immun 48: 378–383

Norgren M, Normark S, Lark D, O'Hanley P, Schoolink G, Falkow S, Svanbor-Eden C, Baga M, Uhlin BE (1984) Mutations in *E. coli* cistrons affecting adhesion to human cells do not abolish Pap pili fiber formation. EMBO J 3: 1159–1165

Ollier JL, Girardeau JP (1983) Structures et fontions hémagglutinantes des pili K99 dont la biosynthèse est glucose-dépendante ou constitutive et des pili F41. Ann Microbiol 134A: 247–254

Orndorff PE, Falkow S (1984a) Organization and expression of genes responsible for type 1 piliation in *Escherichia coli*. J Bacteriol 159: 736–744

Orndorff PE, Falkow S (1984b) Identification and characterization of a gene product that regulates type 1 piliation in *Escherichia coli*. J Bacteriol 160: 61–66

Orskov I, Orskov F (1966) Episome-carried surface antigen K88 of *Escherichia coli*. I. Transmission of the determinant of the K88 antigen and influence on the transfer of chromosomal markers. J. Bacteriol 91: 69–75

Pabo CO, Sauer RT (1984) Protein-DNA interactions. Annu Rev Biochem 53: 293–321

Penaranda ME, Mann MB, Evans DG, Evans DJ (1980) Transfer of an ST:LT:CFA/II plasmids into *Escherichia coli* K-12 strain RR1 by cotransformation with PSC 301 plasmid DNA. FEMS Microbiol Lett 8: 251–254

Penaranda ME, Evans DG, Murray BE, Evans DJ (1983) ST:LT:CFA/II plasmids in enterotoxigenic *Escherichia coli* belonging to serogroups O6, O8, O80, O85 and O139. J Bacteriol 154: 980–983

Plasterk RHA, Brinkman AD, van der Putte P (1983) DNA inversions in the chromosome of *Escherichia coli* and in bacteriophage Mu: relationships to other site-specific recombination systems. Proc Natl Acad Sci USA 80: 5355–5358

Platt T (1986) Transcription termination and the regulation of gene expression. Annu Rev Biochem 55: 339–372

Pohl P, Lintermans P, van Muylen K, Schotte M (1982) Colibacilles entérotoxigènes du veau possédant in antigène d'attachment différent de l'antigène K99. Annu Med Vet 126: 569–571

Reis MHL, Affonso MHT, Trabulski LR, Mazaitis AJ, Maas R, Maas WK (1980) Transfer of a CFA/I-ST plasmid promoted by a conjugative plasmid in a strain of *Escherichia coli* of serotype O128ac:H12. Infect Immun 29:140–143

Rhen M, Mäkelä PH, Korhonen TK (1983) P-fimbriae of *Escherichia coli* are subject to phase variation. FEMS Microbiol Lett 19: 267–271.

Roosendaal B (1987) A molecular study on the biogenesis of K99 fimbriae. Thesis, Tessel Offset, Utrecht

Roosendaal E, Gaastra W, De Graaf FK (1984) The nucleotide sequence of the gene encoding the K99 subunit of enterotoxigenic *Escherichia coli*. FEMS Microbiol Lett 22: 253–258

Roosendaal E, Boots M, De Graaf FK (1987a) Two novel genes, *fanA*, and *fanB*, involved in the biogenesis of K99 fimbriae. Nucleic Acids Res 15:5973–5984

Roosendaal E, Jacobs AAC, Rathman P, Sondermeyer C, Stegehuis F, Oudega B, De Graaf FK (1987b) Primary structure and subcellular localization of two fimbrial subunit-like proteins involved in the biosynthesis of K99 fibrillae. Mol Microbiol 1: 211–217

Roosendaal B, Damoiseaux J, Jordi W, de Graaf FK (1988) Transcriptional organization of the DNA region controlling expression of the K99 gene cluster, MGG (in press)

Scaletsky ICA, Milani SR, Trabulsi LR, Travassos LR (1988) Isolation and characterization of the localized adherence factor of enteropathogenic *Escherichia coli*. Infect Immun 56: 2979–2983

Schmitt R, Mattes R, Schmid K, Altenbuchner J (1979) Raf plasmids in strains of *Escherichia coli* and their possible role in enteropathogeny. In: Timmis KN, Pühler A (eds) Plasmids of medical, environmental and commercial importance.Elsevier, North-Holland Biomedical, New York, pp 199–210

Scotland SM, McConnell MM, Willshaw GA, Rowe B, Field AM (1985) Properties of wild-type strains of enterotoxigenic *Escherichia coli* which produce colonization factor antigen II, and belonging to serogroups other than O6. J Gen Microbiol 131: 2327–2333

Shipley PL, Gyles CL, Falkow S (1978) Characterization of plasmids that encode for the K88 colonization antigen. Infect Immun 20: 559–566

Shipley PL, Dougan G, Falkow S (1981) Identification and cloning of the genetic determinant that encodes for the K88ac adherence antigen. J Bacteriol 145: 920–925

Smith HR, Cravioto A, Willshaw GA, McConnell MM, Scotland SM, Gross RJ, Rowe B (1979) A plasmid coding for the production of colonisation factor antigen I and heat-stable enterotoxin in strains of *Escherichia coli* of serogroup O78. FEMS Microbiol Lett 6: 255–260

Smith HR, Willshaw GA, Rowe B (1982) Mapping of a plasmid, coding for colonization factor antigen I and heat-stable enterotoxin production, isolated from an enterotoxigenic strain of *Escherichia coli*. J Bacteriol 149: 264–275

Smith HR, Scotland SM, Rowe B (1983) Plasmids that code for production of colonization factor antigen II and enterotoxin production in strains of *Escherichia coli*. Infect Immun 40: 1236–1239

Smith HW, Linggood MA (1972) Further observations on *Escherichia coli* enterotoxins with particular regard to those produced by a typical piglet strains and by calf and lamb strains. The transmissible nature of these enterotoxins and of a K antigen possessed by calf and lamb strains. J Med Microbiol 5: 243–250

Smith HW, Parsell Z (1975) Transmissible substrate-utilising ability in enterobacteria. J Gen Microbiol 87: 129–140

Smyth CJ (1982) Two mannose-resistant haemagglutinins on enterotoxigenic *Escherichia coli* of serotype O6:K15:H16 or H-isolated from travellers' and infantile diarrhoea. J Gen Microbiol 128: 2081–2096

Smyth CJ (1984) Serologically distinct fimbriae on enterotoxigenic *Escherichia coli* of serotype O6:K15:H16 or H-. FEMS Microbiol Lett 21: 51–57

So M, Crandall JF, Crosa JH, Falkow S (1974) Extrachromosomal determinants which contribute to bacterial pathogenicity. In: Schlessinger D (ed) Microbiology 1974. American Society for Microbiology, Washington, pp 16–26

Swaney LM, Liu Y-P, Ippen-Ihler K, Brinton CC (1977) Genetic complementation analysis of *Escherichia coli* type 1 somatic pilus mutants. J Bacteriol 130: 506–511

Tacket CO, Maneval DR, Levine MM (1987) Purification, morphology, and genetics of a new fimbrial putative colonization factor of enterotoxigenic *Escherichia coli* O159:H4. Infect Immun 55: 1063–1069

Thomas LV, Cravioto A, Scotland SM, Rowe B (1982) New fimbrial antigenic type (E8775) that may represent a colonization factor in enterotoxigenic *Escherichia coli* in humans. Infect Immun 35: 1119–1124

Thomas LV, McConnell MM, Rowe B, Field AM (1985) The possession of three novel coli surface antigens by enterotoxigenic *Escherichia coli* strains positive for the putative colonization factor PCF 8775. J Gen Microbiol 131: 2319–2326

Thomas LV, Rowe B, McConnell MM (1987) In strains of *Escherichia coli* O167 a single plasmid encodes for the coli surface antigens CS5 and CS6 of putative colonization factor PCF8775, heat-stable enterotoxin, and colicin Ia. Infect Immun 55: 1929–1931

Van Doorn J, Oudega B, Mooi FR, De Graaf FK (1982) Subcellular localization of polypeptides involved in the biosynthesis of the K88ab fimbriae. FEMS Microbiol Lett 13: 99–104

Van Embden JDA, De Graaf FK, Schouls LM, Teppema JS (1980) Cloning and expression of a deoxyribonucleic acid fragment that encodes for the adhesive antigen K99. Infect Immun 29: 1125–1133

Van Verseveld HW, Bakker P, Van der Woude T, Terleth C, De Graaf FK (1985) Production of fimbrial adhesins K99 and F41 by enterotoxigenic *Escherichia coli* as a function of growth-rate domain. Infect Immun 49: 159–163

Williams PH, Knutton S, Brown MGM, Candy DCA, McNeish AS (1984) Characterization of nonfimbrial mannose-resistant protein hemagglutinins of two *Escherichia coli* strains isolated from infants with enteritis. Infect Immun 44: 592–598

Willshaw GA, Smith HR, McConnell MM, Barclay EA, Krnjulac J, Rowe B (1982) Genetic and molecular studies of plasmids coding for colonization factor antigen I and heat-stable enterotoxin in several *Escherichia coli* serotypes. Infect Immun 37: 858–868

Willshaw GA, Smith HR, Rowe R (1983) Cloning of regions encoding colonisation factor antigen I and heat-stable enterotoxin in *Escherichia coli*. FEMS Microbiol Lett 16: 101–106

Willshaw GA, Smith HR, McConnell MM, Rowe B (1985) Expression of cloned plasmid regions encoding colonization factor antigen I (CFA/I) in *Escherichia coli*. Plasmid 13: 8–16

Zieg J, Simon M (1980) Analysis of the nucleotide sequence of an invertible controlling element. Proc. Natl Acad Sci USA 77: 4196–4201

Nature and Organization of Adhesins

K. Jann and H. Hoschützky

1 Introduction

The adherence of pathogenic bacteria to epithelial surfaces is an important early even in their interaction with the host, leading to colonization and infection (Ofek and Beachey 1980; D.C. Savage 1985). This process provides the bacteria with several advantages, such as resistance to the rinsing and cleaning action of body fluids, evasion of host defenses, and improved acquisition of nutrients from their environment. It may also permit an optimal delivery of exotoxins. In extraintestinal infections the specific adhesion to certain tissues may result in an organ tropism of infectious bacteria (Nowicki et al. 1986).

The term adhesion, as used in this article, describes a specific interaction of surface-attached bacterial recognition proteins (adhesins) with the carbohydrate moiety of glycoproteins or glycolipids on mammalian cells. Since the receptors for bacteria are often also expressed on erythrocytes, hemagglutination is used as a simple in vitro assay for the adhesive properties of bacteria (Leffler and Svanborg-Edén 1986; Salit and Gotschlich 1977). Thus, adhesins may also be termed hemagglutinins but the more general term lectins is also used. It was found early that some agglutinations/adhesions can be inhibited with α-mannosides,which

Max-Planck-Institut für Immunbiologie, Stübeweg 51, D-7800 Freiburg, FRG

Current Topics in Microbiology and Immunology, Vol. 151
© Springer-Verlag Berlin·Heidelberg 1990

were termed mannose sensitive (MS). Others were not inhibitable with α-mannosides and were thus termed mannose resistant (MR) (SHARON and OFEK 1986).

The carbohydrate specificity of several MR adhesins could be defined in molecular terms: -α-Gal-(1,4)-β-Gal- is recognized by P-specific adhesins of uropathogenic *Escherichia coli* (VÄISÄNEN-RHEN et al. 1981), α-NeuNAc-(2,3 or 2,6)-β-Gal is recognized by S-specific adhesins of *E. coli* causing sepsis and neonatal meningitis (PARKKINEN et al. 1983), and β-GlcNAc-(1,3)-β-GlcNAc is recognized by G-specific adhesins of uropathogenic *E. coli* (VÄISINEN-RHEN et al. 1983). Some uropathogenic bacteria adhere to cells exhibiting an as yet unknown structural arrangement of galactose, sialic acid, and serine. Since the receptor for the latter bacteria is associated with the expression of the human M/N blood group system, their adhesins are termed M/N specific (VÄISÄNEN-RHEN et al. 1982). Human enteropathogenic *E. coli* adhere to cell linings of the small intestine with the aid of colonization factor antigens (mainly CFA/I and CFA/II) that interact with mucosal gangliosides: NeuNAc seems to be essential for the interaction (FARIS et al. 1980).

The adhesins which govern the interaction of bacteria with receptors on host cells are subject to regulation by growth conditions. Thus, most adhesins are expressed above a growth temperature of 20°C and not below. The expression of others is regulated by the presence of certain nutrients such as glucose or alanine (DE GRAAF et al. 1980; EISENSTEIN and DODD 1982). The mannose-specific (type 1) adhesins are produced in large amounts by bacteria grown in liquid culture and in only very small amounts in agar-grown bacteria (EISENSTEIN and DODD 1982); the expression of other fimbriae as well as nonfimbrial adhesins is more pronounced in agar-grown bacteria. Bacterial adhesins may be subject to variation between an expressed and a nonexpressed state with the genetic information, however, constantly present (EISENSTEIN 1981; RHEN et al. 1983).

The adhesiveness of bacteria is frequently correlated with the presence of proteinaceous fibrillar structures on the bacterial surface. These appendages, which are distinct from the sex pili and from the flagella, have been termed fimbriae or pili (BRINTON 1965; DUGUID et al. 1955; DUGUID and OLD 1980). Thus, fimbriae were assumed to be the mediators of adhesion. The nature of the actual recognition protein has been elusive for many years and has been described only recently for some fimbrial types (ABRAHAM et al. 1987a; HOSCHÜTZKY et al. 1989b; MANSON and BRINTON 1988; MOCH et al. 1987; ABRAHAM et al. 1987a). The nature of adhesins at large became seemingly more complex when bacteria were found that were adhesive but did not express fimbriae. It was possible to isolate proteins from these bacteria which agglutinated red blood cells and which did not form electron microscopically recognizable structures (DUGUID et al. 1979). It should, however, be pointed out that a distinction between fimbriae-associated (or fimbrial) and nonfimbrial adhesins may be superficial and in fact be due to the limitations of electron microscopic resolution. Thus, the adhesins of certain enteropathogenic *E. coli* have been described as nonfimbrial and some of them were later found to form very fine fimbrillae on the bacterial surfacc (ØRSKOV et al. 1985). We are using the terms fimbrial and nonfimbrial with this reservation.

Bacterial adhesins and their interactions with receptors on host cells have been studied with many bacterial genera (for reviews see MIRELMAN and OFEK 1986; ISAACSON 1985). In this article we restrict ourselves to the chemical and topographic characterization of the adhesins of *E. coli*, with special emphasis on P- and S-specific fimbriae and on (blood group M/N specific) nonfimbrial adhesins.

2 Characterization of Adhesins

2.1 Adhesins Associated with Fimbriae

Fimbriae can be detached in a morphologically and functionally unaltered form from the bacterial surface without damage to the cells. SDS–PAGE analysis revealed that fimbrial preparations consist of more than one protein (JANN et al. 1981; HOSCHÜTZKY et al. 1989a). For a long time it remained unclear whether the observed minor proteins were integral components of the fimbriae, or whether they were contaminating proteins. It was also not known which protein is the actual mediator of the adhesive properties. Genetic studies on P, S, and type 1 fimbriae showed that the phenotypes of fimbriation and adhesiveness are encoded by a complex gene cluster but determined by different genes (LINDBERG et al. 1984; HACKER et al. 1985; KLEMM and CHRISTIANSEN 1987; MINION et al. 1986; VAN DIE et al. 1985). This suggested that the protein mediating the adhesive properties of the bacteria is different from that forming the fimbriae. The first genetic system studied in detail was that of the gene complex encoding F13 (pap) fimbriae of uropathogenic *E. coli*. Expression of the genes in minicells showed that the gene complex directed the synthesis of several peptides (A, H, C, D, E, F, G, X, Y) (LINDBERG et al. 1986; NORGREN et al. 1984; TENNENT et al. 1987). Essential peptides concerning the fimbrial structure itself are encoded by the A, E, F, and G genes, whereas the C protein is an outer membrane protein, essential in the passage of the fimbriae through the outer membrane and their anchorage to the cell surface. The D protein seems to be a periplasmic protein, serving as a shuttle in the passage through the periplasm of all proteins forming adhesive fimbriae (TENNENT et al. 1987). It was found that the A peptide is the structural subunit of the fimbrium. Since F, G, and possibly E negative mutants were not adhesive, it was thought that these proteins are involved in the adhesive process. The question remained, however, of whether these proteins (or one of them) are directly responsible for receptor binding, or whether they modify the fimbrial subunits so that they acquire receptor binding properties (LINDBERG et al. 1986).

In our attempts to verify the genetic data on the protein level we have assumed as a working hypothesis that the major fimbrial subunit (A protein) and the adhesin are distinct proteins and that one of the minor proteins we normally detect in preparations of purified fimbriae is responsible for the adhesive properties. The expectation was that the molar ratio of the proteins must be in the order of one adhesin molecule to several hundred fimbrial subunits. We have developed an isolation procedure for these fimbriae-associated adhesins that in principle consists of the following steps: (a) detachment of fimbriae from the bacterial surface, (b) isolation of purified fimbriae, (c) dissociation of the adhesin from the fimbrial polymer, and (d) final purification of the adhesin by HPLC. This isolation procedure, which is described below, is generally applicable to all fimbriae which carry an adhesive protein different from the major subunit of the polymer (MOCH et al. 1987; HOSCHÜTZKY et al. 1989b).

2.1.1 Isolation of the Fimbriae–Adhesin Complex

The separation of the fimbriae–adhesin complex (FAC) from the bacteria can be achieved by shearing a suspension of fimbriated bacteria in an Omnimixer or a similar device, by their incubation with urea or octyl glucoside, or by heating to 50°–70°C. Since omnimixing concomitantly releases substantial amounts of outer membrane material and gives only a low yield of fimbriae, and detergents or chaotropes interfere with subsequent purification of FACs and their dissociation, heating of bacterial suspensions at low ionic strength to 65°C was the method of choice for the removal of FACs (MOCH et al. 1987; HOSCHÜTZKY et al. 1989a). After cooling, the bacteria are removed by centrifugation and the fimbriae are precipitated from the supernatant with 10% ammonium sulfate or with 500 mM LiCl. The presence of EDTA, glycine, and 5% sucrose prevents coprecipitation of bacterial outer membrane proteins. Lipopolysaccharide (LPS) is then removed from the FAC by incubation with 0.5% sodium desoxycholate at 80°C followed by ultracentrifugation. SDS–PAGE analysis at this point reveals the presence of major fimbrillar subunits and two or three additional faint protein bands. One of these has proved to be the respective adhesin. In the case of the genetically well-characterized F13 fimbriae, the other two proteins had molecular sizes (15.5 kd and 16 kd) (HOSCHÜTZKY et al. 1989b), corresponding to the E and F proteins proposed to be minor components of fimbriae (LINDBERG et al. 1986). In our isolation procedure the adhesin (G protein) and the E protein are isolated in a 1:1 ratio, while the F protein is usually isolated in a five- to tenfold lower concentration.

It should be noted that if the fimbriae are removed from the bacteria with octyl glucoside rather than by heating, a 80-kd protein is coextracted and can be obtained separately (HOSCHÜTZKY and JANN, unpublished). This seems to be the C protein (anchorage protein) proposed by NORMAK to be associated with the outer membrane (NORGREN et al. 1984).

The purified S- and P-specific FACs disperse readily in water at pH 7–8. They adhere to cultured cells of a human kidney cell line and to HeLa cells, as could be demonstrated by immunofluorescence using specific antibodies. Fimbriae isolated from recombinant E. coli strains expressing the adhesive negative phenotype, i.e., lacking the adhesive G protein, were not able to adhere to the tissue culture cells.

The purified S- and P-FACs adhered to different cells in the pH range from 5 to 9, and did so optimally at pH 7–8. However, hemagglutination was not observed with purified FACs at neutral pH. At pH values below 6, as well as in the presence of divalent cations, hemagglutination could be readily demonstrated. We interpret these observations as indicating that near neutrality the isolated FACs are dispersed and monovalent with respect to cell interaction. They can therefore adhere to cells but not agglutinate them. At lower pH as well as in the presence of divalent ions the FACs aggregate either isoelectrically or through ionic interactions. Thus they become polyvalent and able to agglutinate cells (HOSCHÜTZKY et al. 1989b). The adherence/agglutination could be specifically inhibited with the trisaccharide α-galactoside-(1,4)-lactose in the case of P-specificity (LEFFLER and SVANBORG-EDÉN 1986; HOSCHÜTZKY et al. 1989a) and with α-sialyl-(2,3/6)-lactose in the case of S-specificity (MOCH et al. 1987). Thus, the purified FACs had retained their specific receptor recognition.

2.1.2 Isolation of Adhesins from Fimbriae–Adhesin Complexes

During our studies we observed that the use of mild detergents like octyl glucoside released the FAC together with the 80-kd C protein. The latter protein was not detectable in FACs that had been released from the bacteria by mechanical treatment or by heat extraction. We were also able to demonstrate that with the use of detergents like SDS or some of the Zwittergents, fimbriae could be isolated that had lost their recognition protein and thus the ability to adhere to eukaryotic cells. Based on these observations we developed a method to specifically release the adhesin from the FAC without their dissociation into the major fimbrillar subunits. The best way to achieve this was heating of the purified FAC in the presence of 0.5% Zwittergent C-16 at 80°C. SDS also dissociates the FAC into adhesin and fimbriae. Since this detergent is difficult to remove and interferes with subsequent purification steps, it was not used further. After incubation with Zwittergent C-16 the fimbriae were precipitated with 500 mM LiCl. The supernatant contained the adhesin and the fimbrillar subunit in a ratio of 1:1 as estimated by SDS–PAGE. This is a 50- to 100-fold enrichment of the adhesin, as compared to the protein composition of intact FAC. The adhesins were concentrated from the supernatant of the detergent extraction either by precipitation with ice-cold ethanol/1-butanol or by dialysis against 50 mM NH_4HCO_3 followed by lyophilization. Subsequent high resolution ion exchange chromatography (HPLC) on a MonoQ or a PW 5-DEAE column was performed, preferentially after denaturation of the proteins with 8 M guanidine hydrochloride prior to the separation by HPLC. After treatment with guanidine the adhesins purified in our laboratory could be renatured to functional proteins.

2.1.3 Characterization of Fimbriae-Associated Adhesins

We have isolated and purified the adhesins from S fimbriae and from P fimbriae of different serotypes ($F7_1$, F8, F13) (MOCH et al. 1987; HOSCHÜTZKY et al. 1989b). The general properties of the proteins are summarized in Table 1. To demonstrate the adhesive specificity of the isolated adhesins, we used a hemadhesion assay, in which erythrocytes were incubated together with the immobilized adhesin in the presence or absence of specific inhibitors. The extent to which the red cells attached to the ELISA plate was determined by measuring hemoglobin at OD_{405}, after lysis of the specifically bound red cells. With this assay we were able to demonstrate that the purified adhesins had maintained their receptor-recognizing specificities.

The N-terminal sequence of the F13 (pap) adhesin (Table 1; HOSCHÜTZKY et al. 1989b) proved to be identical with the amino acid sequence derived from the DNA sequence (LUND et al. 1987). The N-terminal sequence of the $F7_1$ adhesin is not

Table 1. Properties of adhesins and fimbrillins from S and P fimbriae

FAC		Fimbrillin		Adhesin	
		MW (10^3 d)	pI	MW (10^3 d)	pI
S		16.5	6	14	4.7
F7	(P)	22	5	35	4.8–5
F8	(P)	18.5	5	35	4.8–5
F13	(P)	18	5	35	4.8–5

related to the F13 adhesin but is homologous with the adhesins from the P-specific F7$_2$ and F11 fimbriae (Lund et al. 1988). These data indicate that at least groups of recognition proteins exist; however, since they have the same Gal–Gal receptor binding specificity, they most probably have a conserved receptor binding site. Thus, the P-specific adhesins represent a group of related recognition proteins which have the binding region in common.

For an immunochemical and functional analysis of P and S fimbriae (FAC) and their corresponding adhesins, as well as for diagnostic purposes, we prepared monoclonal antibodies. Several antifimbrial antibodies reacted only with the homologous F type and others cross-reacted more or less extensively with fimbriae of other F types (Hoschützky et al. 1989b). Similarly, several antiadhesin antibodies reacted only with the homologous adhesin and a few reacted with all P or S adhesins isolated from the different FACs (H. Hoschützky et al., unpublished material). There was no immunologic cross-reactivity between fimbriae and their adhesins. The monoclonal antibodies reacting with all P or S adhesins were antiadhesive. First results indicate that the (ELISA) titer of their serologic cross-reactivity parallels their capacity to inhibit adherence/agglutination of target cells. This underlines the above interpretation that the adhesins may contain a conserved region which is the receptor binding site.

The question of the location of adhesins on the bacteria was addressed with the aid of immunogold electron microscopy. The S adhesin and the P adhesins from F7$_1$ and F11 and F13 fimbriae were found to be located at the tips of the fimbriae (Moch et al. 1987; Riegman et al. 1989; Lindberg et al. 1982), as shown in Fig. 1. Interestingly, the adhesin and another minor protein (corresponding to the E protein of pap fimbriae) form a functional complex attached to the fimbrial rod. This seems, however, not to be the only mode of adhesin presentation on fimbriae. Analysis of the P-specific F9-FAC revealed P-specific adhesin not only at the tip but also at the side of the fimbriae, which had a kinked appearance (Riegman et al. 1989). Using the same technique, Beachey and Abraham (1988) obtained a similar picture with an *E. coli* recombinant expressing mannose-specific type 1 fimbriae. The authors showed that the adhesin (H protein) is located alongside the fimbriae as distinct globules which were termed fimbriosomes and which may contain additional minor subunits. Although the clone used by Abraham et al. 1987a overproduced the adhesin and thus may not be representative, the evidence cannot be disregarded; it is possible that in some strains the adhesin may be located on the tips of the fimbriae and in other strains alongside the fimbriae.

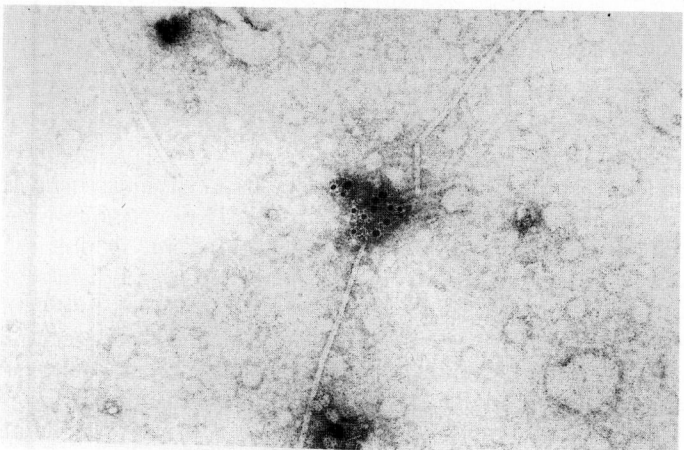

Fig. 1. Immunoelectron micrograph of $F7_1$ fimbriae, double labeled with E- and G-specific antiserum. The larger (11-nm) gold particles adhere to the adhesin; the 7-nm gold particles adhere to the nonadhesive minor subunit E. $F7_2$ and F11 fimbriae show the same labeling pattern. (Courtesy of N. RIEGMAN, Utrecht)

2.1.4 Isolation of adhesins from Outer Bacterial Membranes

Several years ago we found that *E. coli* with mannose-specific fimbriae agglutinated erythrocytes and *S. cerevisiae* even after removal of the fimbriae (ESHDAT et al. 1981). We were also able to show that isolated outer membranes of these bacteria were still agglutinating in an MS way. From the membranes of defimbriated P-specific *E. coli* the adhesin could be solubilized with octyl glucoside or Zwittergent 3-16, but not with Triton X-100 or 4M urea. The P adhesin thus isolated does not associate with the respective fimbriae; it can also be obtained from fimbriated bacteria by direct extraction with octyl glucoside which does not dissociate FAC. Such an extraction leads to a five- to tenfold higher yield of adhesin as compared with the other procedures. The cell wall associated adhesin, which is solubilized with octyl glucoside, does not copurify with the FAC. However, it has the same molecular size, receptor specificity, and serologic specificity as analyzed with monoclonal antibodies (H. HOSCHÜTZKY and K. JANN, in preparation).

These findings, which indicate an FAC-independent adhesin pool in or associated with the outer membrane, are difficult to reconcile with the biogenesis and export assembly of FACs as proposed by NORMARK and co-workers (BAGA et al 1987). These authors postulate that transport of the P adhesin (G protein) is closely linked with that of the other minor subunit proteins (E and F) constituting the functional head of the fimbria and of the major subunit (A protein), which forms the supporting fimbrial rod. Our findings permit the interpretation that the P adhesin is exported independently of the other fimbrial proteins and forms a pool in (or on) the outer bacterial membrane from which the fimbriae-associated adhesin is drawn. Alternatively, the adhesin may be exported by two independent mechanisms, one as proposed by NORMARK et al., and the other using a different export system which channels the protein into the outer membrane. This holds not only for the P adhesin but for the S- and mannose-specific adhesins as well.

2.2 Fimbriae Consisting of Adhesive Subunits

Fimbriae which contain the receptor-recognizing protein (adhesin) as an associated minor subunit are found with many pathogenic *E. coli*, mainly of extraintestinal origin. In contrast, swine pathogenic *E. coli* of O groups 8 and 148 exhibit the fimbrial K88 antigen, which has a different organization. Genetic studies have indicated that the major subunit of the K88 fimbriae is the adhesin (DE GRAAF 1988). Such fimbriae, consisting of adhesive subunits, may be polyvalent if the receptor binding domains of all subunits are available for receptor interaction, or monovalent if only the terminal subunit is available for receptor binding. Like the P and S fimbriae (see Sect. 2.1.1), the K88 fimbriae are monodispers, adhering but not agglutinating at pH 7-8, but aggregating and thus agglutinating at pH 5-6. We therefore assume that the K88 fimbriae consist of adhesive subunits which have their receptor binding sites blocked by their organization into fimbriae.

M fimbriae of uropathogenic *E. coli* seem to be organized in the same way as the K88 fimbriae, while in the SS142 fimbriae (METT et al. 1983) all subunits can interact with the receptor (H. HOSCHÜTZKY and K. JANN, unpublished results).

2.3 Nonfimbrial Adhesins of *E. coli*

The early work of DUGUID has already shown that, not uncommonly, *E. coli* are isolated which show adhesive and agglutinating properties but are not fimbriated (DUGUID et al. 1979). Systematic studies of these bacteria and their respective adhesins have only recently begun in several laboratories (GOLDHAR et al. 1987;

Table 2. Properties of nonfimbrial adhesins of *E. coli*

Nonfimbrial adhesin	MW of subunit (10^3 d)	Receptor	Reference
GV-12	13.4	MR[a]	SHELADIA et al., 1982
444-3	14.5	MR	WILLIAMS et al. 1984
469-3	14	MR	WILLIAMS et al. 1984
O75X	16	MR	VÄISANEN-RHEN 1984
AFA-1	16	MR	WALZ et al. 1985
ZI	14.4	MR	ØRSKOV et al. 1985
M	21	M[b]	RHEN et al. 1986
2230	16	MR	FORESTIER et al. 1987
NFA-1	21	M/N[b]	GOLDHAR et al. 1987
NFA-2	19	M/N	GOLDHAR et al. 1987
NFA-3	17.5	N[b]	GRÜNBERG et al. 1988
NFA-4	28	M	HOSCHÜTZKY et al. 1989b
NFA-5	23	M	HOSCHÜTZKY and K. JANN, unpublished material

[a] MR, mannose resistant;
[b] Blood group specificity

GRUNBERG et al. 1988; HOSCHÜTZKY et al. 1989a; ØRSKOV I et al. 1985; SKELADIA et al. 1982; WALZ et al. 1985; WILLIAMS et al. 1984; VAISÄNEN-RHEN 1984). Strains which express nonfimbrial adhesins are listed in Table 2. These adhesive structures seem to be subject to temperature regulation (restrictive temperatures below 20°C) and to phase variation, as has been described for the fimbrial adhesins. Most of the clinical isolates also express mannose-specific fimbriae, especially when the bacteria are grown in liquid culture. Two of the nonfimbrial adhesins have been cloned (LABIGNE-ROUSSEL et al. 1984, 1985; HALES et al. 1988). Their gene organization was shown to be similar to that of the fimbrial adhesins. The major difference between fimbriated and nonfimbriated strains seems to be the lack of a transcribed gene coding for the major fimbrial subunit. In this context it should be noted that the fimbriae consisting of adhesive subunits (see Sect. 2.2) have a gene organization analogous to that of the nonfimbrial adhesins.

The most intensive genetic studies have been performed on a nonfimbrial (afimbrial) adhesin (NFA) termed AFA-1 (LABIGNE-ROUSSEL et al. 1984, 1985). This or closely related adhesins are expressed in many *E. coli* strains (LABIGNE-ROUSSEL and FALKOW 1988). Amino acid sequence and receptor specificity indicate that AFA-1 is different from the other non-fimbrial adhesins listed in Table 2.

Whereas receptor structures for the fimbrial adhesins are expressed not only on human but also on animal cells, this is not the case with the nonfimbrial adhesins with blood group M/N specificity. The only animal cell found to express an NFA receptor is the chicken erythrocyte (HOSCHÜTZKY et al. 1989b). Such a specificity for chicken as well as human red blood cells is also known for the sialic acid-specific influenza A virus (MARKWELL 1986). The chemical nature of the NFA receptor(s) is not known. Interestingly, most NFAs which we have isolated have a preference for erythrocytes of MM, MN, or NN blood groups (GOLDHAR et al. 1987; GRUNBERG et al. 1988; HOSCHÜTZKY et al. 1989a).

2.3.1 Isolation

The isolation of NFAs was achieved in several laboratories with very similar methods. We have developed a procedure which is similar to that for the isolation of fimbrial-associated adhesin (see Sect. 2.1.1). Crude preparations of NFA are obtained by heating of the agar-grown bacteria to 50–70°C or by the other methods described in Sect. 2.1.1. Temperature, ionic strength, and pH during heat extraction are critical and may vary from one strain to another. After removal of the bacteria by centrifugation, crude NFAs are obtained by differential ammonium sulfate precipitation in the presence of EDTA and glycine. To remove lipid contamination, the crude NFAs may be treated with 50% ethanol. At this stage the solubility of the NFAs is reduced, possibly due to a change in conformation or loss of a compound which stabilizes the proteins in aqueous solution. The NFAs are dissolved in 8M guanidine hydrochloride and, after its exchange for 8 M urea, purified by high resolution ion exchange chromatography. Using this method we have isolated NFAs from six different *E. coli* strains, with yields ranging from 1 to 10 mg per 10 g wet weight.

Table 3. N-terminal sequences of fimbriae-associated and nonfimbrial adhesins

Adhesin	N-terminal sequence					Reference
	1	6	11	16	21	
M	STVTA	THTVE	SDAEF	TI		RHEN et al. 1986
NFA-5	VTVTA	RHTVE	SDAXF	IDXV		HOSCHÜTZKY et al., unpublished material
2230	GNVLS	GGNGT	QTVTM	PVNAA	TXTVS	FORESTIER et al. 1987
NFA-1	DANGL	NTVNA	GDGKN	LGTAA	A	HOSCHÜTZKY et al., unpublished material
AFA-1	NFTSG	GTNGK	VDLIT	TEECR	VTVES	LABIGNE-ROUSSEL et al. 1985
NFA-4	NTTGD	FNGSF	DMNGA	IAADV	YKG	HOSCHÜTZKY et al., 1989a
K88	WMTGD	FNGSV	DIGGS	ITADD	YRQ	GAASTRA et al. 1981
F1 (typge I)	CKTAN	GTAIP	IGGSA	NVYVN	LAPVV	ABRAHAM et al. 1987
F7$_1$	XNNIV	FYSLG	NVNSY	QGG		HOSCHÜTZKY et al. 1989
F7$_2$	WNNIV	FYSLG	NVNSY	QGGNV	VITQR	LUND et al. 1988
F11	WNNIV	FYSLG	DVNSY	QGGNV	VITQR	LUND et al. 1988
F13	GWHNV	MFYAF	NDYLT	TNAGN	FKVID	LUND et al. 1987 HOSCHÜTZKY et al. 1989[b]

2.3.2 Characterization

Data accumulated on NFAs are shown in Table 2 and the known N-terminal amino acid sequences are compared in Table 3. The only sequence homologies seen so far are those between NFA-4 and the fimbrial K88 antigen and between NFA-5 and the M-specific adhesin described by RHEN et al. (1986), HOSCHÜTZKY et al. (1989), and H. HOSCHÜTZKY and K. JANN (unpublished material).

The NFAs have apparent molecular weights in excess of 10^7, as estimated by gel permeation chromatography. SDS–PAGE showed that they consist of noncovalently linked subunits with molecular weights in the range of 15–30 kD, which is

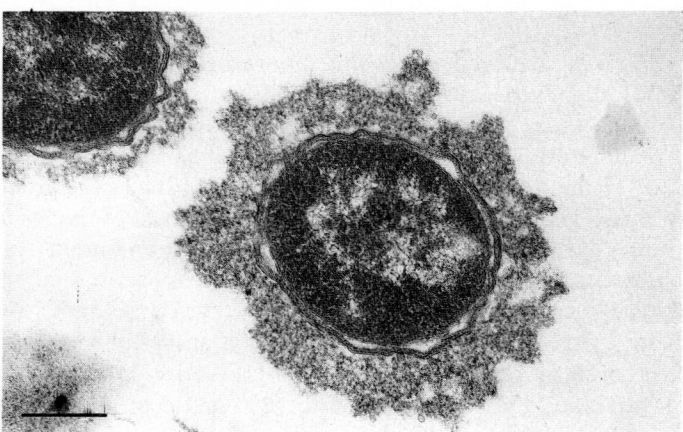

Fig. 2. Electron micrograph of an ultrathin section of *E. coli* 21248 expressing NFA-1. The extracellular adhesin was stabilized with an adhesin-specific monoclonal antibody

comparable to those of fimbrial adhesins and fimbrillins. The NFAs are therefore large polymers consisting of more than 20 subunits. It will be interesting to learn in which way these subunits aggregate and if stable intermediate stages of aggregation exist. No information concerning aggregation states of NFAs in solution and their arrangement on the bacterial cell surface is available at present.

Isolated NFA-1, NFA-2, and NFA-4 were used for the preparation of specific antisera and monoclonal antibodies. Some of the monoclonal antibodies are strongly antiadhesive in the hemadhesion assay. The antibodies were used to elucidate the cellular expression of the NFAs. Electromicroscopic analysis of thin sections from NFA-1, NFA-2, and NFA-4 expressing E. coli after stabilization of the adhesins with homologous antibody revealed that these adhesins surround the bacterial cell like a capsule. This is shown in Fig. 2 with NFA-1 as an example. This type of extracellular presentation was also demonstrated by ØRSKOV et al. (1985) and termed an adhesive capsule. Such an extracellular expression of an NFA may be more common that hitherto known.

3 Coexpression of Polysaccharide Capsules and Adhesins

Escherichia coli causing extraintestinal infections frequently express K antigens, which surround the cells as a polysaccharide capsule (JANN and JANN, this volume). With the aid of these capsules the bacteria are able to overcome the nonspecific defense mechanisms of the host, mostly the action of complement and phagocytes. Since several such strains also express adhesive proteins, it is interesting to know whether or not in these extraintestinal bacteria polysaccharide capsules and adhesins are expressed by the same or by different cells of a given population. In immunoelectron microscopic studies of fimbriated and encapsulated E. coli we have found evidence that polysaccharide capsules and fimbriae are coexpressed in many but not all cells of the population. K.D. KRÖNCKE and K. JANN, unpublished material). Many cells were seen which seemed to express either one or other of these virulence determinants. This may be due to phase variations and/or mutations in the bacterial population. In a statistical mean time both virulence factors are probably expressed by all cells. There is one additional factor which hitherto has escaped close examination, i.e., the regulation of fimbrial expression in certain stages of an infection. We do not know when fimbriae are essential for the bacteria and when they impede the progress of infection. For instance, it may be that fimbriae are not expressed on bacteria which are in the process of invading a tissue or which are encountering receptor-expressing phagocytes. In these cases, phase variation may provide a selective advantage to the nonfimbriated cells.

The problem of coexpression of adhesins and polysaccharide capsules becomes especially interesting when the adhesin also forms a capsule. In such cases the possibility exists that the bacteria express two types of capsule. We have analyzed this phenomenon with the uropathogenic E. coli O7:K98:H6, which exhibits the polysaccharide K98 capsule as well as the nonfimbrial adhesin NFA-4 (KRÖNCKE et al., manuscript in preparation). The bacteria were subjected to immunoelectron

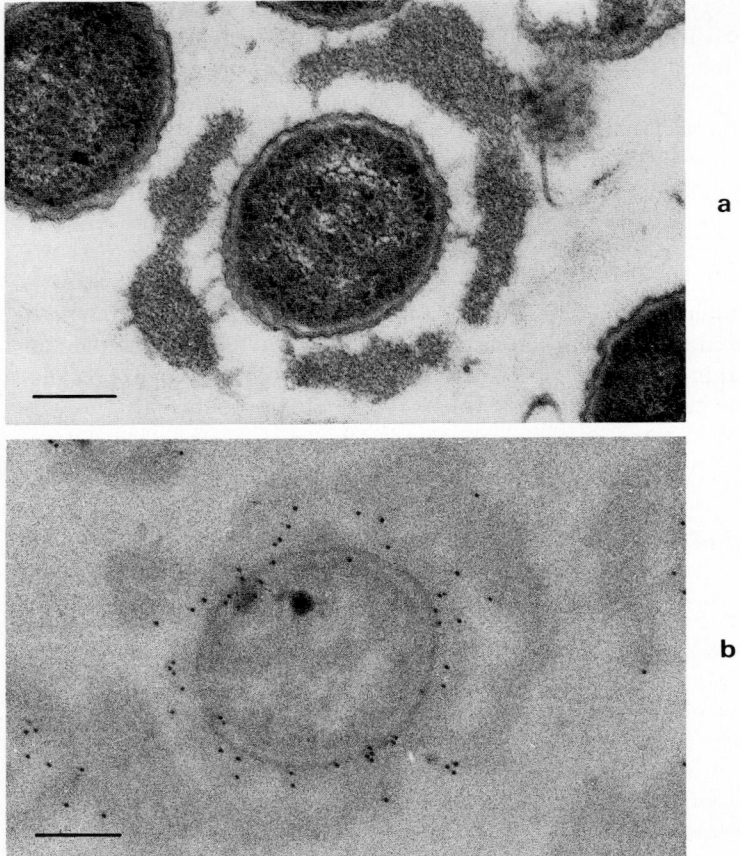

Fig. 3a, b. Coexpression of NFA-4 and the K98 antigen in *E. coli* 21511, a NFA-4 was stabilized with an adhesin-specific monoclonal antibody. **b** NFA-4 was stabilized as in **a**. The 10-nm gold particles indicate the K98-antigen stabilized with a specific polyclonal antiserum

microscopy with a monoclonal anti-NFA-4 antibody, absorbed anti-K98-antiserum, or both. As shown in Fig. 3, the nonfimbrial adhesin NFA-4 surrounds the bacterial cells as a capsule, but without a continuous circumferential contact. There are large gaps between the cell wall and the adhesin, which are bridged in many places. Using only anti K98-antiserum, these structures are not seen. Instead, the bacteria are directly surrounded by a dense layer with the usual appearance of a polysaccharide capsule. Double contrasting with both types of antibody reveals the concentric way in which these extracellular substances are arranged, with the adhesin predominantly on the outside and the polysaccharide more at the cell proximal space. This appearance was seen with practically all cells of the bacterial population. These results show that extraintestinal *E. coli* can express both adhesins and polysaccharide capsules simultaneously. It remains to be analyzed in which way the expression of one or both of these extracellular components is transient during a cell's life span and at different stages of the ineffective process.

4 Conclusions

Within the last two decades our knowledge about the molecular basis of bacterial adhesion and its specificity has increased tremendously. This was to a large measure due to advances in molecular genetics, biochemistry, and immunology as well as to the development of immunoelectron microscopy.

The early interpretation that fimbriae are mediators of bacterial adhesion was corroborated in many cases by genetic and biochemical evidence. It was recently demonstrated that the fimbriae have specific adhesins at their tips. The fimbriae, which were shown to consist of peptide subunits, are more or less rigid structures supporting the adhesins and presenting them to the surrounding of the bacteria.

Other fimbriae, such as the K88 antigen, seem to consist of adhesive subunits. If these block each other in their association to functional fimbriae with only the terminal (tip) one remaining functional, a chemical differentiation between these and the fimbriae with distinct adhesins will not be easy. More recently, adhesins were reported as not being associated with or organized into fimbriae. Although several of them were later shown to be very thin fimbrillae, this demonstration was not possible with others. The latter are operationally termed nonfimbrial. Some of them form adhesive capsules around the bacteria. However, their molecular organization is not known.

It will be interesting to learn why bacterial adhesins serving the same function are so differently expressed on the bacterial surface—ranging from fimbriae to capsules. Perhaps an ancestral adhesin gene has evolved through duplication and variation into a more complex genetic system which expresses the adhesin as part of a functional complex consisting of several different proteins.

Adhesins even of one specificity may in fact be a group of closely related but not identical proteins having a conserved receptor binding domain, as was recently demonstrated for P-specific fimbrial adhesins. This may also be true for several nonfimbrial adhesins, which share the recognition of blood group M and/or N substances.

If one accepts the presence of a conserved receptor binding domain in a group of related adhesins, no matter whether fimbrial or nonfimbrial, a comparison with another group of recognition proteins, namely the antibodies, is interesting. Whereas antibodies have the binding site in a very variable region, the adhesins have a conserved binding site. The structural variability of adhesins outside the binding domain will provide these recognition proteins with a serologic flexibility vis-à-vis the infected host, which in fimbriated bacteria is increased in the structural/serologic variability of the fimbriae.

The specificity or recognition was attributed to complex carbohydrates on host cells and the relevant oligosaccharide structures were elucidated in many cases. Although much knowledge has accumulated in this field, the chemical structures of many receptors, e.g., those associated with bloog group M and/or N substances, remain to be elucidated. A more extended comparative biochemical analysis of adhesins and their receptors will probably reveal that the adhesion of such distinct pathogens as bacteria, viruses, or parasites is governed by their recognition of related, if not identical, host structures—possibly with related recognition systems.

Acknowledgments. The work from the authors' laboratory was in part supported by the Deutsche Forschungsgemeinschaft and the Bundesministerium für Forschung und Technologie/DFVLR.

References

Abraham SN, Goguen JG, Beachey EH (1987a) Analysis of the genetic and structural determinants of the D-mannose sensitive adhesin of *Escherichia coli*. J Cell Biochem 11B: 121

Abraham SN, Goguen JD, Sun D, Klemm P, Beachey EM (1987b) Identification of two ancillary subunits of *Escherichia coli* type 1 fimbriae by using antibodies against synthetic oligopeptides of fim gene products. J Bacteriol 169: 5330–5336

Baga M, Norgren M, Normark S (1987) Biogenesis of *E. coli* pap pili: Pap H, a minor pilin subunit involved in cell anchoring and length modulation. Cell 49: 214–251

Beachey EH, Abraham SN (1988) Biological properties of bacterial surface proteins: type 1 fimbriae of *Escherichia coli*. In: Schrinner E, Richmond PMM, Seibert G, Schwarz U (eds) Surface structures of microorganisms and their interaction with the mammalian host. VCH, Weinheim

Brinton CC (1965) The structure, function, synthesis and genetic control of bacterial pili and a molecular model for DNA and DNA transport in gram-negative bacteria. Trans NY Acad Sci 27: 1003–1054

de Graaf FH (1988) Fimbrial structures of enterotoxigenic *E. coli*. Antonie van Leeuwenhook J Microbiol 54: 395–404

de Graaf FK, Klaasen-Bor P, van Hees JE (1980) Biosynthesis of the K99 surface antigen is repressed by alanin. Infect Immun 30: 125–128

Duguid JP, Old DC (1980) Adhesive properties of Enterobacteriaceae. In: Beachey EH (ed) Bacterial adherence. Chapman and Hall, London, pp 184-217 (Receptors and recognition, series B, vol 6)

Duguid JP, Smith IW, Dempster G, Serol Edmunds PM (1955) Nonflagellar filamentous appendages ("fimbriae") and hemagglutinating acitivity in *Bacterium coli*. J Pathol Bacteriol 70: 335–348

Duguid JP, Clegg S, Wilson MI (1979) The fimbrial and non-fimbrial hemagglutinins of *Escherichia coli*. J Med Microbiol 12: 213–227

Eisenstein BI (1981) Phase variation of type-1 fimbriae in *Escherichia coli* is under transcriptional control. Science 214: 337–339

Eisenstein BI, Dodd D (1982) Pseudocatabolite repression of type 1 fimbriae of *Escherichia coli*. J Bacteriol 151: 1560–1567

Eshdat Y, Speth V, Jann K (1981) Participation of pili and cell wall adhesin in the yeast agglutination activity of *Escherichia coli*. Infect Immun 34: 980–986

Faris A, Lindahl M, Wadström J (1980) GM_2-like glycoconjugate as possible erythrocyte receptor for the CFA/1 and K99 hemagglutinins of enterotoxigenic *Escherichia coli*. FEMS Microbiol Lett 7: 265–269

Forestier C, Welinder HG, Darfeuille-Michaud A, Klemm P (1987) Afimbrial adhesin from *Escherichia coli* strain 2230: purification, characterization and partial covalent structure. FEMS Lett 40: 47–50

Gaastra E, Movi FR, Stnitje AR, and de Graaf FK (1981) The nucleotide sequence of the gene encoding the K88ab protein subunit of porcine enterotoxigenic *Escherichia coli*. FEMS Microbiol Lett 12: 41–46

Goldhar J, Perry R, Ofek I (1984) Extraction and properties of nonfimbrial mannose-resistant hemagglutinin from a urinary isolate of *Escherichia coli*. Curr Microbiol 11: 49–54

Goldhar J, Perry R, Golecki J, Hoschützky H, Jann B, Jann K (1987) Nonfimbrial, mannose-resistant adhesins from uropathogenic *Escherichia coli*. O83:K1:H4 and O14:K:H11. Infect Immun 55: 1837–1842

Grunberg J, Perry R, Hoschützky H, Jann B, Jann K, Goldhar J (1988) Nonfimbrial blood group N-specific adhesin (NFA-3) from *Escherichia coli* O20:X104:H⁻, causing systemic infection. FEMS Microbiol Lett 56: 241–246

Hacker J, Schmidt G, Hughes G, Knapp S, Marget M, Goebel W (1985) Cloning and characterization of genes involved in production of mannose-resistant, neuraminidase-susceptible (X) fimbriae from a uropathogenic O6:K15:H31 *Escherichia coli* strain. Infect Immun 47: 434–440

Hales BA, Beverly-Clarke H, High NJ, Jann K, Perry R, Goldhar J, Boulnois GJ (1988) Molecular cloning and characterization of the genes for a non-fimbrial adhesin from *Escjerichia coli*. Microbiol Pathogen 5: 9–17

Hoschützky H, Lottspeich F, Nimmich W, Jann K (1989a) Isolation and characterization of the nonfimbrial adhesin NFA-4 from uropathogenic *Escherichia coli* O7:K98:H6. Microbiol Pathogen 6: 351–359

Hoschützky H, Lottspeich F, Jann K (1989b) Isolation and characterization of the αGal-(1,4)-βGal (P) specific adhesin from fimbriated *Escherichia coli*. Infect Immun 57: 76–87

Isaacson RC (1985) Pilus adhesins In: Savage DC, Fletcher H (ed) Bacterial adhesion. Mechanisms and physiological consequences. Plenum, New York, pp 307–336

Jann K, Jann B, Schmidt G (1981) SDS polyacrylamide gel electrophoresis and serological analysis of pili from *Escherichia coli* of different pathogenic origin. FEMS Microbiol Lett 11: 21–25

Klemm P, Christiansen G (1987) Three fim genes required for the regulation of length and mediation of adhesion of *Escherichia coli* type 1 fimbriae. MGG 208: 439–445

Knutton S, Williams PH, Llyod DR, Candy DCA, McNeish AS (1984) Ultrastructural study of adherence to and penetration of cultured cells by two invasive *Escherichia coli* strains isolated from infants with enteritis. Infect Immun 44: 599–608

Labigne-Roussel AF, Lark D, Schoolnik G, Falkow S (1984) Cloning and expression of an afimbrial adhesin (AFA-1) responsible for P-blood group-independent mannose-resistant hemagglutination from a pyelo-nephritic *Escherichia coli* strain. Infect Immun 46: 251–259

Labigne-Roussel AF, Schmidt MA, Walz W, Falkow S (1985) Genetic organization of the *afa* operon and nucleotide sequence from a uropathogenic *Escherichia coli* gene encoding an afmbrial adhesin (AFA-1). J Bacteriol 162: 1285–1292

Labigne-Roussel AF, Falkow S (1988) Distribution and degree of heterogeneity of the afimbrial-adhesin-encoding operon (*afa*) among uropathogeneic *Escherichia coli* isolates. Infect Immun 56: 640–648

Leffler H, Svanborg-Edén C (1986) Glycolipids as receptors for *Escherichia coli* lectins or adhesins. In: Mirelman D (ed) Microbiol lectins and agglutinins: properties and biological activity. Wiley, New York, pp 83–112

Lindberg FP, Lund B, Normark S (1984) Genes of pyelonephritogenic *E. coli* required for digalactoside-specific agglutination of human cells. EMBO J 3: 1167-1173

Lindberg F, Lund B, Normark S (1986) Gene products specifying adhesion of uropathogenic *Escherichia coli* are minor components of pili. Proc Natl Acad Sci USA 83: 1891–1895

Lindberg F, Lund B, Johannson L, Normark S (1987) Localization of the receptor-binding protein adhesin at the tip of the bacterial pilus. Nature 328: 84–87.

Lund B, Lindberg G, Marklund B-I, Normark S (1987) The papG protein is the α-D-galactopyranosyl-(2-4)-β-D-galactopyranose-binding adhesion of uropathogenic *Escherichia coli*. Proc Natl Acad Sci USA 84: 5898–5902

Lund B, Lindberg F, Normark S (1988) Structure and antigenic properties of the tip-located P pilus proteins of uropathogenic *Escherichia coli*. J Bacteriol 170: 1887–1894

Manson MS, Brinton CC (1988) Identification and characterization of *E. coli* type-1 pilus tip adhesion protein. Nature 332: 265–268

Markwell MAK (1986) Viruses as hemagglutinins and lectins. In: Mirelman D (ed) Microbiol lectins and agglutinins. Properties and biological activity. Wiley, New York, pp 21–53

Mett H, Kloetzlen L, Vosbeck K (1983) Properties of pili from *Escherichia coli* SS142 that mediate mannose-resistant adhesion to mammalian cells. J Bacteriol 153: 1038–1044

Minion FC, Abraham SN, Beachey EH, Goguen JD (1986) The genetic determinant of adhesive function in type 1 fimbriae of *Escherichia coli* is distinct from the gene encoding the fimbrial subunit. J Bacteriol 165: 1033–1036

Mirelman D, Ofek I (1986) Introduction to microbial lectins and agglutinins. In: Mirelman D (ed) Microbial lectins and agglutinins: properties and biological activity. Wiley, New York, pp 1–19

Moch T, Hoschützky H, Hacker J, Kröncke KD, Jann K (1987) Isolation and characterization of the α-sialyl-β-2,3-galactosyl-specific adhesin from fimbriated *Escherichia coli*. Proc Natl Acad Sci USA 84: 3462–3466

Norgren M, Normark S, Lark D, O'Hanley P, Schoolnik G, Falkow S, Svanborg-Edén C, Baga M, Uhlin BE (1984) Mutations in *E. coli* cistrons affecting adhesion to human cells do not abolish pap pili Hiber formation, EMBO J 3:1159–1165

Nowicki B, Holthöfer H, Saraneva T, Rhen M, Väisanen-Rhen V, Korhonen TK (1986) Location of adhesion sites for P-fimbriate and for O75X-positive *Escherichia coli* in the human kidney. Microb Pathogen 1: 169–180

Ørskov I, Birch-Anderson A, Duguid JP, Stenderup J, Ørskov F (1985) An adhesive protein capsule of *Escherichia coli*. Infect Immun 47: 191–200

Ofek I, Beachey EH (1980) General concepts and principles of bacterial adhesion. In: Beachey EH (ed) Bacterial adherence. Chapman and Hall, London, pp 1–29 (Receptors and recognition, series A, vol 6)

Parkkinen J, Finne J, Achtmann M, Väisanen V, Korhonen TK (1983) *Escherichia coli* strains binding neuraminyl α2-3 galactosides. Biochem Biophys Res Commun 111: 456–461

Rhen M, Mäkelä PH, Korhonen TK (1983) P-fimbriae of *Escherichia coli* are subject to phase variation. FEMS Lett 19: 267–271

Rhen M, Klemm P, Korhonen TK (1986) Identification of two new hemagglutinins of *Escherichia coli*, *N*-acetyl-D-glucosamine-specific fimbriae and a blood group M-specific agglutinin, by cloning the corresponding genes in *Escherichia coli* K12. J Bacteriol 168: 1234-1242

Riegman N, Hoschützky H, Jann K, Bergmans H, Hoekstra W (1989) Immunocytochemical analysis of P-fimbrial structure: localization of the minor subunits and the influence of the minor subunit FSOE on the biogenesis of the adhesion. Mol Microbiol in press

Salit IE, Gotschlich EC (1977) Hemagglutination by purified type 1 *Escherichia coli* pili. J Exp Med 146: 1169–1181

Savage DC (1985) Effects on host animals of bacteria adhering to epithelial surfaces. In: Savage DC, Fletcher M (eds) Bacterial adhesion. Mechanisms and physiological significance. Plenum, New York, pp 437–464

Sharon N, Ofek I (1986) Mannose specific bacterial surface lectins. In: Mirelman D (ed) Microbial lectins and agglutinins: properties and biological activity. Wiley, New York, pp 55–82

Sheladia VL, Chambers JP, Guevara J, Evans DJ (1982) Isolation, purification, and partial characterization of type V–A hemagglutinin from *Escherichia coli* CV-12, O1:H⁻. J Bacterial 152: 757–761

Tennent JM, Baga M, Fossman K, Göransson M, Lindberg F, Lund B, Marklund BI, Norgren M, Uhlin BE, Normaks (1987) Protein interactions essential for the biosynthesis of P pili of uropathogenic *Escherichia coli* In: Schrinner E et al. (eds) Surface structures of microorganisms and their interactions with the mammalian host. Workshop conferences, vol 18 VCH Weinheim

Väisanen-Rhen V (1984) Fimbriae-like hemagglutinin of *Escherichia coli* O75 strains. Infect Immun 46: 401–407

Väisanen-Rhen V, Elo J, Tallgren LG, Siitonen A, Mäkelä PH, Svanborg Edén C, Källenius G, Svenson SB, Hultberg H, Korhonen TK (1981) Mannose-resistant hemagglutination and P-antigen-recognition are characteristic of *Escherichia coli* causing primary pyelonephritis. Lancet ii: 1366–1369

Väisanen-Rhen V, Korhonen TK, Jokinen M, Gahmberg CG, Ehnholm C (1982) Blood group M specific hemagglutinin in pyelonephritogenic *Escherichia coli*. Lancet i: 1192

Väisanen-Rhen V, Korhonen TK, Finne J (1983) Novel cell-binding activity specific for *N*-acetyl-D-glucosamine in an *Escherichia coli* strain. FEBS Lett 159: 233–236

Van Die I, Spierings G, Van Megen I, Zuidweg E, Hoekstra W, Bergmans H (1985) Cloning and genetic organization of the gene cluster encoding F7₁ fimbriae of a uropathogenic *Escherichia coli* and comparison with the F7₂ gene cluster. FEMS Microbiol Lett 28: 329–334

Walz W, Schmidt A, Labigne-Roussel A, Falkow S, Schoolnik G (1985) AFA-1, a cloned afimbrial X-type adhesin from a human pyelonephritic *Escherichia coli* strain. Eur J Biochem 152: 315–321

Williams PH, Knutton S, Brown MGM, Candy DCA, McNeish AS (1984) Characterization of nonfimbrial mannose-resistant protein hemagglutinins of two *Escherichia coli* strains isolated from infants with enteritis. Infect Immun 44: 592–598

Serologic Classification of Fimbriae

I. Ørskov and F. Ørskov

1 Introduction

Many bacteria have the ability to attach to epithelial cells at various sites in man and animal by means of adhesins. The serologically best studied adhesins are those of gram-negative bacteria, particularly Enterobacteriaceae and Neisseria gonorrhoeae. They are proteins (e.g., JONES and ISAACSON 1983) and are commonly structured as filaments of various thickness; if the diameter is wide enough, they are called fimbriae (DUGUID et al. 1955) (synonym: pili (BRINTON 1959)), while very thin filaments are often referred to as fibrils, but no rules exist for the use of these names. Recently adhesive *E. coli* surface proteins have been described consisting of extremely fine filaments which can often be visualized only with great difficulty (HINSON et al. 1987; LEVINE et al. 1984; ØRSKOV et al. 1985).

Fimbriae and adhesins have been classified according to morphology and ability to agglutinate erythrocytes from various animals (DUGUID et al. 1966), and furthermore according to serology (DUGUID et al. 1979; GILLIES and DUGUID 1958). A more recent way of characterizing fimbriae is by their chemical composition, i.e., the sequence of amino acids, or by the receptors to which they adhere (KLEMM 1985). The receptors described to date are carbohydrate constituents of glycoconjugates

International Escherichia and Klebsiella Centre (WHO), Statens Seruminstitut, Amager Boulevard 80, DK-2300 Copenhagen S, Denmark

Table 1. *E. coli* fimbriae groups

Fimbriae group	Antigen types	F antigen number[a]	Ref. to group name
Type 1	Different types	F1A and variants of this	BRINTON 1965; DUGUID et al. 1966
P	Different types	F7–F16 and variants of these	KÄLLENIUS et al. 1981
S	Highly related types[b]		KORHONEN et al. 1984
G	?		RHEN et al. 1986b
M	?		RHEN et al. 1986b
F (= Sex)	Different types		BRINTON et al. 1964
Fib	?		FARIS 1985

[a]ØRSKOV and ØRSKOV (1983a);
[b]Serologically related to F1C, see Table 3 (OTT et al. 1986)

(JONES and ISAACSON 1983; SMYTH 1986). For example, P fimbriae of *E. coli*, particularly from urinary tract infections, bind to the disaccharide Galα(1-4)Gal present on human erythrocytes of the P blood group system and on epithelial cells of the urinary tract (KÄLLENIUS et al. 1980; LEFFLER and SVANBORG-EDEN 1980, 1981). Recent examples of naming fimbriae in *E. coli* according to the receptors are the S fimbriae, which adhere to sialic acid-containing receptors (KORHONEN et al. 1984; PARKKINEN et al. 1986), and the Fib fimbriae, which bind to fibronectin (FARIS 1985) (see Table 1). In this chapter we shall deal with serologic classification of fimbriae and adhesins in some Enterobacteriaceae, particularly *E. coli*. The fimbriae of gonococcal strains are being intensively studied (e.g., SWANSON et al. 1987); antigenically they are extremely heterogeneous, and the fimbriae of each strain can undergo both antigenic variation and phase variation (SPARLING et al. 1986); no serologic classification as such exists; these fimbriae will not be treated here.

2 Organization of Fimbriae and Adhesins

Bacterial adhesion to erythrocytes and cell surfaces is frequently associated with fimbriae (DUGUID et al. 1955; DUGUID and GILLIES 1957; DUGUID and OLD 1980; BRINTON 1965). The adhesion may be sensitive or resistant to inhibition by D-mannose and is referred to as MS or MR, respectively. Type 1 fimbriae are MS. Fimbriae lacking adhering capacity also exist. It has been suggested that the adhesin protein may be distinct from the major subunit of the fimbriae (NORMARK et al. 1983), and recently this has been confirmed regarding type 1, P, and S fimbriae. Gene clusters coding for production and adhesive capacity of these fimbriae have been cloned from *E. coli*, and genetic analyses of the clones have shown that distinct genes code for fimbriation and adhesiveness (HACKER et al. 1985; MAURER and ORNDORFF 1985; MINION et al. 1986; NORGREN et al. 1984; RHEN et al. 1986a; VAN DIE et al. 1986). The MS adhesion of type 1 fimbriae has been suggested to be provided by the product of the *fimH* and either the *fimF* or *fimG* genes (KLEMM and CHRISTIANSEN 1987).

In P fimbriae, a minor protein termed PapG located at the tip of the fimbria is the protein which binds to the Gal-Gal receptor; the major PapA protein is not required for binding (LINDBERG et al. 1987).

Recently the S-specific and P-specific adhesive proteins have been isolated (MOCH et al. 1987; HOSCHÜTZKY et al. 1989). Immunoelectron microscopy (IEM) findings suggested that the S adhesin is located at the tip of the fimbria (MOCH et al. 1987).

Evidence for the presence of an adhesive minor subunit different from a major fimbrial subunit has hitherto not been obtained in K88 *E. coli* fimbriae, which have adhesin molecules as the major component (JACOBS 1987). For K99 fimbriae the situation is not yet clear (JACOBS 1987).

There seem to be two categories of fimbrial structure. The first category, in this chapter referred to as fimbriae I, comprises rigid fimbriae, of about 7 nm in diameter (BRINTON 1965; DUGUID et al. 1966; LEVINE et al. 1984; ØRSKOV et al. 1980). It may be that in all these fimbriae the fimbrial and adhesive subunits are distinct, although forming a functional complex in which the fimbrial subunit is the major protein.

The other category of fimbriae, in this chapter referred to as fimbriae II, comprises flexible thin fibrils or filaments of a diameter well below 7 nm, and sometimes as narrow as 2 nm. They often surround the bacteria as a fur coat, e.g., K88 (STIRM et al. 1967b), which by immunolabeling would look like a capsule. Fimbriae II may consist exclusively of adhesive molecules.

Both fimbriae I and fimbriae II have been serotyped. The major fimbrial protein constitutes the bulk of fimbriae I, thus > 99% of each P fimbria (LUND et al. 1988a), and it is in general believed that only antibodies against the major protein will be demonstrable in polyclonal antisera against fimbriae. The serotypes described to date are thus considered to reflect the antigenic properties of the major fimbrial protein. However, by selective absorption of a polyclonal antiserum against F13 fimbriae, antibodies against minor adhesin protein were demonstrated by a radio-immunoassay (LUND et al. 1988a). Serotyping of fimbriae as carried out today is complicated; if in the future serotyping should include examination of minor proteins present in the fimbria, we can only speculate on the complexity of such an analysis.

The serology dealt with in the following will involve reactions of the major fimbrial protein of fimbriae I and the fimbrial-adhesin protein of fimbriae II.

3 Some Results Obtained by Different Methods

3.1 Introduction

When it appeared that fimbriae in, for example, enterotoxigenic and uropathogenic *E. coli* played a role in attachment of the bacteria to mucosal surfaces, it became of great interest to know whether antigenically different fimbriae existed. In animal husbandry vaccines against K88 and K99 and 987 are in use to avoid diarrhea in young animals (for a review see KLEMM 1985), and it is the aim of many workers to develop a fimbria-based or partly fimbria-based vaccine against pyelonephritis and traveller's diarrhea in man. With this idea in mind serologic differentiation is important.

3.2 Bacterial Agglutination

Fimbriae are good antigens, and fimbrial antisera are easily obtained that can be used for demonstration of the presence of fimbriae and for serologic differentiation of fimbriae. GILLIES and DUGUID (1958) were the first to examine serologic relatedness between fimbriae by bacterial agglutination. Their report and a later one by DUGUID and CAMPBELL (1969) showed that type 1 fimbrial antigens are shared among *Escherichia*, *Shigella*, and to a lesser extent *Klebsiella*, and similarly among *Salmonella* and *Citrobacter*. There was no sharing of antigens between the two groups (see Sect. 3.3 and Table 2).

The agglutination test is a sensitive method and easy to carry out. It is a reliable test if proper antisera are used, i.e., properly absorbed polyclonal antisera against whole cells or purified fimbriae, or monoclonal antibodies (MAbs). In many cases a single strain will produce fimbriae with different receptors and/or antigenically different fimbriae with the same receptor (DUGUID and OLD 1980; KLEMM et al. 1982; ØRSKOV et al. 1982a, b). In such cases the agglutination technique will only distinguish between different antigens if monospecific antisera are used. The K88 and its variants K88ab and K88ac, and K99, were first demonstrated by means of bacterial agglutination, and this method is still used in our laboratory for differentiation of these antigens (ØRSKOV et al. 1961, 1964, 1975) (Table 3).

PARRY et al. (1982) and PARRY and ROOKE (1985) examined the antigens of seven uropathogenic *E. coli* strains with MR hemagglutinins by agglutination tests with cross-absorbed antisera and demonstrated seven antigens designated a–g. All seven strains expressed P fimbriae and four in addition expressed X adhesin, which is an

Table 2. Fimbriae in Enterobacteriaceae serologically examined

Fimbriae classification	Serologic results	Reference
Type 1[a]	Ten distinct groups[b] within Enterobacteriaceae	ADEGBOLA and OLD 1987
Type 2[a]	Serologically similar to type 1, but not hemagglutinating (*Salmonella paratyphi* B, *Salmonella pullorum*)	OLD and PAYNE 1971
Type 3[a]	Nine distinct groups[c] in Enterobacteriaceae (none in *E. coli* or *Shigella*)	OLD and ADEGBOLA 1985
Other types		For a review see CLEGG and GERLACH 1987

[a] Types 1–3 were defined according to morphology and hemagglutination by DUGUID et al. (1966);
[b] See Sects. 3.2 and 3.3;
[c] See Sect. 3.3

Table 3. *E. coli* fimbriae classified as antigens

Fimbria antigen	Different antigen types	Proposed F antigen number[a]	Reference
1C[b]		F1C	KLEMM et al. 1982
CFA/I		F2	EVANS et al. 1975
CFA/II		F3	EVANS and EVANS 1978
	CS1		
	CS2		CRAVIOTO et al. 1982
	CS3		SMYTH 1982
CFA III			HONDA et al. 1984
PCF 8775			THOMAS et al. 1982
= CFA IV			WILLSHAW et al. 1988
	CS4		
	CS5		THOMAS et al. 1985
	CS6		
K88		F4	ØRSKOV et al. 1961
	K88ab		
	K88ac		ØRSKOV et al. 1964
	K88ad		GUINÉE and JANSEN 1979
K99		F5	ØRSKOV et al. 1975
987P		F6	NAGY et al. 1977
F41			MORRIS et al. 1980
F(y)	F(y)		GIRARDEAU et al. 1980
	Att25 etc.		POHL et al. 1982
			MORRIS et al. 1985b
		F17	LINTERMANS et al. 1988
F165			FAIRBROTHER et al. 1986
Z	Z_1Z_2		ØRSKOV et al. 1985

[a] ØRSKOV and ØRSKOV (1983a);
[b] Serologically ralated to S fimbriae; see Table 1 (OTT et al. 1986)

assembly name for adhesins recognizing unknown epitopes on human erythrocytes (VÄISÄNEN-RHEN et al. 1984). The results of PARRY and ROOKE (1985) did not demonstrate which antigen corresponded to P- or to X-specific fimbriae. Table 4 shows the serologic results obtained with these strains by different methods and by different workers. In this table the unnumbered antigen (marked "?") of SP57 and SP88 is related to or identical with a particular X hemagglutinin present in many O75:K5:H⁻ strains and described as a coil-like structure on the basis of electron microscopy (EM) (VÄISÄNEN-RHEN 1984). This hemagglutinin is probably a fimbria II structure.

There are conflicting results in the literature regarding the usefulness of MAbs for bacterial agglutination. MORRIS et al. (1985a) found the reactions of a MAb against K99 to be identical with those of a polyclonal antiserum. DE REE et al. (1986a) experienced that strains which expressed cloned fimbriae but not wild-type strains were agglutinated by MAbs prepared against purified P fimbriae with different F specificities. A MAb against an F8 preparation from a P-fimbriated strain agglutinated this strain but not another one expressing F8, although it bound ¹²⁵I-labeled antibody; EM showed that the latter strain was less fimbriated than the agglutinable one (MOSER et al. 1986). SCHMITZ et al. (1986) reported that MAbs against *E. coli* F antigen 1C did not agglutinate F1C-fimbriated strains although

Table 4. Results of serologic examination of fimbriae in uropathogenic *E. coli* obtained by different methods in different different laboratories

O:K:H serotype[a]	Strain no.[b]	Specificity according to receptor[b]	Antigenic composition		
			PARRY	DE REE	ØRSKOV
			Bacterial aggluti- nation[b]	ELISA (MAbs)[c]	CIE[a]
O18ac:K5:H-	SP7	P + X	c, d, e	F8	F8, F1C
O157:H45	SP57	P + X	a, b	F12 rel. (F16)	F16 + ?[d]
O157:H45	SP88	P + X	a, b	F12 rel. (F16)	F16 + ?[d]
O1:K1:H7	SP101	P + X	b, c, e	F11	F11
rough:K1:H7	SP133	P	c, e	F11	F11
O6:K5:H1	SP144	P	f, g		F14, F1C
O75:K5:H-	SP201	P	g, e		F15, F1C

[a]ØRSKOV I and ØRSKOV F (unpublished results);
[b]Results by PARRY and ROOKE (1985);
[c]Results by DE REE et al. (1986a, b);
[d]An unnumbered adhesin, probably of type fimbriae II

they bound to the bacteria, while PERE et al. (1985) reported that a MAb against 1C agglutinated 44 out of 45 1C-fimbriated strains. The usefulness of MAbs for agglutination probably depends on how richly fimbriated the strains are and how accessible the recognized epitopes are.

3.3 Immunoelectron Microscopy

By IEM the antibody-binding sites become visible by being coated with antibody followed by negative staining. This method of direct labeling was introduced by LAWN (1967), and by use of it LAWN and MEYNELL (1970) concluded that four serotypes were present among sex fimbriae of the F type (present on male bacteria of compatibility group F). This conclusion was based on the different extents to which the sex fimbriae were coated with antibody. Since then other reports have appeared dealing with direct IEM studies of fimbriae (e.g., EVANS et al. 1975; ØRSKOV et al. 1980). The difficult interpretation of agglutination tests with fimbriated cultures caused ADEGBOLA and OLD (1982) and OLD and ADEGBOLA (1983) to use the IEM method in distinguishing different kinds of fimbria formed by multifimbriated strains. By this method the same authors demonstrated nine groups of antigenically distinct type 3 fimbriae which are present in all members of Enterobacteriaceae except *E. coli* and *Shigella* (OLD and ADEGBOLA 1985). Type 3 fimbriae is a name coined by DUGUID et al. (1966) for thin fimbriae which cause agglutination of tannic acid-treated erythrocytes, but not of fresh ones (Table 2). Recently IEM was also used to examine antigenic relationships among type 1 fimbriae of 40 Entero-bacteriaceae strains of 19 species of seven genera, and ten distinct groups were distinguished (ADEGBOLA and OLD 1987) (Table 2). Antigenic relationships were found to be more complex than described in previous reports based on findings from agglutination tests (GILLIES and DUGUID 1958; DUGUID and CAMPBELL 1969; NOWOTARSKA and MULCZYK 1977) (see Sect. 3.2).

Immunoelectron microscopic studies by OLD et al. (1987) of P fimbriae on 37 urinary *E. coli* strains revealed antigenic heterogeneity and marked differences in the extent to which the fimbriae were coated with antibody. Five pure fimbrial antisera were prepared by absorption and used for agglutination tests and IEM, and in this study the degree of activity was reported to be almost identical by both methods. By cross-absorptions of the five antisera, four distinct P antigenic determinants were demonstrated which occurred in different combinations of two in four of the five strains used for preparation of antisera. Unfortunately we cannot compare these results with ours obtained by crossed immunoelectrophoresis (CIE). Our reference strains for F antigens 7, 8, 9, 11, and 12, all antigens of P fimbriae (ØRSKOV and ØRSKOV 1983a, 1985), are included in the material examined, but no results with these strains are reported by OLD et al. (1987).

A fimbriae-containing specimen can, after direct labeling with, for example, rabbit antiserum, be processed further for indirect labeling with anti-rabbit IgG–ferritin, anti-rabbit IgG–gold, or protein A–gold. Indirect labeling is, in contrast to direct labeling, mainly independent of the quality of the negative staining, but it is more time and material consuming (GELDERBLOM et al. 1985).

The indirect immunogold EM method was, for example, used by MOCH et al. (1987) when they showed that the adhesin of an S-fimbriated strain was localized exclusively on the fimbriae with a possible preference for the tips. This method is particularly useful for differentiation when more types of fimbria are present in a culture.

Immuneoelectron microscopy is an excellent method for demonstration of antigenic sites on fimbriae. However, for serologic examinations of fimbriae on many strains the present authors would prefer other methods which are less time consuming and easier to evaluate.

3.4 Electron Microscopy, Immunoelectron Microscopy, and Fimbriae II

For demonstration of very fine filaments (Table 5), i.e., fimbriae II, EM is preferable to IEM, because using IEM, whether direct or indirect, these fimbriae will often appear as a mass since they aggregate and are often present in abundance. When LEVINE et al. (1984) showed that antigen CS3, present on some enterotoxigenic *E. coli* and formerly considered to be nonfimbrial (MULLANY et al. 1983), consisted of thin (2 nm), flexible, "fibrillar" fimbriae, this was primarily seen by improved EM examination and then confirmed by indirect IEM. Similarly, an adhesin produced by *E. coli* 469-3 (021:H⁻) from a case of enteritis was first described as nonfimbrial (WILLIAMS et al. 1984), but after technical improvement in EM procedures the adhesive strain was found to produce fine flexible filaments which were identified as the adhesin by indirect IEM (HINSON et al. 1987). LABIGNE-ROUSSEL et al. (1984) reported about an afimbrial adhesin (AFAI) with MR hemagglutinating ability from a pyelonephritic *E. coli* strain of O group 2. No visible fimbriae of this strain were detected by EM. The protein antigen provisionally termed Z_1(ØRSKOV et al. 1985)

Table 5. Some *E. coli* fine filaments or nonfimbrial adhesins (fimbriae II)

Name of fimbria/adhesin	Demonstrable in O:K:H serotype	Reference to morphology
K88[a]	e.g., O149:H10	STIRM et al. 1967b
F41[a]	e.g., O101:K32:H-	DE GRAAF and ROORDA 1982
CS3[a]	e.g., O6:K15:H16	LEVINE et al. 1984
XO75	O75:K5:H-	VÄISÄNEN-RHEN 1974
Z[a,b]	e.g., O21:K4:H4	ØRSKOV et al. 1985
Fib[c]	Different	FARIS 1985
469-3[b]	O21:H-	HINSON et al. 1987
AFAI[b]	O2	LABIGNE-ROUSSEL et al. 1984
NFAI	O83:K1:H4	GOLDHAR et al. 1984
NFAII	O14:H11	GOLDHAR et al. 1987
2230	O25:H16	DARFEUILLE-MICHAUD et al. 1986
M[c]	O2	RHEN et al. 1986b

[a] See Table 3;
[b] Serologically identical or strongly related (I. Ørskov, unpublished observations);
[c] See Table 1

was found in three strains earlier described as MR hemagglutinating but non-fimbriated by DUGUID et al. (1979) and IP et al. (1981). EM studies of one of the strains, O21:K4, showed that many but not all bacteria were surrounded by a coat consisting of a mesh of exceedingly fine threads. By indirect IEM the bacteria appeared surrounded by a thick capsule-like brim (ØRSKOV et al. 1985). By CIE we have found that AFAI and the 469-3 adhesin mentioned above are antigenically identical with or strongly related to the Z_1 antigen.

Other protein adhesins have been described as nonfimbrial by EM, e.g., one present in *E. coli* strain 2230 (O25:K?:H16) isolated from an infant with diarrhea (FORESTIER et al. 1987; DARFEUILLE-MICHAUD et al. 1986). This adhesin is non-hemagglutinating but adheres to human enterocyte brush borders, not to HEp-2 cells. No antigenic relationship was found by double immunodiffusion to CFA/I or CFA/II. Two other nonfimbrial protein adhesins, termed NFA, have been demonstrated on MR hemagglutinating *E. coli* of serotypes O83:K1:H4 and O14:K?:H11 isolated from urine (GOLDHAR et al. 1984, 1987). Direct IEM with monoclonal as well as polyclonal antisera against purified adhesin followed by embedding showed a capsule-like material around one of the strains and patchy material on the other. The strains and the purified adhesins adhered to cultured human kidney cells as tested by an ELISA type procedure, and inhibition studies of this adhesion showed that the adhesins from the two strains were serologically related but could be differentiated with MAbs. No cross-reactions were found by GOLDHAR et al. (1987) to the Z antigen and the fimbrial antigens F7–F14.

Since it appears from recent literature (HINSON et al. 1987) that demonstration of very fine threads by EM requires improved technique, it cannot be excluded that new studies would reveal that at least some of the adhesins described as nonfimbrial may also consist of delicate filaments, provided that the thickness is above the size visible by EM.

3.5 Enzyme-Linked Immunosorbent Assay

One of the main methods used today for the study of fimbriae serology is ELISA (ENGVALL and PERLMANN 1972); this was used, for example, by KORHONEN et al. (1982, 1984). The method is sensitive and has the advantage that the result is read in an objective way and that both whole bacteria suspensions and purified fimbrial preparations can be used as antigens. In many cases, however, it has the same disadvantage as the agglutination test, i.e., it does not differentiate between different antigens or different antibodies when such are present in an antigen preparation or a polyclonal antiserum, respectively.

When using MAbs, ELISA is in general the method of choice, e.g., DE REE et al. (1986a) found ELISA better than the agglutination test. McCONELL et al. (1985) examined more than 100 enterotoxigenic *E. coli* strains for presence of fimbrial antigens CFAI and CS1 to CS6 (Table 3). These antigens could be detected in more strains by ELISA than by the double immunodiffusion in gel test, which was normally used (e.g., THOMAS and ROWE 1982).

3.6 SDS–PAGE and Immunoblotting (Western Blotting)

In 1981 JANN et al. examined fimbriae preparations of ten *E. coli* strains of different O:K:H types by sodium dodecyl sulfate–polyacrylamide gel electrophoresis (SDS–PAGE) and found that they all exhibited different patterns of peptide bands, and that there was always more than one band corresponding to each strain. Similar multiple bands were found by KORHONEN et al. (1982) and HANLEY et al. (1985). The molecular weights of the subunits of fimbrial preparations from 14 strains ranged from 14 000 to 19 500. Very limited cross-reactivity between the strains was found by either an ELISA inhibition test or Western blotting (HANLEY et al. 1985).

The Western blotting technique (TOWBIN et al. 1979) is most useful when only a few antigens are to be examined with one or a few antisera, e.g., the examinations of CS1 and CS3 (LEVINE et al. 1984) and of the Z antigens (ØRSKOV et al. 1985) (fimbriae II), or when the reaction of several fimbrial preparations, whether originating from wild-type or recombinant strains, are to be examined with one antiserum (OTT et al. 1986).

3.7 SDS–PAGE of Immunoprecipitates

By immunoprecipitation RHEN et al. (1983a) examined whether the multiple peptides found in an SDS–PAGE analysis of a fimbrial preparation originated from different fimbriae or from the same fimbriae composed of two or three different subunits (KORHONEN et al. 1982). Two preparations of antigenically cross-reacting fimbriae derived from two strains were first precipitated with a nonhomologous antiserum reacting with only some of the peptides, and the supernatants were further precipitated with homologous antisera giving a second precipitate. Both precipitates were analyzed by SDS–PAGE, where they formed only distinct bands (RHEN et al. 1983a). By EM it was shown that the preparation from the second precipitate consisted of fimbriae, demonstrating that the different bands originally seen in SDS–PAGE originated from different fimbriae. Serologic examination of both P and S fimbriae was later undertaken by SDS–PAGE analysis of immunoprecipitates obtained with different fimbrial antisera, whereby it was demonstrated that S fimbriae of various strains (O18:K1:H7, O6:K$^+$, rough K1:H33, and O2:K2) showed great immunologic cross-reactivity (KORHONEN et al. 1984); in contrast, the serology of P fimbriae was highly complex (PERE 1986; PERE et al. 1986, 1988).

3.8 Gel Precipitation Tests

Double diffusion in gel precipitation tests (OUCHTERLONY 1949) and immunoelectrophoresis (SCHEIDEGGER 1955) have to a great extent been used for serologic analysis of fimbriae, particularly from enterotoxigenic *E. coli*, e.g., STIRM et al. (1967a) (K88), GUINÉE et al. (1976) (K99), and EVANS et al. (1975) (CFA/I). Both CFA/II (EVANS and EVANS 1978) and the colonization factor PCF 8775 (THOMAS et al. 1982) were resolved into three components in double diffusion, representing different fimbriae called coli surface antigens CS1–3 (CRAVIOTO et al. 1982; SMYTH 1982) and CS4–6 (THOMAS et al. 1985), respectively. CS3 are thin filaments (LEVINE et al. 1984) here named fimbriae II, but the nature of CS6 has not been reported (THOMAS et al. 1985). The other CS fimbriae are fimbriae I.

Another gel precipitation method which has been highly valuable for studies on fimbria serology is CIE because of its high resolving power (AXELSEN 1973, AXELSEN et al. 1973; WEEKE 1973). We used antisera from rabbits immunized with whole bacterial culture subsequently absorbed with culture of the same strain grown at 18°C, at which temperature type 1 fimbriae are expressed, in contrast to most other fimbriae (ØRSKOV et al. 1982b; ØRSKOV and ØRSKOV 1983a). In cases where a nonfimbriated mutant of the strain used for immunization was at hand, the serum was made fimbria-specific by absorption with the mutant grown at 37°C instead of the homologous culture grown at 18°C.

We prepared the fimbrial extract for CIE by heating bacterial suspensions at 60°C for 20 min followed by treatment in an Ultra Turrax apparatus for 2 × 1 min. By experience it was learned at which positions in the gels the different fimbria lines are situated. The relationship between fimbria antigens was examined by crossed line immunoelectrophoresis (CLIE) (KRÖLL 1973), i.e., incorporation of heterologous extract in an intermediate gel of a fimbrial reference system, i.e., a reference fimbrial extract run against its homologous fimbria antiserum. If antigens are related, the precipitation line will be somewhat lifted, i.e., moved anodically. If two antigens are identical, the line from the extract in the intermediate gel will fuse completely with the line or peak of the reference system, and this combined line will be moved upward to an extent depending on the antigen concentration in the intermediate gel.

However, a serologic relationship is more evident when it is examined in the less specific but more sensitive way by incorporation of a heterologous antiserum in the intermediate gel of a reference system (CLOSS et al. 1975). The relationship will be demonstrated by the formation of the precipitation line or peak at a lower position than when saline is incorporated in the intermediate gel. In a sense this method is an examination of one antigen against two antisera in a competitive way. Even if only some of the antigenic determinants, or epitopes, of the fimbrial protein in the bottom gel fit the antibody or antibodies in the intermediate gel, all the fimbrial molecules will participate in formation of the precipitate, and none will be left to react with the homologous antiserum in the upper gel, provided that only one fimbrial protein is present in the reference antigen in the bottom gel.

For general serotyping of fimbriae we examine fimbrial extracts by rocket immunoelectrophoresis. For confirmation of positive reactions, the extracts are incorporated in the intermediate gels (CLIE) of those reference systems indicated by the reactions in rocket immunoelectrophoresis. Only if the precipitation line of the extract under examination fuses with the reference precipitation line is a certain fimbria serotype reported.

Examination of *E. coli* from urinary tract infections (UTIs), particularly pyelonephritis, by CIE and IEM has shown the presence of different fimbrial antigens which have been termed F antigens and numbered F1B, F1C, and F7 to F16 (ØRSKOV et al. 1980; ØRSKOV and ØRSKOV 1983a and I. ØRSKOV and F. ØRSKOV, unpublished results). Serologically these antigens are easily distinguished from each other if studied by CLIE, while some relationships between F1C and F1B and between F7 to F16 are apparent when heterologous antiserum is incorporated in the intermediate gel. F1C, and probably also F1B, does not cause any agglutination of erythrocytes or adhesion to urinary epithelial cells. F7 to F13, F15, and F16 belong to the group of fimbriae (P) binding to gal-gal receptors.

Two striking features have appeared from the F antigen study: (a) an association between F antigens and O:K:H serotype, a finding which suggests nature's selection of certain virulent clones, and (b) a diversity of fimbriae not only among different O:K:H serotypes but also in the same strain (ØRSKOV et al. 1982b). As examples of these features mention can be made of O6:K2:H1 strains, which in our studies characteristically have shown the presence of either $F7_1$, $F7_2$, and F1C or $F7_2$ and F1C at the same time (ØRSKOV et al. 1981a), and O4:K12:H5 strains, which have mainly been found with the following F antigens: F1C, F13, F14, and F16.

At this juncture we should comment on the numbering of F antigens in the O4:K12:H5 strain C134-73 (Table 6). These antigens were originally given other designations, thus F13 was FC134.3, F14 was FC134.4 or Fy, and F16 was FC134.2 or F12 rel. The change to F13 took place when the FC134.3 antigen by CIE was found to be identical with the antigen of the pap fimbria in strain J96 (O4:K6) cloned by HULL et al. (1981). FC134.4 was difficult to separate from the other F antigens in

Table 6. Fimbriae of *E. coli* C134-73 (O4:K12:H5)

FC134 1 = pseudotype 1 = F1C
    2 = F12 rel. = F16
    3 = F13
    4 = Fy = F14

strain C134 by CIE and was therefore provisionally termed Fy (ØRSKOV and ØRSKOV 1983a), which has no relationship to the F(y) antigen found in strains from calves and first described by GIRARDEAU et al. (1980).

A MAb against a fimbrial preparation of strain 20025 of serotype $O4:K12:H^-$, similar to strain C134, was examined in ELISA, immunoblot, and rocket immunoelectrophoresis with various fimbrial preparations, and was shown to react with strains expressing Fy or an Fy-related antigen or the $F7_1$, $F7_2$ antigens (ABE et al. 1987). The fimbrial antigen reacting with this MAb was designated F14, well knowing that it was highly related to but not identical with $F7_2$. F14 was cloned by BOULNOIS (to be published) but its state as a P or an S fimbria is not yet clear.

By CIE the FC134.2 antigen had been found to be related to F12 and was therefore termed F12 rel. An antigen identical with it was detected, also by CIE, in two *E. coli* UTI strains of serotype O157:H45 (Table 4). By use of ELISA DE REE et al. (1986b) similarly demonstrated a difference between MAbs against F12 rel. and F12 (Table 7). The two O157:H45 strains reacted with the MAb against F12 rel. but not with the one against F12. The two MAbs thus seem to recognize different epitopes. The designation F16 was introduced instead of F12 rel. for this fimbrial antigen, which had first been demonstrated in the C134-73 strain.

In $O75:K5:H^-$ strains also from UTIs, another F12 rel. fimbrial antigen was found. This antigen was denoted F15. The relationship to F12 is different from that of F16.

At the beginning of this section it was stated that an antiserum against whole cells can be used for analysis of fimbrial antigens if it is absorbed with the same strain as is used for antiserum production but cultured at 18°C. Such an absorbed antiserum may also contain antibodies against other surface antigens; of particular importance in this respect are antibodies against those of the *E. coli* polysaccharide K antigens which are not expressed at 18°C (ØRSKOV et al. 1984). In CIE such antibodies can be recognized because the strongly anodic position of the precipitates is characteristic. An example is given in Fig. 1, which shows the K15, CS1, and CS3 precipitates. The form of the CS1 line is characteristic of tight compact fimbriae (fimbriae I). If the serum is diluted or the antigen concentrated, this line will be

Table 7. Comparison between two *E. coli* MAbs in ELISA with whole bacteria. Results from Table 2 of DE REE et al. (1986b)

Strain	Serotype	MAbs against fimbriae	
		F12 rel.[a] (F16) in strain 20025 (O4:K12:H-)	F12[b] in strain C1979 (O16:K1:H-)
C1979	O16:K1:H-	+	+
C134	O4:K12:H5	+	−
SP57	O157:H45	+	−
SP88	O157:H45	+	−

[a] MAb prepared by ABE et al. (1987);
[b] MAb prepared by DE REE et al. (1986b);
See also Table 4

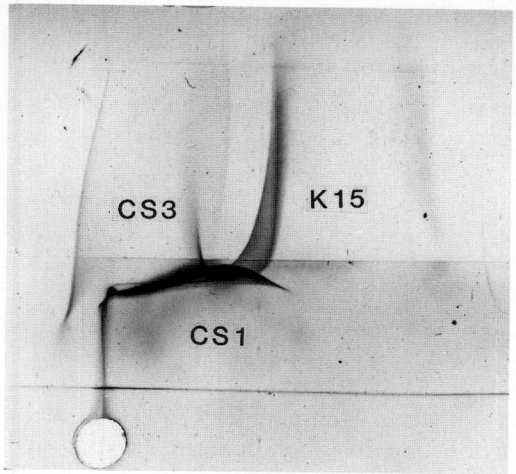

Fig. 1. CIE showing precipitation reactions of strain E1392-75 (06:K15:H16). *Lower gel*: extract E1392; *intermediate gel*: saline; *upper gel*: antiserum E1392 absorbed with culture 1392 grown at 18°C. The high peak to the *left* is the CS3 peak, the high peak to the *right* is the K15 peak, and the almost *horizontal line* is the CS1 line. The anode is to the right in the first dimension and at the top in the second dimension

elevated and form a peak which, however, does not look like the peak of CS3, which is characteristic of fine flexible filaments (fimbriae II). The more anodic position of a fimbriae I line than of a fimbriae II line is also characteristic.

4 Phase Variation

Fimbriae phase variation is a regulatory mechanism affecting phenotypic expression of fimbriae, which means that cells can vary between two alternative forms: fimbriated and nonfimbriated (BRINTON et al. 1954). The role of this change is influenced by cultural conditions (DUGUID and GILLIES 1957; DUGUID et al. 1955).

Based on the hemagglutinating capacity of strains we some years ago separated differently fimbriated organisms of the same strain and were surprised that fimbriae of different antigenic structures could be present on different cells and, although rarely, on the same cell, and that not all cells were fimbriated in a fimbriated culture (ØRSKOV et al. 1982b). Shortly afterward RHEN et al. (1983b) reported that there was a phase variation between P-fimbriated and 1C-fimbriated cells. Differently fimbriated cells were fractionated with specific antifimbria sera, and cells remaining in suspension after this serum treatment were examined for their agglutination properties with the various antisera. The phase variation was very rapid compared with the on-off variation of the *E. coli* type 1 fimbriae (EISENSTEIN 1981). By immunofluorescence microscopy with differently fluorochrome-labeled antibodies specific for either P, type F1C, or type 1 fimbriae it was shown that in a broth culture of the examined strain the different fimbrial types mostly occurred on different cells. Only 9% carried more than one fimbrial type (NOWICKI et al. 1984). A similar fast phase variation was described between S and type 1 fimbriae by the same technique.

The cells of the S-fimbriated strain 3040 (O18ac:K1:H7) changed more rapidly from S-fimbriated or type 1-fimbriated cells into nonfimbriated cells than they shifted from one phase to another (NOWICKI et al. 1985a). Since these phase changes are very fast, overnight culture of fractionated subpopulations carrying only one fimbrial type produces colonies that are heterogeneous in respect of fimbrial antigens (NOWICKI et al. 1985b).

5 Conclusions

What we in the beginning supposed to be distinct antigens of P fimbriae of *E. coli*, i.e., F7–F12, have turned out to be not so distinct. Multiple copies of the pap operon have been found in a high percentage of *E. coli* from the urinary tract (HULL et al. 1985), so it is no wonder that, for example, some strains of serotype O6:K2:H1 express P fimbriae antigen $F7_1$ or $F7_2$, and others $F7_1$ and $F7_2$, etc. (ØRSKOV and ØRSKOV 1983a; ØRSKOV et al. 1982a). By CLIE we can tell when two fimbrial antigens are identical, and when we examine strains of O:K:H serotypes which are frequent in cases of pyelonephritis we find that a high percentage of the strains have F antigens identical with those of our reference strains; such strains often represent clones, i.e., strains of the same bioserotype, electrotype, outer membrane pattern, etc. (ØRSKOV and ØRSKOV 1983b; NIMMICH et al. 1984; NIMMICH and ZINGLER 1984). The F antigens of UTI strains not belonging to these most frequent clones may, upon CLIE examination, show only a relationship to, and not identity with, one or more of the known F antigens. Studies of F7–F12 by CIE with heterologous antiserum in the intermediate gel showed that several of these antigens were related although they were distinct when examined by CLIE (ØRSKOV and ØRSKOV 1983a), a finding which reflects the fact that more than one epitope is present in each F antigen. Results obtained by SDS–PAGE examination of immunoprecipitates have similarly suggested that more antigenic determinants can occur in different P fimbriae in variable combinations, and use of the term "F-specific antigenic determinant" rather than "specific F antigen" was proposed (PERE et al. 1988). Antigenic complexity of *E. coli* fimbrial antigens is reminiscent of the complexity of the Enterobacteriaceae O antigens, where the polysaccharide side chain of the lipopolysaccharide from different O groups often contains similar chemical structures (epitopes) in combinations characteristic for the single O group. Exactly defining which antigenic factors, determinants, or epitopes compose the fimbrial antigens is an important scientific question, and it may be important for the development of a vaccine, independently of what kind of fimbriae we are dealing with. It should be mentioned here that PERE et al. (1988) found that 82 of 84 P-fimbriated *E. coli* reacted in immunoprecipitation with three anti-P-fimbriae sera, and concluded that the antigens used for preparing the antisera might form the basis of a broadly cross-reacting P-fimbrial vaccine. However, recent data have indicated that purified adhesin proteins might be superior vaccines to those prepared from intact fimbriae (LUND et al. 1988a), and that the receptor-binding site of these minor proteins is conserved despite differences in their primary structures (HOSCHÜTZKY et al. 1989). Furthermore, not only gene clusters determining F13 fimbriae together with the Galα(1-4)Gal binding adhesin can be cloned from strain J96, as LUND et al. (1988b) recently cloned the genes for F13 fimbriae and a GalNAcα(1-3)GalNAc binding adhesin from J96. Strains with P fimbriae agglutinate sheep erythrocytes to various degrees, and this ability has been ascribed to the Forssman antigen, which is present on sheep erythrocytes but also on human erythrocytes. However, the Forssman antigen contains both the Forssman-specific GalNAcα(1-3)GalNAc moiety and the Galα(1-4)Gal moiety, and this antigen has recently been isolated from the human kidney (BREIMER 1985; LUND et al. 1988b). Thus the adhesin binding the Forssman-specific determinant may also have an important function in colonization of the upper urinary tract, and its possible antigenic variability should be examined.

VAN DIE et al. (1988) questioned in a recent paper whether determination of F specificity is sufficient for serologic characterization of fimbriae. The answer is probably that some sort of simplified F typing will suffice for most epidemiologic examinations because of the apparent stability of the main F character in the different pathogenic clones. Of course, for other purposes a more detailed analysis may be necessary. Such a detailed analysis could be a combination of an examination by SDS–PAGE of fimbrial preparations and of immunoprecipitates, and an analysis by CIE, particularly CLIE; the fimbria-specific antisera should include cross-absorbed antisera, and production of many MAbs should be considered.

References

Abe C, Schmitz S, Moser I, Boulnois G, High NJ, Ørskov I, Ørskov F, Jann B, Jann K (1987) Monoclonal antibodies with fimbrial F1C, F12, F13, and F14 specificities obtained with fimbriae from *E. coli* O4:K12:H⁻. Microb Pathogen 2: 71–77

Adegbola RA, Old DC (1982) New fimbrial hemagglutinin in *Serratia* species. Infect Immun 38: 306–315

Adegbola RA, Old DC (1987) Antigenic relationships among type-1 fimbriae of Enterobacteriaceae revealed by immuno-electronmicroscopy. J Med Microbiol 24: 21–28

Axelsen NH (1973) Intermediate gel in crossed and in fused rocket immunoelectrophoresis. Scand J Immunol 2 (Suppl 1): 71–77

Axelsen NH, Bock E, Kröll J (1973) Comparison of antigens: the reaction of identity. Scand J Immunol 2 (Suppl 1): 91–94

Breimer ME (1985) Chemical and immunological identification of the Forssman pentaglycosylceramide in human kidneys. Glycoconjugate 2: 375–385

Brinton CC (1959) non-flagellar appendages of bacteria. Nature 183: 782–786

Brinton CC (1965) The structure, function, synthesis and genetic control of bacterial pili and a molecular model for DNA and RNA transport in gram negative bacteria. Trans NY Acad Sci (Series II) 27: 1003–1054

Brinton CC, Buzell A, Lauffer MA (1954) Electrophoresis and phage susceptibility studies on a filament-producing variant of the *E. coli* B bacterium. Biochim Biophys Acta 15: 533–542

Brinton CC, Gemski P, Camahan J (1964) A new type of bacterial pilus genetically controlled by the fertility factor of *E. coli* K12 and its role in chromosome transfer. Proc Natl Acad Sci USA 52: 776–783

Clegg S, Gerlach GF (1987) Enterobacterial fimbriae. J Bacteriol 169: 934–938

Closs O, Harboe M, Wassum Am (1975) Cross reactions between mycobacteria. Scand J Immunol 2 (Suppl 2): 173–185

Cravioto A, Scotland SM, Rowe B (1982) Hemagglutination activity and colonization factor antigens I and II in enterotoxigenic and non-enterotoxigenic strains of *Escherichia coli* isolated from humans. Infect Immun 36: 189–197

Darfeuille-Michaud A, Forestier C, Joly B, Cluzel R (1986) Identification of a nonfimbrial adhesive factor of an enterotoxigenic *Escherichia coli* strain. Infect Immun 52: 468–475

de Graaf FK, Roorda I (1982) Production, purification, and characterization of the fimbrial adhesive antigen F41 isolated from calf enteropathogenic *Escherichia coli* strain B41M. Infect Immun 36: 751–758

de Ree JM, Schwillens P, van den Bosch JF (1986a) Monoclonal antibodies for serotyping the P fimbriae of uropathogenic *Escherichia coli*. J Clin Microbiol 24: 121–125

de Ree JM, Jann K, van de Bosch JF (1986b) Characterization of the P fimbriae F8 and F12 from uropathogenic *Escherichia coli* with monoclonal antibodies. In: de Ree JM (thesis) Molecular and serological characterization of P fimbriae from uropathogenic *Escherichia coli*. Krips Repro, Meppel

Duguid JP, Campbell I (1969) Antigens of the type-1 fimbriae of salmonellae and other enterobacteria. J Med Microbiol 2: 535–553

Duguid JP, Gillies RR (1957) Fimbriae and adhesive properties in dysentery bacilli. J Pathol Bacteriol 74: 397–411

Duguid JP, Old DC (1980) Adhesive properties of Enterobacteriaceae. In: Beachy EH (ed) Bacterial adherence. Chapman and Hall, London, Chap 7 (Receptors and recognition, series B, vol 6)

Duguid JP, Smith IW, Dempster G, Edmunds PN (1955) Non-flagellar filamentous appendages ("fimbriae") and haemagglutinating activity in *Bacterium coli*. J Pathol Bacteriol 70: 335–348

Duguid JP, Anderson ES, Campbell I (1966) Fimbriae and adhesive properties in salmonellae. J Pathol Bacteriol 92: 107–138

Duguid JP, Clegg S, Wilson MI (1979) The fimbrial and non-fimbrial haemagglutinins of *Escherichia coli*. J Med Microbiol 12: 213–227

Eisenstein BI (1981) Phase variation of type 1 fimbriae in *Escherichia coli* is under transcriptional control. Science 214: 337–339

Engvall E, Perlmann P (1972) Enzyme-linked immunosorbent assay, ELISA. III. Quantitation of specific antibodies by enzyme-labelled anti-immunoglobulin in antigen-coated tubes. J Immunol 109: 129–135

Evans DG, Evans DJ (1978) New surface-associated heat-labile colonization factor antigen (CFA/II) produced by enterotoxigenic *Escherichia coli* of serogroups O6 and O8. Infect Immun 21: 638–647

Evans DG, Silver RP, Evans DJ, Chase DG, Gorbach SL (1975) Plasmid-controlled colonization factor associated with virulence in *Escherichia coli* enterotoxigenic for humans. Infect Immun 12: 656–667.

Fairbrother JM, Larivière S, Lallier R (1986) New fimbrial antigen F165 from *Escherichia coli* serogroup O115 strains isolated from piglets with diarrhea. Infect Immun 51: 10–55

Faris A (1985) Adhesive and hydrophobic adsorptive properties of enterotoxigenic and bovine mastitis *Escherichia coli*. Identification of fibronectin-binding fimbriae (Fib fimbriae). Thesis, Swedish University of Agricultural Sciences, Uppsala

Forestier C, Welinder KG, Darfeuille-Michaud A, Klemm P (1987) Afimbrial adhesin from *Escherichia coli* strain 2230: purification, characterization and partial covalent structure. FEMS Microbiol Lett 40: 47–50

Gelderblom H, Beutin L, Hadjiyiannis D, Reupke H (1985) Rapid typing of pili of pathogenic *Escherichia coli* by dispersive immunoelectron microscopy. In: Habermehl K-O (ed) Rapid methods and automation in microbiology and immunology. Springer, Berlin Heidelberg New York Tokyo

Gillies RR, Duguid JP (1958) The fimbrial antigens of *Shigella flexneri*. J Hyg (Lond) 56: 303–318

Girardeau JP, Doubourguier HC, Contrepois M (1980) Attachment des *Escherichia coli* entéro-pathogènes à la muqueuse intestinale. Bull Groupe Techn Vet 80-4-B-190: 49–60

Goldhar J, Perry R, Ofek I (1984) Extraction and properties of nonfimbrial mannose-resistant hemagglutinin from a urinary isolate of *Escherichia coli*. Curr Microbiol 11:49–54

Goldhar J, Perry R, Golecki JR, Hoschützky H, Jann B, Jann K (1987) Nonfimbrial, mannose-resistant adhesins from uropathogenic *Escherichia coli* O83:K1:H4 and O14:K?: H11. Infect Immun 55: 1837–1842

Guinée PAM, Jansen WH (1979) Behavior of *Escherichia coli* K antigens K88ab, K88ac, and K88ad in immunoelectrophoresis, double diffusion, and hemagglutination. Infect Immun 23: 700–705

Guinée PAM, Jansen WH, Agterberg CM (1976) Detection of the K99 antigen by means of agglutination and immunoelectrophoresis in *Escherichia coli* isolates from calves and its correlation with enterotoxigenicity. Infect Immun 13: 1369–1377

Hacker J, Schmidt G, Hughes C, Knapp S, Marget M, Goebel W (1985) Cloning and characterization of genes involved in production of mannose-resistant, neuraminidase-susceptible (X) fimbriae from a uropathogenic O6: K15: H31 *Escherichia coli* strain. Infect Immun 47: 434–440

Hanley J, Salit IE, Hofmann T (1985) Immunochemical characterization of P pili from invasive *Escherichia coli*. Infect Immun 49: 581–586

Hinson G, Knutton S, Lam-Po-Tang MK-L, McNeish AS, Williams PH (1987) Adherence to human colonocytes of an *Escherichia coli* strain isolated from severe infantile enteritis: molecular and ultrastructural studies of a fibrillar adhesin. Infect Immun 55: 393–402

Honda T, Arita M, Miwatani T (1984) Characterization of new hydrophobic pili of human enterotoxigenic *Escherichia coli*: a possible new colonization factor. Infect Immun 43: 959–965

Hoschützky H, Lottspeich F, Jann K (1989) Isolation and characterization of the α-galactosyl-1,4-β-galactosyl-specific adhesin (P adhesin) from fimbriated *Escherichia coli*. Infect Immun 57: 76–81

Hull RA, Gill RE, Hsu P, Minshew BH, Falkow S (1981) Construction and expression of recombinant plasmids encoding type 1, or D-mannose-resistant pili from a urinary tract infection *Escherichia coli* isolate. Infect Immun 33: 933–938

Hull S, Clegg S, Svanborg Eden, C, Hull R (1985) Multiple forms of genes in pyelonephritogenic *Escherichia coli* encoding adhesins binding globoseries glycolipid receptors. Infect Immun 47: 80–83

Ip SM, Chrichton PB, Old DC, Duguid JP (1981) Mannose-resistant and eluting haemagglutinins and fimbriae in *Escherichia coli*. J Med Microbiol 14: 223–226

Jacobs AAC (1987) Structure-function relationships of the K88 and K99 fibrillar adhesins and enterotoxigenic *Escherichia coli*. Thesis, Free University, Amsterdam

Jann K, Jann B, Schmidt G (1981) SDS polyacrylamide gel electrophoresis and serological analysis of pili from *Escherichia coli* of different pathogenic origin. FEMS Microbiol Lett 11: 21–25

Jones GW, Isaacson RE (1983) Proteinaceous bacterial adhesins and their receptors. CRC Crit Rev Microbiol 10: 229–260

Källenius G, Möllby R, Svenson SB, Winberg J, Hultberg H (1980) Identification of a carbohydrate receptor recognized by uropathogenic *Escherichia coli*. Infection 8 (Suppl 3): S228–S293

Källenius G, Möllby R, Hultberg H, Svenson SB, Cedergren B, Winberg J (1981) Structure of carbohydrate part of receptor on human uroepithelial cells for pyelonephritogenic *Escherichia coli*. Lancet II: 604–606

Klemm P (1985) Fimbrial adhesins of *Escherichia coli*. Rev Infect Dis 7: 321–340

Klemm P, Christiansen G (1987) Three *fim* genes required for the regulation of length and mediation of adhesion of *Escherichia coli* type 1 fimbriae. MGG 208: 439–445

Klemm P, Ørskov I, Ørskov F (1982) F7 and type 1-like fimbriae from three *Escherichia coli* strains isolated from urinary tract infections: protein chemical and immunological aspects. Infect Immum 36: 462–468

Korhonen TK, Väisänen V, Saxén H, Hultberg H, Svenson SB (1982) P-antigen-recognizing fimbriae from human uropathogenic *Escherichia coli* strains. Infect Immun 37: 286–291

Korhonen TK, Väisänen-Rhen V, Rhen M, Pere A, Parkkinen J, Finne J (1984) *Escherichia coli* fimbriae recognizing sialyl galactosides. J Bacteriol 159: 762–766

Krøll J (1983) Crossed line-immunoelectrophoresis. Scand J Immunol 2 (Suppl 1): 79–81

Labigne-Roussel AF, Lark D, Schoolnik G, Falkow S (1984) Cloning and expression of an afimbrial adhesin (AFA-I) responsible for P blood group-independent, mannose-resistant hemagglutination from a pyelonephritic *Escherichia coli* strain. Infect Immun 46: 251–259

Lawn AM (1967) Simple immunological labelling method for electron microscopy and its application to the study of filamentous appendages of bacteria. Nature 214: 1151–1152

Lawn AM, Meynell E (1970) Serotypes of sex pili. J Hyg 68: 683–684

Leffler H, Svanborg-Edén C (1980) Chemical identification of a glycosphingolipid receptor for *Escherichia coli* attaching to human urinary tract epithelial cells and agglutinating human erythrocytes. FEMS Microbiol Lett 8: 127–134

Leffler H, Svanborg-Edén C (1981) Glycolipid receptors for uropathogenic *Escherichia coli* on human erythrocytes and uroepithelial cells. Infect Immun 34: 920–929

Levine MM, Ristaino P, Marley G, Smyth C, Knutton S, Boedeker E, Black R, Young C, Clements ML, Cheney C, Patnaik R (1984) Coli surface antigens 1 and 3 of colonization factor antigen II-positive enterotoxigenic *Escherichia coli*: morphology, purification and immune responses in humans. Infect Immun 44: 409–420

Lindberg F, Lund B, Johansson L, Normark S (1987) Localization of the receptor-binding protein adhesin at the tip of the bacterial pilus. Nature 328: 84–87

Lintermans PF, Pohl P, Bertels A, Charlier G, Vandekerckhove J, Vandamme J, Schoup J, Schlicker C, Korhonen T, de Greve H, van Montagu M (1989) Characterization and purification of the F17 adhesin present on the surface of bovine enteropathogenic and septicemic *Escherichia coli*. Am J Vet Res

Lund B, Lindberg F, Marklund B-I, Normark S (1988a) Tip proteins of pili associated with pyelonephritis: new candidates for vaccine development. Vaccine 6: 110–112

Lund B, Marklund B-I, Strömberg N, Lindberg F, Karlsson K-A, Normark S (1988b) Uropathogenic *Escherichia coli* express serologically identical pili with different receptor binding specificities. Mol Microbiol 2: 255–265

Maurer L, Orndorff PE (1985) A new locus, *pilE*, required for the binding of type 1 piliated *Escherichia coli* to erythrocytes. FEMS Microbiol Lett 30: 59–66

McConnell MM, Thomas LV, Day NP, Rowe B (1985) Enzyme-linked immunosorbent assays for the detection of adhesin factor antigens of enterotoxigenic *Escherichia coli*. J Infect Dis 152: 1120–1127

Minion FC, Abraham SN, Beachey EH, Goguen JD (1986) The genetic determinant of adhesive function in type 1 fimbriae of *Escherichia coli* is distinct from the gene encoding the fimbrial subunit. J Bacteriol 165: 1033–1036

Moch T, Hoschützky H, Hacker J, Kröncke K-D, Jann K (1987) Isolation and characterization of the α-sialyl-β-2,3-galactosyl-specific adhesin from fimbriated *Escherichia coli*. Proc Natl Acad Sci USA 84: 3462–3466

Morris JA, Thorns CJ, Sojka WJ (1980) Evidence for two adhesive antigens on the K99 reference strain *Escherichia coli* B41. J Gen Microbiol 118: 107–113

Morris JA, Thorns CJ, Boarer C (1985a) Evaluation of a monoclonal antibody to the K99 fimbrial adhesin produced by *Escherichia coli* enterotoxigenic for calves, lambs and piglets. Res Vet Sci 39: 75–79

Morris JA, Sojka WJ, Ready RA (1985b) Serological comparison of the *Escherichia coli* prototype strains for the F(Y) and Att 25 adhesins implicated in neonatal diarrhoea in calves. Res Vet Sci 38: 246–247

Moser I, Ørskov I, Hacker J, Jann K (1986) Characterization of a monoclonal antibody against the fimbrial F8 antigen of *Escherichia coli*. FEMS Microbiol Lett 34: 329–334

Mullany P, Field AM, McConnell MM, Scotland SM, Smith HR, Rowe B (1983) Expression of plasmids coding for colonization factor antigen II (CFA/II) and enterotoxin production in *Escherichia coli*. J Gen Microbiol 129: 3591–3601

Nagy B, Moon HW, Isaacson RE (1977) Colonization of porcine intestine by enterotoxigenic *Escherichia coli*: selection of piliated forms in vivo, adhesion of piliated forms to epithelial cells in vitro, and incidence of a pilus antigen among porcine enteropathogenic *E. coli* Infect Immun 16: 344–352

Nimmich W, Zingler G (1984) Biochemical characteristics, phage patterns, and O1 factor analysis of *Escherichia coli* O1:K1:H7:F11 and O1:K1:H⁻:F9 strains isolated from patients with urinary tract infections. Med Microbiol Immunol 173: 75–85

Nimmich W, Zingler G, Ørskov I (1984) Fimbrial antigens of *Escherichia coli* O1:K1:H7 and O1:K1:H⁻ strains isolated from patients with urinary tract infections. Zentralbl Bakteriol Mikrobiol Hyg [A] 258: 104–111

Norgen M, Normark S, Lark D, O'Hanley P, Schoolnik G, Falkow S, Båga M, Uhlin BE (1984) Mutations in *E. coli* cistrons affecting adhesion to human cells do not abolish Pap pili fiber formation. EMBO J 3: 1159–1165

Normark S, Lark D, Hull R, Norgren M, Båga M, O'Hanley P, Schoolnik G, Falkow S (1983) Genetics of digalactocide-binding adhesin from a uropathogenic *Escherichia coli* strain. Infect Imun 41: 942–949

Nowicki B, Rhen M, Väisänen-Rhen V, Pere A, Korhonen TK (1984) Immunofluorescence study of fimbrial phase variation in *Escherichia coli* KS71. J Bacteriol 160: 691–695

Nowicki B, Rhen M, Väisänen-Rhen V, Pere A, Korhonen TK (1985a) Kinetics of phase variation between S and type-1 fimbriae of *Escherichia coli*. FEMS Microbiol Lett 28: 237–242

Nowicki B, Rhen M, Väisänen-Rhen V, Pere A, Korhonen TK (1985b) Organization of fimbriate cells in colonies of *Escherichia coli* strain 3040. J Gen Microbiol 131: 1263–1266

Nowotarska M, Mulczyk M (1977) Serological relationship of fimbriae among Enterobacteriaceae. Arch Immunol Ther Exp (Warsz) 25: 7–46

Old DC, Adegbola RA (1983) A new mannose-resistant haemagglutinin in *Klebsiella*. J Appl Bacteriol 55: 165–172

Old DC, Adegbola RA (1985) Antigenic relationships among type-3 fimbriae of Enterobacteriaceae revealed by immunoelectronmicroscopy. J Med Microbiol 20: 113–121

Old DC, Payne SB (1971) Antigens of the type-2 fimbriae of salmonellae: "cross-reacting material" (CRM) of type-1 fimbriae. J Med Microbiol 4: 215–225

Old DC, Yakubu DE, Chrichton PB (1987) Demonstration by immunoelectronmicroscopy of antigenic heterogeneity among P fimbriae of strains of *Escherichia coli*. J Med Microbiol 23: 247–253

Ørskov I, Ørskov F (1983a) Serology of *Escherichia coli* fimbriae. In: Hanson LÅ, Kallós P, Westphal O (eds) Host parasite relationships in gram-negative infections. Prog Allergy 33: 80–105

Ørskov F, Ørskov I (1983b) Summary of a workshop on the clone concept in the epidemiology, taxonomy, and evolution of the Enterobacteriaceae and other bacteria. J Infect Dis 148: 346–357

Ørskov I, Ørskov F (1985) *Escherichia coli* in extra-intestinal infections. J Hyg 95: 551–575

Ørskov I, Ørskov F, Sojka WJ, Leach JM (1961) Simultaneous occurrence of *E. coli* B and L antigens in strains from diseased swine. Acta Pathol Microbiol Scand 53: 404–422

Ørskov I, Ørskov F, Sojka WJ, Wittig W (1964) K antigens K88ab(L) and K88ac(L) in *E. coli*. A new O antigen: O147 and a new K antigen: K89(B). Acta Pathol Microbiol Scand 62: 439–447

Ørskov I, Ørskov F, Smith HW, Sojka WJ (1975) The establishment of K99, a thermolabile, transmissible *Escherichia coli* K antigen, previously called "Kco", possessed by calf and lamb enteropathogenic strains. Acta Pathol Microbiol Scand [B] 83: 31–36

Ørskov I, Ørskov F, Birch-Andersen A (1980) Comparison of *Escherichia coli* fimbrial antigen F7 with type 1 fimbriae. Infect Immun 27: 657–666

Ørskov I, Ørskov F, Birch-Andersen A, Klemm P, Svanborg-Edén C (1982a) Protein attachment factors: fimbriae in adhering *Escherichia coli* strains. In: Robbins JB, Hill JC, Sadoff JC (eds) Bacterial vaccines. Part 4. Surface antigens: pili. Thieme-Stratton, New York (Seminars in infectious disease, vol 4)

Ørskov I, Ørskov F, Birch-Andersen A, Kanamori M, Svanborg-Edén C (1982b) O, K, H and fimbrial antigens in *Escherichia coli* serotypes associated with pyelonephritis and cystitis. Scand J Infect Dis [Suppl] 33: 18–25

Ørskov F, Sharma V, Ørskov I (1984) Influence of growth temperature on the development of *Escherichia coli* polysaccharide K antigens. J Gen Microbiol 130: 2681–2684

Ørskov I, Birch-Andersen A, Duguid JP, Stenderup J, Ørskov F (1985) An adhesive protein capsule of *Escherichia coli*. Infect Immun 47: 191–200

Ott M, Hacker J, Schmoll T, Jarchau T, Korhonen TK, Goebel W (1986) Analysis of the genetic determinants coding for the S-fimbrial adhesin (*sfa*) in different *Escherica coli* strains causing meningitis or urinary tract infections. Infect Immun 54: 646–653

Ouchterlony O (1949) Antigen-antibody reactions in gel. Acta Pathol Microbiol Scand 26: 507–515

Parkkinen J, Rogers GN, Korhonen T, Dahr W, Finne J (1986) Identification of the O-linked sialyloligosaccharides of glycophorin A as the erythrocyte receptors for S-fimbriated *Escherichia coli*. Infect Immun 54: 37–42

Parry SH, Rooke DM (1985) Adhesins and colonization factors of *Escherichia coli*. In: Sussman M (ed) The virulence of *Escherichia coli*. Academic, London (Special publications of the Society for General Microbiology, 13)

Parry SH, Abraham SN, Sussman M (1982) The biological and serological properties of adhesion determinants of *Escherichia coli* isolated from urinary tract infections in children. In: Schulte-Wisserman H (ed) Clinical, bacteriological and immunological aspects of urinary tract infections in children. Thieme, Stuttgart

Pere A (1986) P fimbriae on uropathogenic *Escherichia coli* O16:K1 and O18 strains. FEMS Microbiol Lett 37: 19–26

Pere A, Leinonen M, Väisänen-Rhen V, Rhen M, Korhonen TK (1985) Occurrence of type-1C fimbriae on *Escherichia coli* strains isolated from human extraintestinal infections. J Gen Microbiol 131: 1705–1711

Pere A, Väisänen-Rhen V, Rhen M, Tenhunen J, Korhonen TK (1986) Analysis of P fimbriae on *Escherichia coli* O2, O4, and O6 strains by immunoprecipitation. Infect Immun 51: 618–625

Pere A, Selander RK, Korhonen TK (1988) P fimbriae on O1, O7, O75, rough and nontypable strains of *Escherichia coli*. Infect Immun 56: 1288–1294

Pohl P, Lintermans P, Van Muylen K, Schotte M (1982) Colibacillis entérotoxigènes de veau possedant un antigène d'attachment différent de l'antigène K99. Ann Med Vet 126: 569–571

Rhen M, Wahlström E, Korhonen TK (1983a) P-fimbriae of *Escherichia coli*: fractionation by immune precipitation. FEMS Microbiol Lett 18: 227–232

Rhen M, Mäkelä PH, Korhonen TK (1983b) P-fimbriae of *Escherichia coli* are subject to phase variation. FEMS Microbiol Lett 19: 267–271

Rhen M, Tenhunen J, Väisänen-Rhen V, Pere A, Båga M, Korhonen TK (1986a) Fimbriation and P-antigen recognition of *Escherichia coli* strains harbouring mutated recombinant plasmids encoding fimbrial adhesins of the uropathogenic *E. coli* strains KS71. J Gen Microbiol 132: 71–77

Rhen M, Klemm P, Korhonen TK (1986b) Identification of two new hemagglutinins of *Escherichia coli*, *N*-acetyl-D-glucosamine-specific fimbriae and a blood group M-specific agglutinin, by cloning the corresponding genes in *Escherichia coli* K-12. J Bacteriol 168: 1234–1242

Scheidegger J-J (1955) Une micro-méthode de l'immuno-électrophorèse. Int Arch Allergy 7: 103–110

Schmitz S, Abe C, Moser I, Ørskov I, Ørskov F, Jann B, Jann K (1986) Monoclonal antibodies against the nonhemagglutinating fimbrial antigen 1C (pseudotype 1) of *Escherichia coli*. Infect Immun 51: 54–59

Smyth CJ (1982) Two mannose-resistant haemagglutinins on enterotoxigenic *Escherichia coli* of serotype O6:K15:H16 or H- isolated from travellers' and infantile diarrhoea. J Gen Microbiol 128: 2081–2096

Smyth CJ (1986) Fimbrial variation in *Escherichia coli*. In: Birkbeck TH, Penn CW (eds) Antigenic variation in infectious diseases. IRL, Oxford (Special publications of the Society for General Microbiology, 19)

Sparling PF, Cannon JG, So M (1986) Phase and antigenic variation of pili and outer membrane. J Infect Dis 153: 196–201

Stirm S, Ørskov F, Ørskov I, Mansa B (1967a) Episome-carried surface antigen K88 of *Escherichia coli*. II. Isolation and chemical analysis. J Bacteriol 93: 731–739

Stirm S, Ørskov F, Ørskov I, Birch-Andersen A (1967b) Episome-carried surface antigen K88 of *Escherichia coli*. III. Morphology. J Bacteriol 93: 740–748

Swanson J, Robbins K, Barrera O, Koomey JM (1987) Gene conversion variations generate structurally distinct pilin polypeptides in *Neisseria gonorrhoeae*. J Exp Med 165: 1016–1025

Thomas LV, Rowe B (1982) The occurrence of colonisation factors (CFA/I, CFA/II and E8775) in enterotoxigenic *Escherichia coli* from various countries in South East Asia. Med Microbiol Immunol 171: 85–90

Thomas LV, Cravioto A, Scotland, SM, Rowe B (1982) New fimbrial antigenic type (E8775) that may represent a colonization factor in enterotoxigenic *Escherichia coli* in humans. Infect Immun 35: 1119–1124

Thomas LV, McConnell MM, Rowe B, Field AM (1985) The possession of three novel coli surface antigens by enterotoxigenic *Escherichia coli* strains positive for the putative colonization factor PCF8775. J Gen Microbiol 131: 2319–2326

Towbin H, Stachelin T, Gordon J (1979) Electrophoretic transfer of proteins from polyacrylamide gels to nitrocellulose sheets: procedure and some applications. Proc Natl Acad Sci USA 76: 4350–4354

van Die I, Zuidweg E, Hoekstra W, Bergmans H (1986) The role of fimbriae of uropathogenic *Escherichia coli* as carriers of the adhesin involved in mannose-resistant hemagglutination. Microb Pathogen 1: 51–56

van Die I, Riegman N, Gaykema O, van Megen I, Hoekstra W, Bergmans H, de Ree H, van den Bosch H (1988) Localization of antigenic determinants on P-fimbriae of uropathogenic *Escherichia coli*. FEMS Microbiol Lett 49: 95–100

Väisänen-Rhen V (1984) Fimbria-like hemagglutinin of *Escherichia coli* O75 strains. Infect Immun 46: 401–407

Väisänen-Rhen V, Elo J, Väisänen E, Siitonen A, Ørskov I, Ørskov F, Svenson SB, Mäkelä PH, Korhonen TK (1984) P-fimbriated clones among uropathogenic *Escherichia coli* strains. Infect Immun 43: 149–155.

Weeke B (1973) General remarks on principles, equipment, reagents and procedure. Scand J Immunol 2 (Suppl 1): 15–35

Williams PH, Knutton S, Brown MGM, Candy DCA, McNeish AS (1984) Characterization of nonfimbrial mannose-resistant protein hemagglutinins of two *Escherichia coli* strains isolated from infants with enteritis. Infect Immun 44: 592–598

Willshaw GA, Smith HR, McConnell MM, Rowe B (1988) Cloning of genes encoding coli-surface (CS) antigens in enterotoxigenic *Eschericia coli*. FEMS Microbiol Lett 49: 473–478

Adhesins as Lectins: Specificity and Role in Infection

I. Ofek[1] and N. Sharon[2]

1 Introduction

The idea that proteins with lectin-like properties on bacterial surfaces serve as adhesins, which bind the organisms to animal cells, was first suggested in 1977 for *Escherichia coli* expressing type 1 fimbriae (or pili) specific for mannose (OFEK et al. 1977). Since then the adhesins of many bacteria studied have been found to be carbohydrate-binding proteins, akin to lectins (Table 1). These lectins have become the focus of interest in research because of the mounting evidence for the crucial role they, like other adhesins, play in infectious disease (SHARON 1987; OFEK and SHARON 1988).

The bacterial surface lectins serve as molecules of recognition in cell–cell interactions. For many of them, only limited information is available on their

---

[1] Department of Human Microbiology, Sackler School of Medicine, Tel Aviv University, Tel Aviv, and
[2] Department of Biophysics, The Weizmann Institute of Science, Rehovot, Israel

Current Topics in Microbiology and Immunology, Vol. 151
© Springer-Verlag Berlin·Heidelberg 1990

Table 1. Sugar specificity of bacterial surface lectins[a]

Monosaccharide	Bacteria[b]	Fimbriae[c]
Mannose	*Citrobacter freundii, Enterobacter* spp., *Erwinia carotovora, Escherichia coli, Klebsiella aerogenes, Klebsiella pneumoniae, Salmonella* spp., *Serratia marcescens, Shigella flexneri* (1)	Type 1
Galactose	*Escherichia coli* (2), *Fusobacterium nucleatum* (3)	
L-Fucose	*Vibrio cholerae* (4)	
L-Rhamnose, D-fucose	*Capnocytophaga ochracea* (5)	
N-Acetylgalactosamine	*Eikenella corrodens* (6), *Escherichia coli* (7)	
N-Acetylglucosamine	*Escherichia coli* (8)	Type G
Galα4Galβ-	*Escherichia coli* (9)	Type P
Galβ3GalNAc	*Actinomyces naeslundii, Actinomyces viscosus* (10, 11), *Bacteroides* spp. (12), *Clostridia* (12), *Lactobacillus* (12), *Propionibacterium* (12)	Type 2
Galβ4GlcNAc	*Erwinia rhapontici* (13), *Staphylococcus saprophyticus* (14)	
GalNAcβ4Gal	*Haemophilus influenzae* (15), *Klebsiella pneumoniae* (15), *Neisseria gonorrhoeae* (16), *Pseudomonas aeruginosa* (17), *Pseudomonas cepacia* (17)	
GlcNAcβ3Gal	*Streptococcus pneumoniae* (18, 19)	
NeuGcα2-3Galβ4Glc-	*Escherichia coli* (20)	K99
NeuAcα2-3Galβ	*Escherichia coli* (21, 22)	Type S
	Bordetella bronchiseptica (23), *Campylobacter pylori* (24), *Mycoplasma gallisepticum* (25), *Mycoplasma pneumoniae* (26), *Streptococcus sanguis* (27)	
(Glcα6)$_{5-8}$	*Streptococcus cricetus* (28), *Streptococcus sobrinus* (29)	

[a] For a more complete listing, as well as for agglutinins of viruses, fungi, and protozoa, the reader is referred to MIRELMAN and OFEK (1986);
[b] References to literature are numbered in parentheses as follows: (1) DUGUID and OLD 1980; (2) NILSSON et al. 1983; (3) MURRAY et al. 1988; (4) JONES and FRETER 1976; (5) WEISS et al. 1987; (6) YAMAZAKI et al. 1981; (7) Faris et al. 1980; (8) VÄISÄNEN-RHEN et al. 1983; (9) LEFFLER and SVANBORG EDÉN 1986; (10) Cisar 1986; (11) BRENNAN et al. 1987; (12) HANSSON et al. 1983; (13) KORHONEN et al. 1988; (14) GUNNARSSON et al. 1984; (15) KRIVAN et al. 1988a; (16) STROMBERG et al. 1988; (17) KRIVAN et al. 1988b; (18) ANDERSSON et al. 1983; (19) PULVERER et al. 1987; (20) SMIT et al. 1984; (21) PARKKINEN et al. 1986; (22) KORHONEN et al. 1984; (23) ISHIKAWA and ISAYAMA 1987; (24) EVANS et al. 1988; (25) GLASGOW and HILL 1980; (26) LOOMES et al. 1984; (27) MURRAY et al. 1982; (28) DRAKE et al. 1988; (29) LANDALE and MCCABE 1987;
[c] Type of fimbria is given only for those microorganisms for which a lectin activity has been ascribed.

properties and sugar specificity. Several of the bacterial surface lectins have, however, been studied to a considerable extent with respect to not only their detailed carbohydrate specificity but also to their role in infection. The knowledge obtained has provided insights into various features of the sugar-combining site of the lectins and the nature of the interaction between bacteria and cell surfaces. Eventually, it may lead to the design of highly effective inhibitors that will prevent adhesion, colonization of mucosal surfaces, and infections in humans.

   In this chapter we summarize the knowledge about the sugar specificity and role in infection of the better characterized bacterial surface lectins. For a more thorough coverage of bacterial lectins, as well as of lectins or agglutinins whose function as adhesins has not yet been defined, the reader is referred to a recent book on the subject (MIRELMAN 1986).

2 Sugar Specificity

The approaches used to elucidate the sugar specificity of the bacterial lectins are based on procedures commonly used in studies of lectins in general, and of lectin–cell interactions in particular (SHARON et al. 1981; OFEK et al. 1985). The first step involves the testing of the effect of a panel of sugars (monosaccharides, simple glycosides, and oligosaccharides) on bacterial adherence to, or agglutination of, erythrocytes or other cells. This step is based on the well-known hapten inhibition techniques used in the study of the specificity of antibodies and plant lectins (KABAT 1978).

The specificity of the lectin is usually defined by the sugar which is the best inhibitor of the adherence or agglutination reaction (Table 1). Alternatively, agglutination of cells, e.g., yeasts, or synthetic particles, e.g., latex beads carrying defined carbohydrates, may be employed (OFEK and BEACHEY 1978; DE MAN et al. 1987). Important information about the structure of the carbohydrates which serve as attachment sites on cell surfaces for a bacterial lectin, i.e., the lectin receptors, may be obtained by examining the effect on agglutination or adherence of treatment of the cells with enzymes (e.g., glycosidases or galactose oxidase) and chemical reagents (e.g., periodate) which modify carbohydrates (reviewed in OFEK et al. 1985).

For further identification of the receptors, they must be isolated and characterized. In general, the receptors that carry the sugar structure(s) for which a bacterial lectin is specific are glycoproteins and/or glycolipids. For receptors that are glycoproteins, techniques using affinity chromatography on immobilized bacterial lectins (RODRIGUEZ-ORTEGA et al. 1987; GIAMPAPA et al. 1988), or lectin overlays of blots of electrochromatograms of cell membrane glycoproteins, are used (PRAKOBPHOL et al. 1987). For receptors that are glycolipids, membrane glycolipids are separated on thin layer silica gel plates and the chromatograms are probed with the lectin-carrying bacteria (HANSSON et al. 1985). Although the number of bacterial lectin receptors known is still very small, it is already clear that they may differ among the various cell types which bind the same bacterial lectin (KARLSSON 1986); the term "isoreceptors" was suggested to describe such a family of molecules. This has been shown for the P fimbrial lectin, whose sugar specificity appears to be absolute for Galα4Galβ (galabiose). It is also true for the mannose-specific type 1 fimbrial lectin, which binds to three distinct glycoproteins derived from the cell membrane of human polymorphonuclear leukocytes (RODRIGUEZ-ORTEGA et al. 1987) and to a glycoprotein from guinea pig erythrocytes (GIAMPAPA et al. 1988) which is different from the former ones. The accessibility of such isoreceptors may vary with the type of tissue and, as a consequence, the affinity of different bacterial strains carring the same lectin to various tissues or cells may differ too (SHERMAN et al. 1985; OFEK et al. 1982).

Bacteria which carry the gene coding for a certain lectin may not always express it on their surface. In particular, many lectin-producing bacteria undergo phase variation, i.e., they switch readily back and forth from the phenotype which expresses the lectin to one that does not. Outgrowth of one phenotype over the other may occur in a bacterial population growing under specific conditions. In addition, bacterial cultures frequently contain two or more types of lectin. This may be the result of either coexpression on the cell surface of multiple types of lectin or the presence of heterogeneous populations of cells, each bearing a different lectin. In such cases, combinations of sugars are more effective inhibitors than individual ones. In addition, the percentage of the total test population of bacteria which express the different lectin may vary (MAAYAN et al. 1985; GOOCHEE et al. 1987).

2.1 Surface Lectins of Enterobacteriaceae

Much of our knowledge of the surface lectins of Enterobacteriaceae is based on the pioneering studies performed in the 1950s and 1960s in the laboratories of DUGUID and of BRINTON (reviewed by DUGUID and OLD 1980; BRINTON 1965). DUGUID was the first to study the agglutination properties of enteric bacteria systematically. He observed that many of them agglutinated erythrocytes of different animal species and classified the agglutinating bacteria into two groups: (a) mannose sensitive (MS) i.e., those that agglutinated guinea pig erythrocytes very strongly and human erythrocytes of all blood groups (A, B, O) moderately to strongly, and whose agglutination activity was inhibited by small concentrations (0.01%–0.5% w/v) of mannose, methyl α-mannoside, and yeast mannan; (b) mannose resistant (MR), i.e., those that agglutinated human erythrocytes strongly and were not inhibited by mannose. It is now well established that many of the latter strains exhibit distinct sugar specificities (Table 1).

Most bacterial strains are genotypically capable of producing more than one type of lectin. This is particularly pronounced in the case of pathogenic enterobacteria which possess the ability to produce type 1 fimbriae as well as one or more additional lectins (usually in the form of fimbriae) or hemagglutinins which may not be sugar specific (GOLDHAR et al. 1984).

In this chapter we focus on those fimbrial lectins of the enteric bacteria for which information is available on sugar specificity.

2.2 Type 1 Fimbriae

Most of the information on the carbohydrate specificity of type 1 fimbriae is based on studies of the inhibitory effect of a variety of glycosides and oligosaccharides of mannose on the agglutination of yeasts by type 1 fimbriated *E. coli*, *Klebsiella pneumoniae*, *Salmonellae* species, and, in the case of *E. coli*, also type 1 fimbriae isolated from this organism (FIRON et al. 1983, 1984) (Table 2). A similar pattern of specificity was subsequently found by NEESER et al. (1986), who used guinea pig erythrocyte instead of yeasts as indicator cells. We have postulated that the combining site of the type 1 fimbrial lectin corresponds to the size of a trisaccharide (Fig. 1) and that it is in the form of a depression or pocket on the surface of the lectin. Extended carbohydrate binding sites have been described for enzymes (e.g., lysozyme), several lectins, and antibodies (KABAT 1978). In the case of the *E. coli*

Table 2. Relative inhibition by mannose derivatives of yeast agglutination by *E. coli* 346 and by the type 1 fimbriae isolated from this organism (Firon et al. 1984)

Sugar	E. coli 346	Isolated fimbriae
MeαMan	1.0	1.0
Mannose	0.8	–
Manα6Man	0.5	–
Manα2Man	1.3	1.8
Manα3Man	1.2	–
Manα2Manα2Man	1.4	–
Manβ4GlcNAc	<0.4	–
Manα3Manβ4GlcNAc	21.0	30.0
Manα6Manβ4GlcNAc	0.7	–
Manα2Manα3Manβ4GlcNAc	0.7	5.5
Manα2Manα2Manα3Manβ4GlcNAc	0.7	–

Manα6
    Manα6
    Manα3 ⟩ManαOMe
Manα3
	3.5	–

Manα2Manα6 Manα6 Manα3 ⟩ManαOMe Manα3	4.7	4.8
Manα6 ⟩ManαOMe Manα3	10.5	–

Manα6
    Manα6
Manα3
    Manα3 ⟩ManαOMe
	30.0	–

Manα6
    Manα6
Manα3
    Manα3 ⟩ManαOMe
 Manα2Manα3
	30.0	36.0

pNPαMan	30.0	48.0
pNPβMan	1.2	–
Galβ4GlcNAcβ2Manα6Manβ4GlcNAc	0.25	–
Galβ4GlcNAcβ2Manα6 Manβ4GlcNAc Galβ4GlcNAcβ2Manα3	0.6	–

pNP = *p*-nitrophenyl

lectin, there are probably three adjacent subsites, each of which accommodates a monosaccharide residue.

The presence of a hydrophobic binding region adjacent to the binding site of type 1 fimbriae is indicated by the finding that aromatic α-mannosides are powerful inhibitors of the agglutination of yeasts by *E. coli* and of the adherence of the bacteria to guinea pig ileal epithelial cells (Firon et al. 1984, 1987). In both systems, the best inhibitors were 4-methylumbelliferyl α-mannoside and *p*-nitro-*o*-chlorophenyl α-mannoside (500–1000 times more inhibitory than methyl α-mannoside). 4-Methylumbelliferyl α-mannoside was also more effective than methyl α-mannoside in removing adherent *E. coli* from ileal epithelial cells.

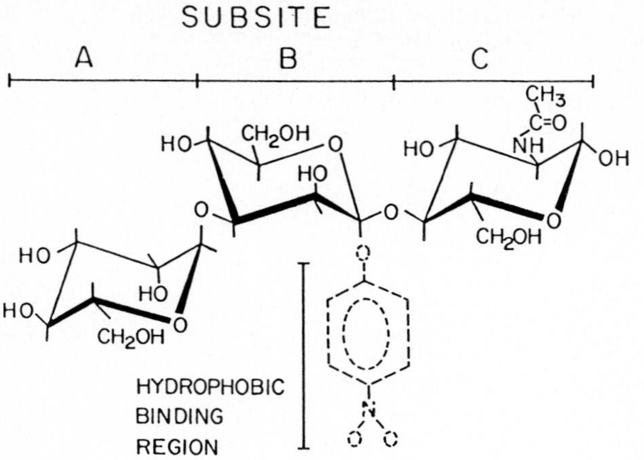

SUBSITE

Fig. 1. Scheme of combining site of the mannose-specific lectin (type 1 fimbriae) of *E. coli* (Sharon 1987)

The combining site of type 1 fimbriae appears to differ from that of concanavalin A, a plant lectin with a closely related sugar specificity. The agglutinating activity of concanavalin A is inhibited by both mannose and glucose, whereas that of type 1 fimbriated *E. coli* is not inhibited by glucose. In addition, Manα2Man is a considerably better inhibitor of concanavalin A than is methyl α-mannoside, in contrast to what is found with the *E. coli* lectin.

Although all strains of *E. coli* examined, as well as *K. pneumoniae*, exhibited essentially the same pattern of specificity this is not the case with other entero-bacteria (FIRON et al. 1984). For example, with several *Salmonella* species examined, aromatic α-mannosides, as well as the trisaccharide Manα3Manβ4GlcNAc, were weaker inhibitors than methyl α-mannoside. The combining site of different *Salmonellae* species is probably smaller than that of *E. coli* or *K. pneumoniae*, and it is devoid of a hydrophobic region. Although classified under the general term mannose specific (or mannose sensitive), the fimbrial lectins of different genera and species differ in their sugar specificity. Within a given genus, however, all strains tested exhibited the same specificity.

In general, it appears that mannose-specific bacteria preferentially bind structures found in short oligomannose chains (and in hybrid units, see below) of N-linked glycoproteins. Such structures are common constituents of many eukaryotic cell surfaces (SHARON and LIS 1982), which accounts for the fact that mannose-specific bacteria bind to a wide variety of cells. Membrane glycolipids are unlikely to serve as receptors for mannose-specific bacteria, since they are devoid of mannose residues.

These conclusions are supported by studies of the binding of mannose-specific *E. coli* to mammalian cells that differ in the level of oligomannose units on their surfaces. Mutants of baby hamster kidney (BHK) cells with increased levels of N-linked oligomannose or hybrid units in their glycoproteins bound considerably larger numbers of mannose-specific *E. coli* (FIRON et al. 1985) (Fig. 2). These cells were also more sensitive to agglutination by these bacteria than the parental wild-type cells. The best example is a mutant (Ric^{R}14) that lacks the enzyme N-

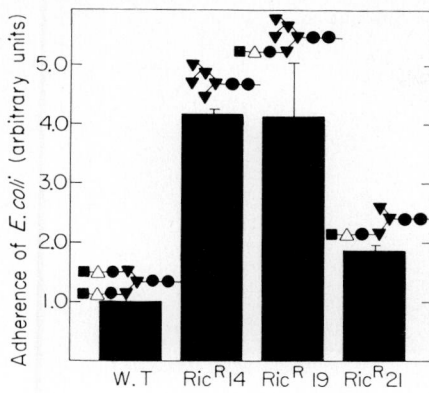

Fig. 2. Mannose-specific binding of type 1 fimbriated *E. coli* to wild-type and ricin-resistant (Ric) BHK cells. The symbols above each bar denote the predominant N-linked oligosaccharide expressed on the surface of the BHK cells: ▼, mannose; ●, *N*-acetylglucosamine; △, galactose; ■, sialic acid. (Sharon 1987; data from Firon et al. 1985)

acetylglucosaminyltransferase I, which catalyzes the first step in the conversion of oligomannose units into complex ones. This mutant bound four times more type 1 fimbriated *E. coli* and was agglutinated at a rate more than ten times faster than the parental BHK cells. Preference for binding to hybrid units was indicated from results of experiments with mutants which express high levels of such units on their surface. The *N*-linked oligomannosides specific for type 1 fimbrial lectin probably reside on more than one type of molecule on the animal cell surface. For example, type 1 fimbriae bound to three glycoproteins derived from the surface membrane of human polymorphonuclear leukocytes (RODRIGUEZ-ORTEGA et al. 1987).

2.3 P Fimbriae

P fimbriae are specific for the disaccharide Galα4Galβ, whether located in a terminal nonreducing portion or internally (LEFFLER and SVANBORG EDÉN 1980, 1986). Compounds containing this disaccharide were the best inhibitors of agglutination of human erythrocytes by P fimbriated *E. coli* and of adhesion of these bacteria to epithelial cells (Table 3). Monosaccharides did not inhibit the attachment of the bacteria to the cells. Globotetraose (GalNAcβ3Galα4Galβ4Glc) blocked the agglutination of human type P erythrocytes by the isolated fimbriae.

The receptor for P-specific *E. coli* on epithelial cells appears to be globotetraosyl-ceramide. Thus, such bacteria did not bind to epithelial cells (e.g., urinary tract cells of p⁻ individuals) lacking this substance. Also, erythrocytes not agglutinated by the bacteria or by the isolated P fimbriae became susceptible to agglutination after they had been coated with globotetraosylceramide (or trihexosylceramide) (LEFFLER and SVANBORG EDÉN 1980, 1986; KORHONEN et al. 1982). Latex beads coated with Galα4Gal, or suitable derivatives of this disaccharide, have been used to detect the presence of P fimbriae in clinical isolates of *E. coli* (DE MAN et al. 1987). Additional information on the specificity of P fimbriae has been obtained in experiments in which the binding of the bacteria to glycolipids separated by thin layer chromatography has been examined (BOCK et al. 1985; KARLSSON 1986).

Table 3. Inhibition of adherence of P fimbriated *E. coli* 3669 to human urinary epithelial cells[a]

Compound tested	Inhibtory concentration[b]
GalNAcβ3Galα4Galβ4Glcβ1-Cer (globotetraosylceramide)	0.1
Galα4Galβ4GlcβO-(CH$_2$)$_2$-S-(CH$_2$)$_{17}$-CH$_3$	0.2
GalNAcβ3Galα4Galβ4Glc	0.2
Galα4Galβ4GlcNAcβOEt	1.3
Galα4GalβOEt	2
Galα4Galβ4GlcβOEt	2
Galα4Gal	11
GalNAcβ3GalαOMe	> 50
Galα4GlcβOEt	> 50

[a] Data from LEFFLER and SVANBORG EDÉN (1986);
[b] Concentration (mM) required for complete inhibition of the adherence of 50 bacteria/cell.

Computer-based calculations and molecular models of the preferred conformation of the Galα4Gal-containing glycolipids revealed that the disaccharide forms a bend, or knee, in the chain, on which there is a continuous nonpolar surface surrounded by polar oxygens. The sugar-combining site of the P fimbrial lectin seems to bind to the latter side of the disaccharide (BOCK et al. 1985; KARLSSON 1986).

2.4 Fimbriae Specific for Sialic Acid

Enterotoxigenic strains of *E. coli* isolated from human and farm animals express fimbrial hemagglutinins specific for glycoconjugates containing sialic acids. This conclusion was originally based on the observation that hemagglutination caused by fimbriated bacteria was decreased or completely abolished after sialidase treatment of the erythrocytes. Several fimbrial lectins with this type of specificity are known. They include: K99, the adhesin expressed by enterotoxigenic strains isolated from piglets, calves, and lambs suffering from diarrhea; CFA/I and CFA/II expressed by human enterotoxigenic isolates; and type S expressed by strains isolated from newborn infants suffering from sepsis and meningitis.

2.4.1 K99 Fimbriae

The detailed sugar specificity of K99 fimbrial lectin was recently studied by LINDAHL et al. (1987), who examined the effect of various sialic acid-containing oligosaccharides on hemagglutination caused by bacteria carrying K99 fimbriae. The affinity of K99 fimbriae for *N*-glycolylneuraminic acid was found to be twice that of *N*-acetylneuraminic acid, and the monosaccharides were found to be less than one-tenth as effective as inhibitors. SMIT et al. (1984) have isolated, from horse erythrocytes, glycolipids that contain the attachment site for binding the K99 fimbrial lectin. The receptor was identified as the glycolipid NeuGcα2-

3Galβ4Glcβ1-ceramide. The purified glycolipid inhibited hemagglutination caused by the whole bacteria and by the purified fimbrial lectin, as well as adherence of the bacteria to intestinal cells.

2.4.2 CFA/I and CFA/II Fimbriae

The effect of various naturally occurring glycopeptides and oligosaccharides on hemagglutination caused by bacteria carrying CFA/I and CFA/II fimbriae was examined by NEESER et al. (1988). Although simple sugars were not employed to define precisely the specificity of the lectins, several interesting findings emerged. First, only complex-type N-linked or human milk oligosaccharides were inhibitory and this activity was dependent on the presence of sialic acids in the compounds. Second, there was a close analogy between the carbohydrate specificities of the CFA/I and CFA/II fimbrial lectins. Finally, the combining site of the lectins seemed to best fit oligosaccharide structures of glycopeptides derived from human red blood cells, in spite of the fact that CFA/II expressing bacteria do not agglutinate human erythrocytes. Thus, the results obtained emphasize the notion that even though the target cells contain glycoconjugates for which the bacterial lectin is specific, the cells may not be agglutinated by intact bacteria expressing the lectin, probably due to inaccessibility of the receptors on the cells or the lectins on the bacterial surface (OFEK et al. 1982).

2.4.3 Type S Fimbriae of E. coli Causing Sepsis and Meningitis

The specificity of S-fimbrial lectins was investigated using the overlay method in which erythrocyte cell membrane glycoproteins are electrophoresed, blotted, and incubated with radiolabeled E. coli bacteria expressing these fimbriae (PARKKINEN et al. 1986). The only band which bound the bacteria was glycophorin A, suggesting that the latter contains an attachment site for the bacterial lectin. Binding of bacteria to glycophorin A-coated plates was inhibited by several sialylated oligosaccharides, the most active of which was NeuAcα2-3Galβ3GalNAc, also a constituent of glycophorin A.

2.5 Lectins of *Vibrio cholerae* Specific for L-Fucose

Cholera vibrios produce a variety of hemagglutinins (FINKELSTEIN and HANNE 1982). Agglutination of human group O erythrocytes by *V. cholerae* and adhesion of the organism to brush borders was specifically inhibited by L-fucose and various glycosides of L-fucose, and to a lesser extent by mannose (JONES and FRETER 1976). The bacteria adhered specifically to agarose beads that carried covalently linked L-fucose on their surfaces. It has been suggested that structures containing L-fucose on eukaryotic cell surfaces may function as receptors for the vibrio hemagglutinins and may therefore be an important determinant of host susceptibility to these bacteria.

HANNE and FINKELSTEIN (1982) reported that *V. cholerae* is capable of producing four different hemagglutinins. The El-Tor biotype produced a cell-associated mannose-specific hemagglutinin that was inhibited by fructose. This lectin was active on all human (A, B, O) and all chicken erythrocytes tested. A lectin specific for L-fucose was detected transiently in early log-phase growth in several strains of *V. cholerae*, including a number of mutants that were not inhibited by mannose.

2.6 *Mycoplasma pneumoniae* and *Mycoplasma gallisepticum*

Mycoplasma pneumoniae is specific for N-acetylneuraminic acid attached by $\alpha2 \to 3$ linkage to the terminal galactose residue of the poly-N-acetyllactosamine sequences of blood type I/i antigens (LOOMES et al. 1984, 1985). This conclusion is based in part on measurements of the adherence of sialidase-treated human erythrocytes that have been resialylated by specific sialyltransferases. Highest levels of binding were observed with erythrocytes having the sequence NeuAcα2-3Galβ3/4GlcNAc on their surfaces.

The specificity of the *M. pneumoniae*–erythrocyte interaction was also investigated using oligosaccharides and glycoproteins as inhibitors of binding. A preference for sialic acid $\alpha2 \to 3$ linked rather than $\alpha2 \to 6$ linked to galactose was found. In its strong preference for sialic acid $\alpha2 \to 3$ linked to galactose residues, *M. pneumoniae* differs from *M. gallisepticum*. Since glycophorin is the major sialoglycoprotein of human erythrocyte membranes, it is not surprising that it binds readily to *M. gallisepticum*, and that asialoglycophorin binds poorly (BANAI et al. 1978; GLASGOW and HILL 1980). Other sialoglycoproteins (such as ovine and bovine submaxillary mucins, α_1-acid glycoprotein, and fetuin) bind poorly to *M. gallisepticum* although they inhibit, to varying extents, the binding of glycophorin to the organisms (GLASGOW and HILL 1980). Sialic acid binds equally well to the organisms when linked either $\alpha2 \to 3$ to galactose or $\alpha2 \to 6$ to N-acetylgalactosamine.

Clustering of sialic acid residues (like in glycophorin) appears to be required for effective binding to *M. gallisepticum* (GLASGOW and HILL 1980). Thus, among the glycoproteins and glycolipids tested, the most potent inhibitors were the polyglycosyl peptides of human erythrocytes (derived mainly from bands 3 and 4.5) and the bovine erythrocyte glycoprotein GP-2, both of which are rich in branched poly-N-acetyllactosamine sequences of type I antigen. Human glycophorin A and α_1-acid glycoprotein were substantially less active and fetuin was inactive, in accordance with their lack of poly-N-acetyllactosamine sequences. Inhibition experiments with glycolipids have also indicated that sialylated poly-N-acetyllactosamine sequences are the preferred sequences for *M. pneumoniae* binding (LOOMES et al. 1984).

The availability of the variant erythrocytes of i-blood type provided the strongest evidence for the importance of sialylated poly-N-acetyllactosamine sequences as receptors for *M. pneumoniae*. Cells of this blood type have a high content of linear poly-N-acetyllactosamine sequences which are more susceptible to digestion with endo-β-galactosidase than the corresponding branched sequences on erythrocytes of type I. The binding of these cells to *M. pneumoniae* was decreased by 85% following treatment with endo-β-galactosidase although only 5% of the total erythrocyte sialic acid is released by this glycosidase (LOOMES et al. 1985). Thus, minor sialylated oligosaccharides of the poly-N-acetyllactosamine series (such as those carried by glycoprotein bands 3 and 4.5) and glycolipids are the main receptors for *M. pneumoniae*, rather than the carbohydrate chains of the major sialogly-coproteins, which are not susceptible to digestion by endo-β-galactosidase.

2.7 *Streptococcus pneumoniae*

Among the bacterial species listed in Table 1, only three gram-positive ones were found to express agglutinins with defined sugar activity. Convincing evidence for adhesive function was obtained for the surface lectin of *Streptococcus pneumoniae*, although it has not yet been isolated. N-Acetylglucosamine inhibited binding of the organisms to tissue cells only at relatively high concentrations (25 mM) (BEUTH et al. 1987). The most active inhibitory compound was found to be GlcNAcβ3Gal (ANDERSSON et al. 1983), and the receptor is likely to belong to the neolacto- and lacto-series of glycolipids containing this disaccharide unit. Whole bacteria bind to glycoconjugates bearing structures with the disaccharide. For example, coating of guinea pig erythrocytes, normally not agglutinated by the organisms, with glycolipids containing GlcNAcβ3Gal made them susceptible to agglutination by the streptococci.

The pneumococcal lectin is inhibited strongly by the tetrasaccharide Galβ4GlcNAcβ3Galβ4Glc. This tetrasaccharide is also present in human milk and it has been suggested that it may protect breast-feeding infants from pneumococcal infections (ANDERSSON et al. 1986).

3 Role in Infection

The best experimental approach to assess the role of bacterial lectins in infections caused by lectin-carrying bacteria is by testing whether injection into animals of the organisms together with suitable inhibitory sugars results in lower rates of infection than are caused by the bacteria alone. Another direct approach is to examine the effect on infectivity of antibodies specific for the carbohydrate of the receptors of the lectin. A different approach is to compare the infectivity of variants or phenotypes that differ in their expression of the fimbrial lectins.

The bacterial lectins may influence the infectious process in several ways. First, and foremost, is the ability of the lectins to mediate attachment to epithelial cells of mucosal tissues in order to withstand the cleansing mechanisms which may remove the bacteria at the very early stage of infection before they have had the chance to proliferate (BEACHEY 1981; BEACHEY et al. 1982). Second, the lectin-mediated adherence may confer growth advantage and enhanced toxicity to the adherent pathogen, due to restricted diffusion of products secreted by the target cell and bacteria, respectively (ZAFRIRI et al. 1987). Third, the bacteria may bind via their surface lectins to glycoconjugates on phagocytic cells with resultant killing of the bacteria once the latter invade deep tissue by a process termed lectinophagocytosis

(OFEK and SHARON 1988). Fourth, the bacterial lectins may interact with soluble glycoproteins containing sugar residues specific for the lectin in body fluids, resulting in inhibition of the bacteria–host cell interaction (ØRSKOV et al. 1980; KURIYAMA and SILVERBLATT 1986).

In the following we summarize what is known about the role of surface lectins in infections caused by enteric and oral bacteria, since these have been investigated in considerable detail and they serve as a paradigm for bacterial infections that are initiated on mucosal and smooth surfaces, respectively.

3.1 Adherence to and Colonization of Mucosal Surfaces

Organisms belonging to the family of enterobacteria cause various types of infections, all of which are initiated at mucosal surfaces. Subsequent to colonization, the organisms may cause intestinal and extraintestinal infections by spreading to other sites, such as the urinary tract, blood, meninges, and other tissues, or they may secrete toxins which cause symptomatic infections. The role of enterobacterial lectins in these infectious processes has been best studied in gastrointestinal and urinary tract infections, in particular those caused by *E. coli* (GAASTRA and DE GRAAF 1982; DE GRAAF and MOOI 1986; CLEGG and GERLACH 1987; REID and SOBEL 1987). Almost all of these studies were carried out with mannose or Galα4Gal-specific fimbrial lectins.

In three separate systems, each employing different type 1 fimbriated enterobacterial species, it has been demonstrated that mannose or methyl α-mannoside inhibited experimental infection or mucosal colonization by the organisms (Table 4). In each of these, glucose (or methyl α-glucoside), which is not an inhibitor of type 1 fimbriae, did not affect the infectivity when injected together with the bacteria. Similarly, Galα4Gal-containing sugars were shown to prevent urinary tract infection

Table 4. Inhibitors of sugar-specific adherence that prevent infection

Organism	Animal, site of infection	Inhibitor	Reference
Type 1 fimbriated			
Escherichia coli	Mice, UT	MeαMan	ARONSON et al. 1979
	Mice, GIT	Mannose	GOLDHAR et al. 1986
	Mice, UT	Anti-Man antibody	ABRAHAM et al. 1985
Klebsiella pneumoniae	Rats, UT	MeαMan	FADER and DAVIS 1980
Shigella flexnerii	Guinea pigs, eye	Mannose	ANDRADE 1980
P fimbriated			
Escherichia coli	Mice, UT	Globotetraose	SVANBORG EDÉN et al. 1982b
	Monkeys, UT	Galα4GalβOMe	ROBERTS et al. 1984

UT = urinary tract; GIT = gastrointestinal tract

in both mice and monkeys by P fimbriated *E. coli*. In addition, antimannose antibodies also prevented urinary tract infection in mice by type 1 fimbriated *E. coli*. These findings provide the most convincing evidence for the central role of bacterial lectin in infection, and particularly in mucosal colonization. They also illustrate the great potential of simple sugar in the prevention of infections caused by bacteria that express surface lectins. For example, the studies on the sugars specificity of type 1 fimbrial lectin suggest that aromatic glycosides such as 4-methylumbelliferyl α-mannoside and *p*-nitro-*o*-chlorophenyl α-mannoside can provide a basis for the design of therapeutic agents that may prevent adherence in vivo and infection by *E. coli* strains that express this lectin (FIRON et al. 1987).

Another experimental approach to assess the role of enterobacterial lectins in colonization of mucosal surfaces included the use of animals that were passively or actively immunized against the corresponding lectins. This approach was employed in the case of type 1 and P fimbriated *E. coli*. In all cases infection was prevented in the immunized animals challenged with strains expressing the fimbrial lectin (Table 5). It should be pointed out that, although the antifimbrial antibodies inhibit adherence in vitro (SILVERBLATT and COHEN 1979; SVANBORG EDÉN et al. 1982a), there is no evidence that these antibodies are directed against the sugar-combining site of the fimbrial lectin.

Comparison between fimbriated phenotypes and nonfimbriated ones showed in all cases that the infectivity of the former was significantly higher. This was found for strains expressing type 1 fimbriae in gastrointestinal infections induced by *Salmonella typhimurium* in mice (DUGUID et al. 1976), as well as in urinary tract infection induced by *K. pneumoniae* in rats (FADER and DAVIS 1982) or in mice (MAAYAN et al. 1985) and by *E. coli* in mice for type 1 fimbriated (IWAHI et al. 1983; HULTGREN et al. 1985), P fimbriated (HAGBERG et al. 1983a, b), and S fimbriated (MARRE and HACKER 1987) strains.

In an interesting study (HAGBERG et al. 1983a, b), mice were injected intraurethrally with mixtures of fimbriated and nonfimbriated, as well as with mixtures of type 1 and P fimbriated isogens. Examination of the bladder 1 day after challenge showed a marked preponderance of type 1 fimbriated bacteria, whereas in the kidney there

Table 5. Prevention of infection caused by enterobacteria carrying fimbrial lectin by antifimbrial immunity

Type of immunity	Fimbrial antigen	Animal, site of infection	Reference
Passive	Type 1	Mice, UT	ABRAHAM et al. 1985
		Pigs, GIT	JAYAPPA et al. 1985
Active	Type 1	Rats, UT	SILVERBLATT and COHEN 1979
		Rats, GIT	GUERINA et al. 1983
		Pigs, GIT	JAYAPPA et al. 1985
	Type P	Monkeys, UT	KAACK et al. 1988
			ROBERTS et al. 1984
		Mice, UT	O'HANLEY et al. 1985
			SCHMIDT et al. 1988
	Type K99	Pigs, GIT	ISAACSON et al. 1980

was a preponderance of P fimbriated bacteria. It was concluded that the type 1 fimbrial lectin confers an advantage to organisms during growth in the bladder. In contrast, in deep tissue (e.g., kidney) the P fimbrial lectin seems advantageous for infectivity.

3.2 Growth Advantage and Enhanced Toxicity

Once the organisms adhere to mucosal cells they start to multiply in order to colonize these tissues. The growing organisms may elaborate toxins that cause tissue damage. Recently ZAFRIRI et al. (1987) examined in tissue culture the growth and toxicity induced by the heat-labile toxin of *E. coli* using tissue culture cells and a pair of bacterial isogens, one of which was nonfimbriated and the other type 1 fimbriated. Growth advantage and enhanced toxicity were observed with the isogen adherent via type 1 fimbriae to the target tissue cells. It was concluded that this adherence-dependent activity is due to restricted diffusion of products secreted by the cells. Other investigators also found growth advantage of type 1 fimbriated phenotypes in vivo (ALKAN et al. 1986; TANAKA 1982).

3.3 Lectinophagocytosis

Following colonization, the organisms may invade and multiply in deep tissues. While the adherence to cells of these tissues may confer growth advantage to the bacteria, the latter have at the same time to confront the phagocytic cells also present at such sites. If these cells possess receptors for the bacterial lectins, they may bind, engulf, and kill the lectin-carrying bacteria. This process of lectin-mediated phagocytosis was termed "lectinophagocytosis" by us (OFEK and SHARON 1988). There is ample evidence showing that type 1 fimbriae are mediators in lectino-phagocytosis of *E. coli* expressing these fimbriae by a variety of phagocytic cells. In contrast, P fimbriated bacteria do not undergo phagocytosis by human polymor-phonuclear leukocytes since receptors for these fimbriae are absent from the phagocytic cells (SVANBORG EDÉN et al. 1984). Lectinophagocytosis of the bacteria, however, was observed by leukocytes that were coated with Galα4Gal-containing glycolipids.

The evidence that the recognition of type 1 fimbriated *E. coli* by phagocytes is mediated by interaction of the fimbrial lectin with mannose-containing glyco-proteins on the surface of the phagocyte is based on various findings: (a) the specificity pattern of inhibition of bacteria–host cell interaction observed is the same for phagocytic and nonphagocytic target cells (FIRON et al. 1983, 1985; BONER et al. 1987); (b) whenever sugars other than mannose or its derivatives were examined, they inhibited poorly (if at all) the interaction of type 1 fimbriated *E. coli* with mouse peritoneal macrophages or human polymorphonuclear leukocytes (BAR-SHAVIT et al. 1977; SILVERBLATT et al. 1979); (c) a very good correlation was found between the mannose-binding activity of the bacteria and the extent of their attachment to mouse peritoneal macrophages (BAR-SHAVIT et al. 1980); (d) the finding that pretreatment of type 1 fimbriated bacteria with yeast mannan inhibited their attachment to mouse and human phagocytes, whereas pretreatment of the phagocytes did not have such an effect (BAR-SHAVIT et al. 1977), shows that the receptor for the bacterial lectin is on the surface of the phagocytes; and (e) latex particles coated with purified type 1 fimbriae stimulated human polymorphonuclear leukocytes and this activity was inhibited by mannose (GOETZ and SILVERBLATT 1987).

The lectin-mediated binding of the bacteria to phagocytic cells described above leads to ingestion, stimulation of antimicrobial systems (e.g., oxygen burst, degranulation), and killing of the bacteria, as summarized in Table 6. Maximum stimulation of antimicrobial systems in phagocytes appears to require cross-linking of type 1 fimbriae on the bacterial surfaces (PERRY et al. 1983; SÖDERSTRÖM and ÖHMAN 1984). Such cross-linking may cause aggregation of the receptors on the phagocytic cells. Aggregation of receptors is important for many other membrane-initiated events.

Table 6. Stages of lectinophagocytosis of *E. coli* mediated by type 1 fimbriae

Stage	Temp.	Cell	Assay system	Reference
Attachment	4°C	Macrophage, mouse	CFU	OFEK and SHARON 1988
Ingestion	37°C	PMNL, human	EM	SILVERBLATT et al. 1979; ROTTINI et al. 1979
			FITC-*E. coli*	ÖHMAN et al. 1982, 1985
		Macrophage, mouse	FITC-anti-*E. coli*	BAR-SHAVIT et al. 1980
Stimulation	37°C	PMNL, human	O_2 consumption	GOETZ and SILVERBLATT 1987; ROTTINI et al. 1979
			CL	MANGAN and SNYDER 1979; BJÖRKSTÉN and WADSTRÖM 1982; SÖDERSTRÖM and ÖHMAN 1984
			Lysozyme release	SVANBORG EDÉN et al. 1984; MANGAN and SNYDER 1979
			Protein iodination	PERRY et al. 1983
		Macrophage, human	CL	BONER et al. 1987; MHASHILKAR 1988
		Macrophage, rat	CL	BLUMENSTOCK and JANN 1982
Killing	37°C	PMNL, human	CFU	SILVERBLATT et al. 1979; ÖHMAN et al. 1982
		Macrophage, mouse	CFU	OFEK and SHARON 1988
		Macrophage, rat	CFU	BLUMENSTOCK and JANN 1982
		Macrophage, human	CFU	BONER et al. 1989

Abbreviations: CFU, colony forming units; CL, chemiluminescence; EM, electron microscopy; FITC, fluorescein isothiocyanate; PMNL, polymorphonuclear leukocytes

Whereas the occurrence of lectinophagocytosis mediated by mannose-specific bacterial surface lectins has been established unequivocally in vitro, little is known about its occurrence in vivo. Indirect evidence that it may take place in vivo was obtained in experimental infection with mixed bacterial phenotypes (or isogens), one of which expresses type 1 fimbriae while the other does not. Such experiments revealed that, whenever the organisms reached phagocyte-rich sites, the non-fimbriated phenotype survived, while at phagocyte-poor sites the fimbriated one survived, irrespective of the bacterial species or experimental animal employed or of the route or site of the infection (Table 7). The selective survival of the fimbriated phenotype on mucosal surfaces was interpreted as being due to its ability to bind to the epithelial cells, whereas the selective survival of the nonfimbriated phenotype in deep tissues (e.g., kidney or peritoneal cavity) could be explained by the elimination of the fimbriated phenotype by macrophages and/or polymorphonuclear leukocytes usually present in such deep tissues (OFEK et al. 1981; OFEK 1984). It was suggested that phase variation, a random on–off switching process that allows the cells to alternate between fimbriated and nonfimbriated states, is an important virulence trait of type 1 fimbriated *E. coli* (OFEK and SILVERBLATT 1982; OFEK 1984), as well as of other fimbriated bacterial species (SILVERBLATT and OFEK 1978, 1983; OFEK 1984), in the survival of the organisms in phagocyte-rich sites.

Lectinophagocytosis has recently been demonstrated to occur with type 2 fimbriated *Actinomyces* species. The fimbrial lectin of these bacteria is specific for Galβ3GalNAc (Table 1) and for lactose (Galβ4Glc). Attachment, internalization, and killing of *Actinomyces viscosus* T14V and *Actinomyces naeslundii* WVU45, both carrying type 2 fimbriae, by human polymorphonuclear leukocytes was blocked by methyl β-galactoside and lactose, whereas cellobiose of methyl α-galactoside was ineffective (SANDBERG et al. 1986, 1988). Phagocytosis was markedly enhanced (to more than 95%) by pretreating the polymorphonuclear leukocytes with sialidase, which unmasks galactose residues on oligosaccharide units of glycoproteins and glycolipids. Only bacteria expressing the galactose-specific lectin were phagocytosed in the absence of serum; phagocytosis of mutants of *A. viscosus* T14V and *A. naeslundii* WVU45 deficient in type 2 fimbriae was minimal or absent.

Table 7. Survival of mannose-specific fimbriated and nonfimbriated phenotypes or isogens in phagocyte-rich and -poor sites following infection with mixed strains

Organism	Animal	Route of infection	Site of infection	Level of phagocytes at infection site	Fimbriated phenotype or genotype	Reference
E. coli	Mouse	Intravesicular	Kidney	Rich	−	1, 2
			Bladder	Poor	+	
		Oral	Blood	Rich	−	3
			Oral cavity	Poor	+	
		Peritoneal	Peritoneum	Rich	−	4
	Rat	Intravesicular	Kidney	Rich	−	5
		Peritoneal	Blood	Rich	−	6
			Peritoneum	Rich	−	6
K. pneumoniae	Mouse	Intravesicular	Kidney	Rich	−	7
			Bladder	Poor	+	

Key to references: (1) HAGBERG et al. 1983a, b; (2) SCHAEFFER et al. 1987; (3) GUERINA et al. 1983; (4) ALKAN et al. 1986; (5) MARRE and HACKER 1987; (6) SAUKKONEN et al. 1988; (7) MAAYAN et al. 1985

3.4 Interaction with Soluble Glycoconjugates and Lectins in Body Fluids

Soluble glycoconjugates in body fluids may contain sugars to which bacterial lectins bind. Such glycoconjugates may, therefore, inhibit lectin-mediated interaction of bacteria with host cells. A case in point is Tamm-Horsfall glycoprotein, the most abundant glycoprotein in normal human urine, which contains N-linked oligomannose units (SERAFINI-CESSI et al. 1984; DELL'OLIO et al. 1988). It is not surprising, therefore, that it binds type 1 fimbriae and that this binding is inhibited by methyl α-mannoside. Bacteria coated with Tamm-Horsfall glycoprotein were considerably less susceptible to mannose-specific lectinophagocytosis than were the uncoated bacteria (KURIYAMA and SILVERBLATT 1986). However, the glycoprotein did not affect the phagocytosis of opsonized bacteria. It was suggested that these observations may explain the virulence of *E. coli* in the bladder and kidney of a susceptible host, where local serum activity is low and Tamm-Horsfall glycoprotein reaches high concentrations (KURIYAMA and SILVERBLATT 1986).

In bladder colonization, Tamm-Horsfall glycoprotein seems to play a dual role. On the one hand, it may facilitate the persistence of type 1 fimbriated bacteria since it sticks to the epithelial cells and provides a matrix to which these bacteria can readily adhere (CHICK et al. 1981; ØRSKOV et al. 1980). On the other hand, adherence of the bacteria to aggregated Tamm-Horsfall glycoprotein in suspension would facilitate their elimination during micturition (ØRSKOV et al. 1980). Thus, depending on its concentration and physical state in urine, the glycoprotein may either enhance or act as natural host defense against the development of infection from type 1 bacteria. This hypothesis is supported by the finding that significantly lower concentrations of Tamm-Horsfall glycoprotein are excreted by patients suffering from urinary tract infection as compared to controls (ISRAELE et al. 1987). These studies show that interaction of lectin-expressing bacteria with soluble glycoconjugates carrying suitable sugars may affect the infectious process.

4 Concluding Remarks

Studies on bacterial lectins expressed on the surfaces of enterobacteria and, in particular, those associated with type 1 and P fimbriae, clearly demonstrate that they influence the infectious process in more ways than one. This is by virtue of their ability to serve as adhesins which bind the organisms to host cells and to glycoconjugates in body fluids. It should thus be possible to intervene in the infectious process by administering sugars which inhibit the lectins.

Expression of the bacterial lectins during the infectious process is not always beneficial to the survival of the bacteria. For example, any adverse effect of type 1 fimbriae on the survival of the organisms due to their interaction with phagocytes or mannose-containing soluble glycoproteins, such as Tamm-Horsfall glycoprotein, will not necessarily result in the termination of the infection, since these bacteria may undergo phase variation in vivo (MAAYAN et al. 1985). This would ensure that a nonfimbriated phenotype or a phenotype expressing another surface lectin of the same infecting strain be present to proceed with the infection in an environment hostile to the type 1 fimbriated phenotype. We expect that similar mechanisms of survival may exist in other bacteria that express surface lectins capable of interacting with phagocytes or with soluble glycoconjugates in body fluids.

The bacterial lectins may have additional functions in vivo besides those discussed above. For example, why, among all *E. coli* strains which express the mannose-specific lectin associated with type 1 fimbriae, do only those that also express the Galα4Gal-specific lectin associated with P fimbriae induce pyelonephritis in humans? Clearly, the P fimbrial lectin possesses some role that is unique for the survival of the organisms during the natural course of infection that leads to pyelonephritis. It is also noteworthy that the genes coding for the expression of type 1 fimbriae are conserved among most enteric bacteria, suggesting that this lectin confers advantages not shared by other bacterial lectins. One such advantage may be related to the distribution of receptors on the host tissues. Receptors containing N-linked oligomannosides and hybrid units to which type 1 fimbriae bind are ubiquitous and abundant on many types of cells, whereas those for which other lectins are specific have a more limited distribution. It follows that each bacterial lectin may have several functions, some of which are shared by other types of lectin while others are unique. Further studies are expected to reveal new and exciting functions for these recognition molecules in host–pathogen interactions.

Acknowledgment. We thank Dvorah Ochert for editorial assistance. Part of the work from the authors' laboratories discussed in this review was supported by grant AI23165 from the National Institutes of Health, Bethesda, MD.

References

Abraham SN, Babu JP, Giampapa CS, Hasty DL, Simpson WA, Beachey EH (1985) Protection against *Escherichia coli*-induced urinary tract infections with hybridoma antibodies directed against type 1 fimbriae or complementary D-mannose receptors. Infect Immun 48: 625–628

Alkan ML, Wong L, Silverblatt FJ (1986) Change in degree of type 1 piliation of *Escherichia coli* during experimental peritonitis in the mouse. Infect Immun 52: 549–554

Andersson B, Dahmén J, Frejd T, Leffler H, Magnusson G, Noori G, Svanborg-Edén C (1983) Identification of an active disaccharide unit of a glycoconjugate receptor for pneumococci attaching to human pharyngeal epithelial cells. J Exp Med 158: 559–570

Andersson B, Porras O, Hanson LA, Lagergård T, Svanborg-Edén C (1986) Inhibition of attachment of *Streptococcus pneumoniae* and *Haemophilus influenzae* by human milk and receptor oligosaccharides. J Infect Dis 153: 232–237

Andrade JRC (1980) Role of fimbrial adhesiveness in guinea-pig keratoconjunctivitis by *Shigella flexneri*. Rev Microbiol (S Paulo), 11: 117–125

Aronson M, Medalia O, Schori L, Mirelman D, Sharon N, Ofek I (1979) Prevention of colonization of the urinary tract of mice with *Escherichia coli* by blocking of bacterial adherence with methyl α-D-mannopyranoside. J Infect Dis 139: 329–332

Banai M, Kahane I, Razin S, Bredt W (1978) Adherence of *Mycoplasma gallisepticum* to human erythrocytes. Infect Immun 21: 365–372

Bar-Shavit Z, Ofek I, Goldman R, Mirelman D, Sharon N (1977) Mannose residues on phagocytes as receptors for the attachment of *Escherichia coli* and *Salmonella typhi*. Biochem Biophys Res Commun 78: 455–460

Bar-Shavit Z, Goldman R, Ofek I, Sharon N, Mirelman D (1980) Mannose-binding activity in *Escherichia coli*: a determinant of attachment and ingestion of the bacteria by macrophages. Infect Immun 29: 417–424

Beachey EH (1981) Bacterial cdherence: adhesin-receptor interactions mediating the attachment of bacteria to mucosal surfaces. J Infect Dis 143: 325–345

Beachey EH, Eisenstein BI, Ofek I (1982) Bacterial adherence in infectious diseases. Upjohn, Kalamazoo (Current concepts series)

Beuth J, Ko HL, Uhlenbruk G, Pulverer G (1987) Lectin-mediated bacterial adhesion to human tissue. Eur J Clin Microbiol 6: 591–593

Björkstén B, Wadström T (1982) Interaction of *Escherichia coli* with different fimbriae and polymorphonuclear leukocytes. Infect Immun 38: 298–305

Blumenstock E, Jann K (1982) Adhesion of piliated *Escherichia coli* strains to phagocytes: differences between bacteria with mannose-sensitive pili and those with mannose-resistant pili. Infect Immun 35: 264–269

Bock K, Breimer ME, Brignole A, Hansson GC, Karlsson KA, Larson G, Leffler H, Samuelsson BE, Strömberg N, Svanborg-Edén C, Thurin J (1985) Specificity of binding of a strain of uropathogenic *Escherichia coli* to Galα1 → 4Gal-containing glycosphingolipids. J Biol Chem 260: 8545–8551

Boner G, Mhashilkar AM, Rodriguez-Ortega M, Sharon N (1989) Lectin-mediated, nonopsonic phagocytosis of type 1 *Escherichia coli* by human peritoneal macro-phages of uremic patients treated by peritoneal dialysis. J Leuk Biol 46: 239–245

Brennan MJ, Joralmon RA, Cisar JO, Sandberg AL (1987) Binding of *Actinomyces naeslundii* to glycosphingolipids. Infect Immun 55: 487–489

Brinton CC (1965) The structure, function, synthesis and genetic control of bacterial pili and a molecular model for DNA and RNA transport in gram negative bacteria. Trans NY Acad Sci 27: 1003–1054

Chick S, Harber MJ, MacKenzie R, Asscher AW (1981) Modified method for studying bacterial adhesion to isolated uroepithelial cells and uromucoid. Infect Immun 34: 256–261

Cisar JO (1986) Fimbrial lectins of the oral actinomyces. In: Mirelman D (ed) Microbial lectins and agglutinins: properties and biological activity. Wiley, New York, pp 183–196

Clegg S, Gerlach GF (1987) Enterobacterial fimbriae. J Bacteriol 169: 934–938

de Graaf FK, Mooi FR (1986) The fimbrial adhesins of *Escherichia coli*. Adv Microb Physiol 28: 65–143

Dell'Olio F, de Kanter FJJ, van den Eijnden DH, Serafini-Cessi F (1988) Structural analysis of the preponderant high-mannose oligosaccharide of human Tamm-Horsfall glycoprotein. Carbohydr Res 178: 327–332

de Man P, Cedergren B, Enerbäck S, Larsson A-C, Leffler H, Lundell A-L, Nilsson B, Svanborg-Edén C (1987) Receptor-specific agglutination tests for detection of bacteria that bind globoseries glycolipids. J Clin Microbiol 25: 401–406

Drake D, Taylor KG, Bleiweis AS, Doyle RJ (1988) Specificity of the glucan-binding lectin of *Streptococcus cricetus*. Infect Immun 56: 1864–1872

Duguid JP, Old DC (1980) Adhesive properties of Enterobacteriaceae. In: Beachey EH (ed) Bacterial adherence. Chapman and Hall, London, pp 185–217 (Receptors and recognition, series B, vol 6)

Duguid JP, Darekar MR, Wheater DWF (1976) Fimbriae and infectivity in *Salmonella typhimurium*. J Med Microbiol 9: 459–473

Evans DG, Evans DJ, Moulds JJ, Graham DY (1988) *N*-Acetylneuraminyllactose-binding fibrillar hemagglutinin of *Campylobacter pylori*: a putative colonization factor antigen. Infect Immun 56: 2896–2906

Fader RC, Davis CP (1980) Effect of piliation on *Klebsiella pneumoniae* infection in rat bladders. Infect Immun 30: 554–561

Fader RC, Davis CP (1982) *Klebsiella pneumoniae*- induced experimental pyelitis: the effect of piliation on infectivity. J Urol 128: 197–201

Faris A, Lindahl M, Wadström T (1980) GM$_2$-like glycoconjugate as possible receptor for the CFA/I and K99 hemagglutinins of enterotoxic *Escherichia coli*. FEMS Microbiol Lett 7: 265–269

Finkelstein RA, Hanne LF (1982) Purification and characterization of the soluble hemagglutinin (cholera lectin) produced by *Vibrio cholerae*. Infect Immun 36: 1199–1208

Firon N, Ofek I, Sharon N (1983) Carbohydrate specificity of the surface lectins of *Escherichia coli*, *Klebsiella pneumoniae* and *Salmonella typhimurium*. Carbohydr Res 120: 235–249

Firon N, Ofek I, Sharon N (1984) Carbohydrate-binding sites of the mannose-specific fimbrial lectins of Enterobacteria. Infect Immun 43: 1088–1090

Firon N, Duksin D, Sharon N (1985) Mannose-specific adherence of *Escherichia coli* to BHK cells that differ in their glycosylation patterns. FEMS Microbiol Lett 27: 161–165

Firon N, Ashkenazi S, Mirelman D, Ofek I, Sharon N (1987) Aromatic alpha-glycosides of mannose are powerful inhibitors of the adherence of type 1 fimbriated *Escherichia coli* to yeast and intestinal epithelial cells. Infect Immun 55: 472–476

Gaastra W, de Graaf FK (1982) Host-specific fimbrial adhesins of noninvasive enterotoxigenic *Escherichia coli* strains. Microbiol Rev 46: 129–161

Giampapa CS, Abraham SN, Chiang TM, Beachey EH (1988) Isolation and characterization of a receptor for type 1 fimbriae of *Escherichia coli* from guinea pig erythrocytes. J Biol Chem 263: 5362–5367

Glasgow LR, Hill RL (1980) Interaction of *Mycoplasma gallisepticum* with sialyl glycoproteins. Infect Immun 30: 353–361

Goetz MB, Silverblatt FJ (1987) Stimulation of human polymorphonuclear leukocyte oxidative metabolism by type 1 pili from *Escherichia coli*. Infect Immun 55: 534–540

Goldhar J, Perry R, Ofek I (1984) Extraction and properties of nonfimbrial mannose-resistant hemagglutinin from a urinary isolate of *Escherichia coli*. Curr Microbiol 11: 49–54

Goldhar J, Zilberberg A, Ofek I (1986) Infant mouse model of adherence and colonization of intestinal tissues by enterotoxigenic strains of *Escherichia coli* isolated from humans. Infect Immun 52: 205–208

Goochee CF, Hatch RT, Cadman TW (1987) Some observations on the role of type 1 fimbriae in *Escherichia coli* autoflocculation. Biotech Bioeng 29: 1024–1034

Guerina NG, Kessler TW, Guerina VJ, Neutra MR, Clegg HW, Langermann S, Scannapieco FA, Goldmann DA (1983) The role of pili and capsule in the pathogenesis of neonatal infection with *Escherichia coli* K1. J Infect Dis 148: 395–404

Gunnarsson A, Mårdh PE, Lundblad A, Svensson S (1984) Oligosaccharide structures mediating agglutination of sheep erythrocytes by *Staphylococcus saprophyticus*. Infect Immun 45: 41–46

Hagberg L, Hull R, Hull S, Falkow S, Freter R, Svanborg-Edén C (1983a) Contribution of adhesion to bacterial persistence in the mouse urinary tract. Infect Immun 40: 265–272

Hagberg L, Engberg I, Feter R, Lam J, Olling S, Svanborg-Edén C (1983b) Ascending unobstructed urinary tract infection in mice caused by pyelonephritogenic *Escherichia coli* of human origin. Infect Immun 40: 273–283

Hanne LF, Finkelstein RA (1982) Characterization and distribution of the hemagglutinins produced by *Vibrio cholerae*. Infect Immun 36: 209–214

Hansson GC, Karlsson K-A, Larson G, Lindberg A, Strömberg N, Thurin J (1983) Lactosylceramide is the probable adhesion site for major indigenous bacteria of the gastrointestinal tract. In: Chester MA, Heinegård D, Lundblad A, Svensson S (eds) Glycoconjugates. 7th International symposium on glycoconjugates, Lund Sweden, July 17–23, p 631

Hansson GC, Karlsson D-A, Larson G, Strömberg N, Thurin J (1985) Carbohydrate-specific adhesion of bacteria to thin-layer chromatograms: a rationalized approach to the study of host cell glycolipid receptors. Anal Biochem 146: 158–163

Hultgren SJ, Porter TN, Schaeffer AJ, Duncan JL (1985) Role of type 1 pili and effects of phase variation on lower urinary tract infections produced by *Escherichia coli*. Infect Immun 50: 370–377

Isaacson RE, Dean EA, Morgan RL, Moon HW (1980) Immununization of suckling pigs against enterotoxigenic *Escherichia coli*-induced diarrheal disease by vaccinating dams with purified K99 or 987P pili: antibody production in response to vaccination. Infect Immun 29: 824–826

Ishikawa H, Isayama Y (1987) Evidence for sialyl glycoconjugates as receptors for *Bordetella bronchiseptica* on swine nasal mucosa. Infect Immun 55: 1607–1609

Israele V, Darabi A, McCracken GH (1987) The role of bacterial virulence factors and Tamm-Horsfall protein in the pathogenesis of *Escherichia coli* urinary tract infection in infants. Am J Dis Child 141: 1230–1234

Iwahi T, Abe Y, Nakao M, Imada A, Tsuchiya K (1983) Role of type 1 fimbriae in the pathogenesis of ascending urinary tract infection induced by *Escherichia coli* in mice. Infect Immun 39: 1307–1315

Jayappa HG, Goodnow RA, Geary SJ (1985) Role of *Escherichia coli* type 1 pilus in colonization of porcine ileum and its protective nature as a vaccine antigen in controlling colibacillosis. Infect Immun 48: 350–354

Jones GW, Freter R (1976) Adhesive properties of *Vibrio cholerae*: nature of the interaction with isolated rabbit brush border membranes and human erythrocytes. Infect Immun 14: 240–245

Kaack MB, Roberts JA, Baskin G, Patterson GM (1988) Maternal immunization with P fimbriae for the prevention of neonatal pyelonephritis. Infect Immun 56: 1–6

Kabat EA (1978) Dimensions and specificities of recognition sites on lectins and antibodies. J Supramol Struct 8: 79–88

Karlsson K-A (1986) Animal glycolipids as attachment sites for microbes. Chem Phys Lipids 42: 153–172

Korhonen TK, Väisänen V, Saxén H, Hultberg H, Svenson SB (1982) P-Antigen-recognizing fimbriae from human uropathogenic *Escherichia coli* strains. Infect Immun 37: 286–291

Korhonen TK, Väisänen-Rehn V, Rehn M, Pere A, Parkkinen J, Finne J (1984) *Escherichia coli* fimbriae recognizing sialyl galactosides. J Bacteriol 159: 762–766

Korhonen TK, Haahtela K, Pirkola A, Parkkinen J (1988) A *N*-acetyllactosamine-specific cell-binding activity in a plant pathogen, *Erwinia rhapontici*. FEBS Lett 236: 163–166

Krivan HC, Roberts DD, Ginsburg V (1988a) Many pulmonary pathogenic bacteria bind specifically to the carbohydrate sequence GalNAcβ1-4Gal found in some glycolipids. Proc Natl Acad Sci USA 85: 6157–6161

Krivan HC, Ginsburg V, Roberts DD (1988b) *Pseudomonas aeruginosa* and *Pseudomonas cepacia* isolated from cystic fibrosis patients bind specifically to gangliotetraosylceramide (asialo GM1) and gangliotriaosylceramide (asialo GM2). Arch Biochem Biophys 260: 493–496

Kuriyama SM, Silverblatt FJ (1986) Effect of Tamm-Horsfall urinary glycoprotein on phagocytosis and killing of type I-fimbriated *Escherichia coli*. Infect Immun 51: 193–198

Landale EC, McCabe MM (1987) Characterization by affinity electrophoresis of an α-1,6-glucan-binding protein from *Streptococcus sobrinus*. Infect Immun 55: 3011–3016

Leffler H, Svanborg-Edén C (1980) Chemical identification of a glycosphingolipid receptor for *Escherichia coli* attaching to human urinary tract epithelial cells and agglutinating human erythrocytes. FEMS Microbiol Lett 8: 127–134

Leffler H, Svanborg-Edén C (1986) Glycolipids as receptors for *Escherichia coli* lectins or adhesins. In: Mirelman D (ed) Microbial lectins and agglutinins: properties and biological activity. Wiley, New York, pp 83–111

Lindahl M, Brossmer R, Wadström T (1987) Carbohydrate receptor specificity of K99 fimbriae of enterotoxigenic *Escherichia coli*. Glycoconjugate J 4: 51–58

Loomes LM, Uemura K, Childs RA, Paulson JC, Rogers GN, Scudder PR, Michalski J-C, Hounsell EF, Taylor-Robinson D, Feizi T (1984) Erythrocyte receptors for *Mycoplasma pneumoniae* are sialylated oligosaccharides of Ii antigen type. Nature 307: 560–563

Loomes LM, Uemura K, Feizi T (1985) Interaction of *Mycoplasma pneumoniae* with erythrocyte glycolipids of I and i antigen types. Infect Immun 47: 15–20

Maayan MC, Ofek I, Medalia O, Aronson M (1985) Population shift in mannose-specific fimbriated phase of *Klebsiella pneumoniae* during experimental urinary tract infection in mice. Infect Immun 49: 785–789

Mangan DF, Snyder IS (1979) Mannose-sensitive interaction of *Escherichia coli* with human peripheral leukocytes in vitro. Infect Immun 26: 520–527

Marre R, Hacker J (1987) Role of S- and common-type 1-fimbriae of *Escherichia coli* in experimental upper and lower urinary tract infection. Microb Pathogen 2: 223–226

Mhashilkar A (1988) Interaction of *Escherichia coli*, *Salmonella typhimurium* and *Pseudomonas aeruginosa* with human polymorphonuclear leukocytes and human peritoneal macrophages— comparative study. MSc thesis, Weizmann Institute of Science, Rehovot

Mirelman D (ed) (1986) Microbial lectins and agglutinins: properties and biological activity. Wiley, New York

Mirelman D, Ofek I (1986) Introduction to microbial lectins and agglutinins. In: Mirelman D (ed) Microbiol lectins and agglutinins: properties and biological activity. Wiley, New York, pp 1–19

Murray PA, Levine MJ, Tabak LA, Reddy MS (1982) Specificity of salivary-bacterial interactions. II. Evidence for a lectin on *Streptococcus sanguis* with specificity for a NeuAcα2,3Galβ1,3GalNAc sequence. Biochem Biophys Res Commun 106: 390–396

Murray PA, Kern DG, Winkler JR (1988) Identification of a galactose-binding lectin on *Fusobacterium nucleatum* FN-2. Infect Immun 56: 1314–1319

Neeser J-R, Koellreutter B, Wuersch P (1986) Oligomannoside-type glycopeptides inhibiting adhesion of *Escherichia coli* strains mediated by type 1 pili: preparation of potent inhibitors from plant glycoproteins. Infect Immun 52: 428–436

Neeser J-R, Chambaz A, Hoang KY, Link-Amster H (1988) Screening for complex carbohydrates inhibiting hemagglutinations by CFA/I- and CFA/II-expressing enterotoxigenic *Escherichia coli* strains. FEMS Microbiol Lett 49: 301–307

Nilsson G, Svensson S, Lindberg AA (1983) The role of the carbohydrate portion of glycolipids for the adherence of *Escherichia coli* K88$^+$ to pig intestine. In: Chester MA, Heinegård D, Lundblad A, Svensson S (eds) Glycoconjugates. 7th International symposium on glycoconjugates, Lund, July 17–23 Sweden, pp 637–638

O'Hanley P, Lark D, Falkow S, Schoolnik G (1985) Molecular basis of *Escherichia coli* colonization of the upper urinary tract in BALB/c mice. J Clin Invest 75: 347–360

Ofek I (1984) Adhesin receptor interaction mediating adherence of bacteria to mucosal tissues: importance of phase transition in the expression of bacterial adhesins. In: Falcon G (ed) Bacterial and viral inhibition and modulation of host defenses. Academic, London, pp 7–24

Ofek I, Beachey EH (1978) Mannose binding and epithelial cell adherence of *Escherichia coli*. Infect Immun 22: 247–254

Ofek I, Sharon N (1988) Lectinophagocytosis: a molecular mechanism of recognition between cell surface sugars and lectins in the phagocytosis of bacteria. Infect Immun 56: 539–547

Ofek I, Silverblatt FJ (1982) Bacterial surface structures involved in adhesion to phagocytic and epithelial cells. In: Schlesinger D (ed) Microbiology 1982. American Society for Microbiology, Washington, pp 296–300

Ofek I, Mirelman D, Sharon N (1977) Adherence of *Escherichia coli* to human mucosal cells mediated by mannose receptors. Nature 265: 623–625

Ofek I, Mosek A, Sharon N (1981) Mannose specific adherence of *Escherichia coli* freshly excreted in the urine of patients with urinary tract infections and of isolates subcultured from the infected urine. Infect Immun 34: 708–711

Ofek I, Goldhar J, Eshdat Y, Sharon N (1982) The importance of mannose specific adhesins (lectins) in infections caused by *Escherichia coli*. Scand J Infect Dis [Suppl] 33: 61–67

Ofek I, Lis H, Sharon N (1985) Animal cell surface membranes. In: Savage DC, Fletcher M (eds) Bacterial adhesion: mechanisms and physiological significance. Plenum, New York, pp 71–88

Öhman L, Hed J, Stendahl O (1982) Interaction between human polymorphonuclear leukocytes and two different strains of type 1 fimbriae-bearing *Escherichia coli*. J Infect Dis 146: 751–757

Öhman L, Magnusson K-E, Stendahl O (1985) Mannose-specific and hydrophobic interaction between *Escherichia coli* and polymorphonuclear leukocytes—influence of bacterial culture period. Acta Pathal Microbiol Immunol Scand [B] 93: 125–131

Ørskov I, Ørskov F, Birch-Andersen A (1980) Comparison of *Escherichia coli* fimbrial antigen F7 with type 1 fimbriae. Infect Immun 27: 657–666

Parkkinen J, Rogers GN, Kornhonen T, Dahr W, Finne J (1986) Identification of the O-linked sialyloligosaccharides of glycophorin A as the erythrocyte receptors for S-fimbriated *Escherichia coli*. Infect Immun 54: 37–42

Perry A, Ofek I, Silverblatt FJ (1983) Enhancement of mannose-mediated stimulation of human granulocytes by type 1 fimbriae aggregated with antibodies on *Escherichia coli* surfaces. Infect Immun 39: 1334–1345

Prakobphol A, Murray PA, Fisher SJ (1987) Bacterial adherence on replicas of sodium dodecyl sulfate-polyacrylamide gels. Anal Biochem 164: 5–11

Pulverer G, Beuth J, Ko HL, Sölter J, Uhlenbruck G (1987) Modification of glycosylation by tunicamycin treatment inhibits lectin-mediated adhesion of *Streptococcus pneumoniae* to various tissues. Zentralbl Bakteriol Mikrobiol Hyg [A] 266: 137–144

Reid G, Sobel JD (1987) Bacterial adherence in the pathogenesis of urinary tract infection: a review. Rev Infect Dis 9: 470–487

Roberts JA, Kaack B, Källenius G, Möllby R, Winberg J, Svenson SB (1984) Receptors for pyelonephritogenic *Escherichia coli* in primates. J Urol 131: 163–168

Rodriguez-Ortega M, Ofek I, Sharon N (1987) Membrane glycoproteins of human polymorphonuclear leukocytes that act as receptors for mannose-specific *Escherichia coli*. Infect Immun 55: 968–973

Rottini G, Cian F, Soranzo MR, Albrigo R, Patriarca P (1979) Evidence for the involvement of human polymorphonuclear leucocyte mannose-like receptors in the phagocytosis of *Escherichia coli*. FEBS Lett 105: 307–312

Sandberg AL, Mudrick LL, Cisar JO, Brennan MJ, Mergenhagen SE, Vatter AE (1986) Type 2 fimbrial lectin-mediated phagocytosis of oral *Actinomyces* spp. by polymorphonuclear leukocytes. Infect Immun 54: 472–476

Sandberg AL, Mudrick LL, Cisar JO, Metcalf JA, Malech HL (1988) Stimulation of superoxide and lactoferrin release from polymorphonuclear leukocytes by the type 2 fimbrial lectin *Actinomyces viscosus* T14V. Infect Immun 56: 267–269

Saukkonen KMJ, Nowicki B, Leinonen M (1988) Role of type 1 and S fimbriae in the pathogenesis of *Escherichia coli* O18:K1 bacteremia and meningitis in the infant rat. Infect Immun 56: 892–897

Schaeffer AJ, Schwan WR, Hultgren SJ, Duncan JL (1987) Relationship of type 1 pilus expression in *Escherichia coli* to ascending urinary tract infections in mice. Infect Immun 55: 373–380

Schmidt MA, O'Hanley P, Lark D, Schoolnik GK (1988) Synthetic peptides corresponding to protective epitopes of *Escherichia coli* digalactoside-binding pilin prevent infection in a murine pyelonephritis model. Proc Natl Acad Sci USA 85: 1247–1251

Serafini-Cessi F, Dall'Olio F, Malagolini N (1984) High-mannose oligosaccharides from human Tamm-Horsfall glycoprotein. Biosci Rep 4: 269–274

Sharon N (1987) Bacterial lectins, cell–cell recognition and infectious disease. FEBS Lett 217: 145–157

Sharon N, Lis H (1982) Glycoproteins. In: Neurath H, Hill RL (eds) The proteins, 3rd edn, vol 5. Academic, New York, pp 1–144

Sharon N, Eshdat Y, Silverblatt FJ, Ofek I (1981) Bacterial adherence to cell surface sugars. In: Elliot K, O'Connor M, Whelan J (eds) Adhesion and micro-organism pathogenicity. Pitman Medical, London, pp 119–135 (Ciba symposium series 80)

Sherman PM, Houston WL, Boedeker EC (1985) Functional heterogeneity of intestinal *Escherichia coli* strains expressing type 1 somatic pili (fimbriae): assessment of bacterial adherence to intestinal membranes and surface hydrophobicity. Infect Immun 49: 797–804

Silverblatt FJ, Cohen LS (1979) Antipili antibody affords protection against experimental ascending pyelonephritis. J Clin Invest 64: 333–336

Silverblatt FJ, Ofek I (1978) Influence of pili on the virulence of *Proteus mirabilis* in experimental hematogenous pyelonephritis. J Infect Dis 138: 664–667

Silverblatt FJ, Ofek I (1983) Interaction of bacterial pili and leukocytes. Infection 11: 235–238

Silverblatt FJ, Dreyer JS, Schauer S (1979) Effect of pili susceptibility of *Escherichia coli* to phagocytosis. Infect Immun 24: 218–223

Smit H, Gaastra W, Kamerling JP, Vliegenthart JFG, de Graaf FK (1984) Isolation and structural characterization of the equine erythrocyte receptor for enterotoxigenic *Escherichia coli* K99 fimbrial adhesin. Infect Immun 46: 578–584

Söderström T, Öhman L (1984) The effect of monoclonal antibodies against *Escherichia coli* type 1 pili and capsular polysaccharides on the interaction between bacteria and human granulocytes. Scand J Immunol 20: 299–305

Stromberg N, Deal C, Nyberg G, Normark S, So M, Karlsson K-A (1988) Identification of carbohydrate structures that are possible receptors for *Neisseria gonorrhoeae*. Proc Natl Acad Sci USA 85: 4902–4906

Svanborg Edén C, Marlid S, Korhonen TK (1982a) Adhesion inhibition by antibodies. Scand J Infect Dis [Suppl] 33: 72–78

Svanborg Edén C, Freter R, Hagberg L, Hull R, Hull S, Leffler H, Schoolnik G (1982b) Inhibition of experimental ascending urinary tract infection by an epithelial cell-surface receptor analogue. Nature 298: 560–562

Svanborg Edén C, Bjursten L-M, Hull R, Hull S, Magnusson K-E, Moldovano Z, Leffler H (1984) Influence of adhesins on the interaction of *Escherichia coli* with human phagocytes. Infect Immun 44: 672–680

Tanaka Y (1982) Multiplication of the fimbriate and nonfimbriate *Salmonella typhimurium* organisms in the intestinal mucosa of mice treated with antibodies. Jpn J Vet Sci 44: 523–527

Väisänen-Rhen V, Korhonen TK, Finne J (1983) Novel cell-binding activity specific for *N*-acetyl-D-glucosamine in an *Escherichia coli* strain. FEBS Lett 159: 233–236

Weiss EI, London J, Kolenbrander PE, Kagermeier AS, Andersen RN (1987) Characterization of lectinlike surface components on *Capnocytophaga ochracea* ATCC 33596 that mediate coaggregation with gram-positive oral bacteria. Infect Immun 55: 1198–1202

Yamazaki Y, Ebisu S, Okada H (1981) *Eikenella corrodens* adherence to human buccal epithelial cells. Infect Immun 31: 21–27

Zafriri D, Oron Y, Eisenstein BI, Ofek I (1987) Growth advantage and enhanced toxicity of *Escherichia coli* adherent to tissue culture cells due to restricted diffusion of products secreted by the cells. J Clin Invest 79: 1210–1216

Tissue Tropism of *Escherichia coli* Adhesins in Human Extraintestinal Infections

T. K. Korhonen[1], R. Virkola[1], B. Westurlund[1], H. Holthöfer[2], and
J. Parkkinen[3]

1 Introduction

Mechanical protective factors at epithelial surfaces are an important part of our defenses against invading bacteria. These factors include direct removal by washing, e.g., by the flow of urine in the urinary tract, by peristalsis in the intestine, and by continuous secretion of saliva in the oral cavity. To colonize epithelial surfaces, bacteria have to overcome the washing effect by firmly attaching to the infection site; this is apparently necessary for the bacteria both to cause infectious diseases and to establish themselves as members of our normal flora.

Svanborg-Edén and co-workers (1976) showed that the ability to adhere to epithelial cells of the human urinary tract is a characteristic of *Escherichia coli* strains associated with upper urinary tract infections. Subsequently, a number of epidemiologic (Hagberg et al. 1981: Källenius et al. 1981; Väisänen-Rhen et al. 1984; Johnson et al. 1987) and experimental (Hagberg et al. 1983; Roberts et al. 1984; O'Hanley et al. 1985) studies have led to the conclusion that the ability to adhere to uroepithelium is perhaps the most important single virulence determinant of *E. coli* strains associated with human pyelonephritis and urosepsis. The importance of bacterial adhesion in the pathogenesis of human pyelonephritis is further supported by the fact that susceptibility to *E. coli* adhesion is directly related to an individual's susceptibility to pyelonephritis caused by these bacteria

Departments of General Microbiology[1], Bacteriology and Immunology[2], and Medical Chemistry[3], University of Helsinki, SF-00280, Helsinki, Finland

(KÄLLENIUS and WINBERG 1978). So far the role of bacterial adhesion in the invasive phase of septic infections remains open, but the findings that the *E. coli* O18:K1 strains associated with neonatal meaningitis have their typical adhesins (KORHONEN et al. 1985) that are expressed in vivo during a systemic infection (NOWICKI et al. 1986b) and have receptors in the brain (PARKKINEN et al. 1988a) suggest that bacterial adhesion is involved in the determination of tissue tropism of neonatal meningitis.

The adhesion assays used to demonstrate the correlation between adhesive capacity and uropathogenicity of *E. coli* were performed with exfoliated epithelial cells of human urine (SVANBORG-EDÉN et al. 1977; KÄLLENIUS et al. 1980). These cells, however, are heterogeneous in origin, being derived mainly from the lower regions of the ureter, the urinary bladder, and the urethra. Thus the information about the distribution of bacterial receptors that can be obtained with these assays is rather limited. Recent results have shown that the distribution of cell-surface glycoconjugates in the urinary tract, e.g., in the kidney, is strictly limited to various tissue subcompartments, as shown, for example, by histochemical studies with labeled lectins (HOLTHÖFER et al. 1982; HOLTHÖFER 1983; HENNIGAR et al. 1985). Since these glycoconjugates are potential receptors for bacterial adherence, one could expect that various epithelial surfaces along the urinary tract would also show variable reactivity for bacterial adhesins. Information concerning the bacterial adhesion sites in mammalian tissues is important to increase our understanding of the molecular mechanisms of bacterial pathogenicity. Therefore we recently developed methods for determining the precise sites for bacterial adhesion in frozen tissue sections.

2 Types of Adhesins on *Escherichia coli* Strains Causing Extraintestinal Infections

A number of adhesins have been characterized on *E. coli* strains associated with urinary tract infections or neonatal meningitis (Table 1). These adhesins differ in molecular binding specificities and in serologic properties, and, excepting the type 1 fimbriae, they occur only on certain strain clusters of *E. coli*. Of these adhesins, only the P and S fimbriae are conclusively associated with strains from infectious diseases, the former with strains from cases of pyelonephritis and the latter with strains from cases of neonatal sepsis and meningitis. The type 1C fimbriae are more common on strains from upper urinary tract infections than on fecal strains, although less frequent than the P fimbriae.

An important feature of adhesins on pathogenic *E. coli* strains is their multiplicity. A strain can have two or three different adhesin types, and many of the P-fimbriated strains further possess two or three variants of the P fimbria (PERE et al. 1986). The different fimbriae on a strain are under phase variation: they are mostly expressed by separate cells that can rapidly switch their fimbrial synthesis (NOWICKI et al. 1984). This kind of fimbrial phase variation also takes place in vivo during

Table 1. Adhesins of *E. coli* strains associated with urinary tract infections or neonatal meningitis

Adhesin	Binding characteristics	Occurrence/Remarks	Reference
P fimbria	Blood group P-specific glycolipids; minimal receptor for most is DGalα1-4DGal	O1:K1:H7, O4:K12, O4:K-, O6:K2:H1, O16:K1, O18, and O7:K1:H1 strains associated with pyelonephritis and urosepsis	KÄLLENIUS et al. 1980 VÄISÄNEN-RHEN et al. 1984 ØRSKOV and ØRSKOV 1983 BOCK et al. 1986 JOHNSON et al. 1987
S fimbria	NeuNAcα2-3DGal and NeuNAcα2-3DGalβ1-3-DGalNac sequences in glycoproteins	O18:K1:H7 strains associated with neonatal sepsis and meningitis	PARKKINEN et al. 1983, 1986 KORHONEN et al. 1985 OTT et al. 1986
Type 1 fimbria	Oligomannoside chains in glycoproteins	About 80% of all strains	DUGUID and OLD 1980 FIRON et al. 1982 NEESER et al. 1986
Type 1C fimbria	Molecular details unknown, not hemagglutinating	O4 and O6 strains associated with pyelonephritis; some O22 and O18 strains	PERE et al. 1985
O75X adhesin	Binds to type IV collagen in basement membranes; binding inhibited by chloramphenicol	O75 strains	VÄISÄNEN-RHEN 1984 NOWICKI et al. 1988 WESTERLUND et al. 1989a
M agglutinin	Binds to glycophorin A^M, terminal serine part of receptor	Rare	JOKINEN et al. 1985 RHEN et al. 1986
G fimbria	Terminal DG1cNAc residues	Rare on human pathogens	VÄISÄNEN-RHEN et al. 1983 RHEN et al. 1986

systemic infection in mice (NOWICKI et al. 1986b) and urinary tract infection in man (PERE et al. 1987). Due to phase variation, the infecting *E. coli* population in the urinary tract is heterogeneous, consisting of cells with different adhesins and of nonfimbriate cells.

3 Tissue Specificity of the Adhesins

3.1 Methodologic Aspects

In our approach we have used two parallel methods to identify the receptor-active domains in tissue samples. The first (NOWICKI et al. 1986a) involves staining of frozen tissue sections with fluorochrome-labeled bacteria. Fimbriated bacteria can be directly labeled with fluorescein isothiocyanate (FITC) or tetramethyl rhodamine (TRITC) without damaging the binding activity of the fimbriae. FITC-tagged bacteria can then be incubated at various cell concentration (10^8–10^{10} per ml, the optimal concentration depending on the strain) on frozen tissue sections, washed, and examined by fluorescence microscopy. For identification of tissue structures, sections with bound bacteria can be double stained with specific tissue markers, i.e., TRITC-labeled lectins or antibodies.

Using a suitable bacterial concentration, adhesion sites for the bacteria can be identified by this simple procedure (Fig. 1A). Specificity of the adhesion should always be controlled by testing adhesin-specific Fab fragments (Fig. 1B) or receptor-active carbohydrates for inhibition and also by including adhesin-negative strains as controls (NOWICKI et al. 1986a; KORHONEN et al. 1986a). It should be noted that, due to multiplicity of adhesins, pathogenic wild-type strains are only rarely suitable for adhesion tests, and the use of recombinant strains expressing only one fimbrial type is preferable.

The other method involves indirect immunofluorescent staining of tissue sections with purified adhesins (KORHONEN et al. 1986a, b; WESTERLUND et al. 1987). This method, which has the advantage of a more precise localization of the binding sites, can be combined with double staining with TRITC-labeled lectins to identify

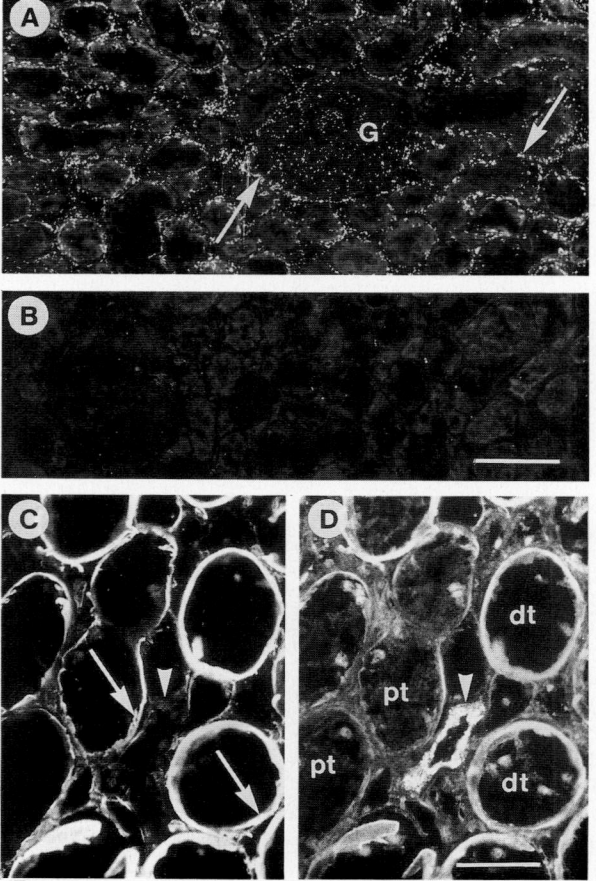

Fig. 1A–C. Binding of the O75X adhesin in kidney. A Adhesion of an O75X-positive strain to frozen sections of kidney; note bacteria on the basement membranes of tubules and on Bowman's capsule (*arrows*). Glomerulus (*G*) is also positive for the bacteria. **B** As in **A**, but in the presence of Fab fragments against the O75X adhesin. Note the almost complete inhibition of the adhesion. **C** and **D** Double staining of the same section with the O75X adhesin (**C**) and FITC-DBA (**D**). The adhesin binds to the basement membranes of both distal (*dt*; positive for FITC-DBA) and proximal (*pt*; negative for FITC-DBA) tubules. *Arrowhead* in **D** indicates a DBA-positive structure, most likely the loop of Henle, that is negative for the adhesin (**C**). *Bars*, 100 μm in **A** and **B**; 40 μm in **C** and **D**

specific tissue elements (Fig. 1C, D). Specificity of the adhesin binding and the common problem of autofluorescence can be controlled by the use of receptor analogs (Fig. 2A, B) and by staining the tissues with the antibodies alone. We have found it necessary to absorb the antibodies with tissue homogenates before using them in the binding tests. In general, a distinct tissue-substructure specificity in fimbrial binding has been observed (Fig. 2), and the results with the recombinant strains and with the corresponding purified adhesins have been in good agreement (Fig. 1).

An alternative approach to localize bacterial receptors in tissue sections is immunofluorescence staining with antibodies raised against receptor-active disaccharides (O'HANLEY et al. 1985) or glycoproteins (DEAN and ISAACSON 1985). This approach is, however, limited by the fact that antibodies and bacterial adhesins can recognize different configurations of the same epitopes, which may lead to different reacitvities at the macromolecular or tissue level. The minimum receptor structure for most P fimbriae is the DGalα1-4DGal disaccharide unit of blood group P-specific glycolipids. P fimbriae recognize this sequence more avidly in an internal

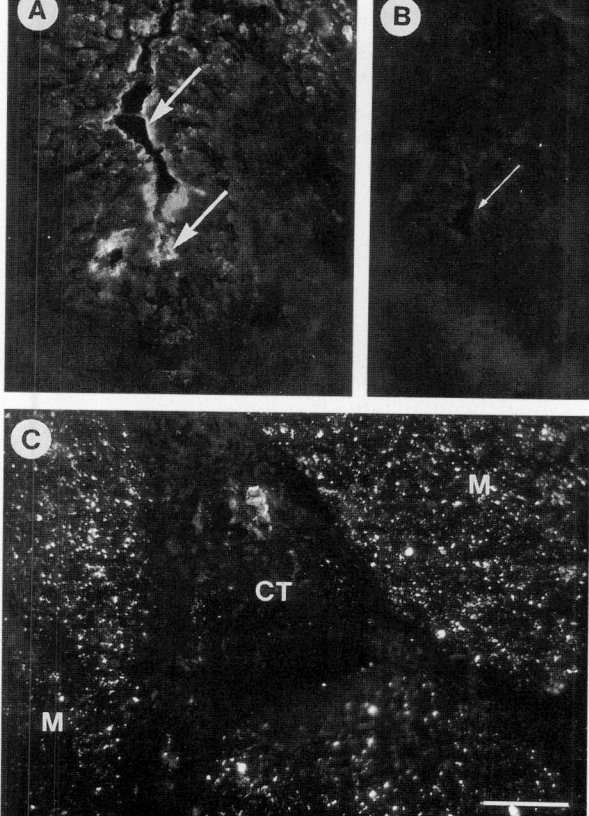

Fig. 2A–C. Fimbrial binding to epithelial cells and to muscular layer. **A** Binding of S fimbriae to epithelial cells of canine urethra (*arrows*). **B** As in **A**, but in the presence of 30 mM sialyl α2-3 lactose. Note complete inhibition of binding (*thin arrow*). **C** Binding of type 1 fimbriae to muscular layer (*M*) of human bladder. Note that connective tissue (*CT*) between the muscle cells is almost negative for the fimbrial binding

position, e.g., in globoside (Bock et al. 1986), but antibodies to the disaccharide bind only to terminal sequences and not to globoside, a major receptor-active glycolipid at the infection site (O'Hanley et al. 1985). Such differences may contribute to the slightly differing distribution of P and type 1 fimbrial receptors in kidneys obtained with antibodies (O'Hanley et al. 1985) or with purified fimbriae and labeled bacteria (Nowicki et al. 1986a; Korhonen et al. 1986b; Virkola 1987). For receptor localization, the methods employing either directly labeled bacteria or indirect staining of tissue sections with purified fimbriae or adhesins are obviously preferable over the use of antibodies raised against receptor-active carbohydrates or glycoproteins.

3.2 Binding Sites in the Urinary Tract

We have used the methods described above to localize binding sites for P, type 1, type 1C, and S fimbriae and for the O75X adhesin in human kidney and bladder (Nowicki et al. 1986a; Korhonen et al. 1986a,b,c; Virkola 1987; Virkola et al. 1988) and along the whole urinary tract of the dog (Westerlund et al. 1987; B. Westerlund, unpublished material). The results have revealed considerable tissue tropism for the *E. coli* adhesins: they bind differently and selectively to restricted tissue domains (Figs. 1, 2) and various epithelial surfaces of the urinary tract show differences in their reactivity with the adhesins (Table 2).

   P and S fimbriae have a closely similar binding pattern in the urinary tract. They bind to epithelial cells of proximal and distal nephron and of urinary bladder, to glomeruli in the kidney, to vascular endothelium in the kidney and bladder, and to exfoliated epithelial cells of human urine (Table 2). The binding of the two fimbrial types seems to differ only with respect to the podocytes in glomeruli (Korhonen et al. 1986a, b) and to connective tissue elements in the bladder. The type 1 fimbriae bind avidly to luminal aspects of proximal tubules but only weakly, or not at all, to other epithelial surfaces in the kidney and bladder. This is in line with their lack of binding to exfoliated uroepithelial cells. Muscular layer of the urinary bladder is strongly stained by the type 1 fimbriae (Fig. 2C). A different pattern was found for the type 1C fimbriae, which bind to epithelial cells of the distal nephron and to vascular endothelium. A strikingly different binding pattern was found for the O75X adhesin, which binds strictly to nonepithelial elements in the kidney (mainly to basement membranes; Fig. 1) but reacts with epithelial cells of the bladder and urine sediment (Table 2).

   The results indicate that P fimbriae bind to most, if not all, epithelial surfaces in the human urinary tract, thus giving a physical basis for the colonization of the upper urinary tract by P-fimbriated *E. coli* strains. Interestingly, however, the P and the S fimbriae, which are associated with different types of infection caused by *E. coli* (Table 1), show similar tissue binding patterns in the urinary tract (Table 2). This indicated to us that, in addition to the presence of binding sites on uroepithelia, other factors also contribute to the specific pathogenetic role of P fimbriae in human upper urinary tract infections. One such factor might be the presence of inhibitory compounds for bacterial binding, which led us to test this hypothesis further.

Table 2. Binding of the *E. coli* adhesins to human kidney and bladder, to sediment cells, and to inhibitors in normal human urine

Tissue site (inhibitor)	Adhesin binding[a]				
	P fimbria	S fimbria	Type 1 fimbria	Type 1C fimbria	O75X adhesin
Kidney:					
Bowman's capsule	+ + +	+ + +	−	−	+ + +[b]
Glomerulus	+ + +	+ + +	−	−	−
Proximal tubule	+ +	+ +	+ + +	−	+ + +[b]
Distal tubule	+ +	+ +	(+)	+ +	+ + +[b]
Collecting duct	+	+ +	(+)	+ +	+ + +[b]
Vessel walls	+ + +[c]	+ + +[c]	+ + +	+ + +[c]	−
Urinary bladder:					
Epithelium	+	+ +	−	−	+
Vessel walls	+ + +[c]	+ + +[c]	+ +	+ + +[c]	−
Muscular layer	+	+	+ + +	+	+
Connective tissue	−	+ +	−	−	+ + +
Sediment cells[d]:	+	+	−	−	+
Inhibitors in urine[e]:					
Oligosaccharides	−	−	+	ND[f]	+ / −[g]
Tann-Horsfall glycoprotein	−	+	(+)	−	−

[a]Binding to tissues is graded from + + + (intense) to − (not detectable); (+) denotes very weak binding. Data for the tissues is from NOWICKI et al. 1986a; KORHONEN et al. 1986a, b, c; VIRKIOLA 1987; VIRKOLA et al. 1988;
[b]To capsular or tubular basement membranes;
[c]Mainly to endothelial cells;
[d]Data for sediment cells are from KORHONEN et al. 1980, 1981, 1986c; ØRSKOV et al. 1980; VIRKOLA et al. 1988. Only positivity (+) or negativity (−) is shown;
[e]Data are from PARKKINEN et al. 1988b. (+), modest interaction; +, intense interaction; −, no interaction;
[f]Not determined;
[g]Inhibitors observed in 2/11 urine samples originating from different individuals

4 Identification of Inhibitors in Human Urine

In order to gain insight into the capacity of the different *E. coli* adhesins to mediate bacterial adhesion to urinary tract epithelia in vivo, we tested urine from different individuals for inhibition of hemagglutination by recombinant strains carrying different fimbriae genes (PARKKINEN et al. 1988b). Urine was found to inhibit the hemagglutination by S and type 1 fimbriae but not that by P fimbriae. Fractionation of the urine showed that the Tamm-Horsfall glycoprotein, the most abundant protein in normal human urine, was mainly responsible for the inhibitory effect for S fimbriae, whereas the major inhibitors for the type 1 fimbrial binding were identified as low molecular weight mannosides. These urine fractions inhibited hemagglutination by S and type 1 fimbriate bacteria at concentrations found in normal urine, indicating that the interactions have physiologic significance, and no significant individual variation in the inhibitory capacity of different urines was found (PARKKINEN et al. 1988b). S and type 1 fimbriae bound also to immobilized Tamm-Horsfall glycoprotein, whereas no such interaction was found for P and type 1C fimbriae or the O75X adhesin (Table 2). Some urine samples were found to contain low molecular weight inhibitors against hemagglutination by the O75X adhesin.

5 Role of the Different Adhesins in Human Urinary Tract Infections

The results summarized in Table 2 provide an insight into the role and function of the *E. coli* adhesins in human urinary tract infections. P fimbriae, which by epidemiologic criteria appear important for pyelonephritis and urosepsis (Table 1), probably bind to all epithelial surfaces in the human urinary tract. The identified binding sites suggest an ascending route for the invasion of P-fimbriated *E. coli* into the kidney and subsequently into the circulatory system. Binding properties of P fimbriae thus support their role as a major bacterial virulence factor in pyelonephritis.

S fimbriae, which only rarely can be found on uropathogenic *E. coli* strains, have a similar binding pattern in the urinary tract to that of the P fimbriae. This indicates that the presence of binding sites on uroepithelia is not enough to explain the pathogenetic function of P fimbriae. The two fimbriae differ in that normal human urine contains inhibitors, mainly Tamm-Horsfall glycoprotein, which at concentrations found in urine can interfere with S fimbriae-mediated adhesion in the urinary tract. Such inhibitors have not been found for the P fimbriae (ORSKOV et al. 1980; O'HANLEY et al. 1985; PARKKINEN et al. 1988b). These results suggest an important principle: P fimbriae are important for pyelonephritis because they readily attach to uroepithelia and their binding is not inhibited by compounds secreted by the host in soluble form into the urine.

We have recently observed a novel interaction of P fimbriae that is independent of the DGalα1-4DGal binding. Both as a purified protein and on bacterial cells, P fimbriae interact with immobilized fibronectin (WESTERLUND et al. 1989b). This interaction is not inhibited by DGalα1-4DGal and is as effective with wild-type P fimbriae as with mutated ones lacking the lectin activity. Insoluble fibronectin is a component of the extracellular matrix and basement membranes and is known to be involved in a number of molecular and cellular binding processes. Our observation suggests a further function for the P fimbriae, i.e., interaction in a lectin-independent manner with a component of the extracellular matrix (WESTERLUND et al. 1988a). Such interaction could be useful for the bacteria at later stages of the infection after epithelial trauma and exposure of connective tissue. It could be that in such conditions the lectin activity of the fimbriae is no longer useful.

Our results do not support a pathogenetic function for type 1 fimbriae in human pyelonephritis or cystitis. These fimbriae do not bind to the distal nephron or to the epithelia of the urinary bladder, which are important surfaces for bacterial ascent. Moreover, low molecular weight α-mannosides in normal human urine would effectively prevent their binding to urinary tract components and to precipitated Tamm-Horsfall glycoprotein occurring in urinary casts and slime (PARKKINEN et al.

1988b). It should be noted that the results suggesting a pathogenetic function for type 1 fimbria in cystitis (e.g., KEITH et al. 1986) have been obtained with uroepithelial cells of mice or rats. The latter are known to differ from epithelial cells of human urine in respect of type 1 fimbrial binding (KORHONEN et al. 1981).

The type 1C fimbriae bound to vascular endothelium and to the epithelial cells of proximal tubules and of collecting ducts in the kidney. This fimbrial type is associated with pyelonephritogenic strains that often also carry P fimbriae (Table 1). Binding of the type 1C fimbriae to the lumen of collecting ducts might increase the invasive potential of *E. coli* as P-fimbriated bacteria adhere to collecting ducts only poorly (NOWICKI et al. 1986b). However, the presence of inhibitors in urine for type 1C fimbrial binding has not been determined so far.

In the kidneys, the O75X adhesin binds only to basement membranes (Fig. 1) and we have recently shown that it strongly interacts with type IV collagen, a major basement membrane glycoprotein (WESTERLUND et al. 1989a). Binding of O75X to basement membranes and to purified type IV collagen is specifically inhibited by chloramphenicol and *N*-acetyltyrosine (NOWICKI et al. 1988; WESTERLUND et al. 1989a), which indicates a protein–protein interaction. The active region in type IV collagen is the aminoterminal 7S domain (WESTERLUND et al. 1989a). The O75X adhesin also binds to bladder epithelium and to exfoliated uroepithelial cells, suggesting that it may facilitate colonization of *E. coli* in epithelia of the lower urinary tract. Moreover, binding to basement membranes may be an important factor after epithelial trauma. The epidemiologic data, however, do not support a significant pathogenetic role for the O75X adhesin, nor does the presence of inhibitors for it in urine.

6 Possible Role of S Fimbriae in Neonatal Meningitis

The distinct association of S fimbriae with *E. coli* O18:K1:H7 strains isolated from neonatal sepsis and meningitis (KORHONEN et al. 1985) suggests that S fimbriae might have a pathogenetic function in these infections. Strains of the O18:K1:H7 serotype predominate among the *E. coli* strains isolated from septic neonatal infections and cause experimental meningitis in newborn rats (BORTOLUSSI et al. 1978), which provides an animal model to study the interactions of S fimbriae during the invasive stage of the infection.

The presence of binding sites for S fimbriae in the different tissues of the newborn rat was recently investigated (PARKKINEN et al. 1988a). In the brain, S fimbriae specifically bind to the luminal surface of the vascular endothelium lining the choroid plexuses and brain ventricles (Fig. 3). Interestingly, there was a distinct decrease in the binding of S-fimbriated bacteria to the choroid plexus of rats older than 2 weeks. Investigation of other organs of the newborn rat indicated that S fimbriae bind to the glomeruli and vascular endothelium in kidney and, more weakly, to vascular endothelium and alveolar epithelium in the lung. No binding was observed in liver and spleen.

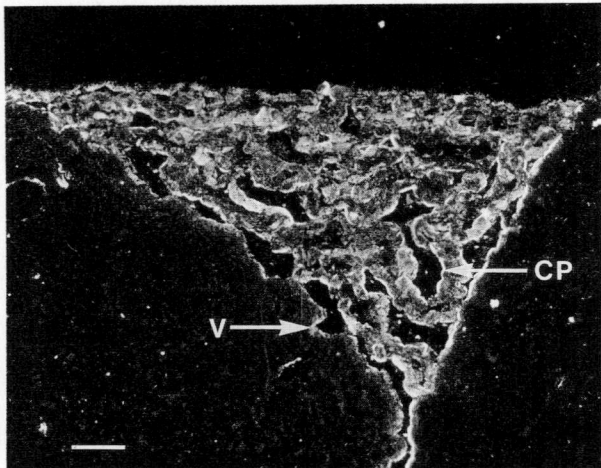

Fig. 3. Binding of S fimbriae to rat brain. The fimbriae bind to the luminal surface of epithelial cells lining the choroid plexuses (*CP*) and brain venticles (*V*). *Bar*, 100 μm

The presence of specific binding sites for S fimbriae in the brain is of particular interest as the bacterial and host factors that, in the pathogenesis on meningitis, result in the invasion of bacteria from circulation into the cerebrospinal fluid are unknown (McGEE 1985). In particular, the intensive binding of S fimbriae to the choroid plexuses may be of pathogenic importance since the choroid plexus is regarded as the site of bacterial invasion into the cerebrospinal fluid (McGEE 1985; LEVINE 1987). During invasion into the cerebrospinal fluid, bacteria must first penetrate the vascular endothelium, and the endothelial receptors for S fimbriae could facilitate this process. On the other hand, in animal models infection of the cerebrospinal fluid compartment is a kinetic process, with bacteria entering from blood and being cleared within the cerebrospinal fluid flow through the arachnoid villi to the cerebral venous sinuses (SCHELD et al. 1979). In view of this, the ability of S-fimbriated bacteria to adhere to the epithelium lining the choroid plexuses and brain ventricles might provide a means for the bacteria to resist mechanical clearance by the cerebrospinal fluid flow (PARKKINEN et al. 1988a).

7 Perspectives

As pointed out earlier, the receptor density for P-fimbriated *E. coli* may determine the susceptibility of certain individuals to pyelonephritis (KÄLLENIUS and WINBERG 1978). There is also evidence suggesting that the expression by calves and rats of sialic acid-containing receptors for *E. coli* is age dependent, perhaps contributing to the susceptibility of newborns to diarrheal and septic infections by *E. coli* (RUNNELS et al. 1980; PARKKINEN et al. 1988a). Determination of the density and localization of tissue receptors could thus be an important step in the identification of individuals who are prone or susceptible to bacterial infections. Our results show that immunofluorescence techniques, either with whole bacteria or purified fimbriae or adhesins (Table 2) or possibly with the receptor-binding components of the adhesins (MOCH et al. 1987), can be used for such purposes. With better understanding of the biochemistry and serologic features of the receptor-active macromolecules, antibodies against them could also be utilized in determining receptor density in selected tissue sites. On the other hand, the obvious ability of naturally occurring inhibitors in urine to prevent bacterial adhesion (PARKKINEN et al. 1988b) encourages the development of synthetic receptor analogues to be used to prevent bacterial adhesion in urinary tracts and in septic metastatic infections.

Acknowledgement. This study was supported by the Academy of Finland.

References

Bock K, Breimer ME, Brignole A, Hansson GG, Karlsson K-A, Keffler HA, Samuelsson Bü-E, Strömberg N, Svanborg-Edén C, Thurin J (1986) Specificity of binding of a strain of uropathogenic *Escherichia coli* to Galα(1-4)Gal-containing glycosphingolipids. J Biol Chem 260: 8545–8551

Bortolussi R, Ferrieri P, Wannamaker LW (1978) Dynamics of *Escherichia coli* infection and meningitis in infant rats. Infect Immun 22: 480–485

Dean EA, Isaacson RE (1985) Location and distribution of a receptor for the 987P pilus of *Escherichia coli* in small intestines. Infect Immun 47: 345–348

Duguid JP, Old DC (1980) Adhesive properties of Enterobacteriaceae. In: Beachey EH (ed) Bacterial adherence. Chapman and Hall, London, pp 185–271 (Receptors and recognition, series B, vol 6)

Firon N, Ofek I, Sharon N (1982) Interaction of mannose-containing oligosaccharides with the fimbrial lectin of *Escherichia coli*. Biochem Biophys Res Commun 105: 1426–1432

Hagberg L, Jodal U, Korhonen TK, Lidin-Janson G, Lindberg U, Svanborg-Edén C (1981) Adhesion, hemagglutination and virulence of *Escherichia coli* causing urinary tract infections. Infect Immun 31: 564–570

Hagberg L, Hull R, Hull S, Falkow S, Freter R, Svanborg-Edén C (1983) Contribution of adhesion to bacterial presistence in the mouse urinary tract. Infect Immun 40: 265–272

Hennigar R, Schulte BA, Spicer SS (1985) Hetreogeneous distribution of glycoconjugates in human kidney tubules. Anat Res 211: 376–390

Holthöfer H (1983) Lectin binding sites in kidney. A comparative study of 14 animal species. J Histochem Cytochem 31: 531–537

Holthöfer H, Virtanen I, Pettersson E, Törnroth I, Alfthan O, Linder E, Miettinen A (1982) Lectins as fluorescence markers for saccharides in the human kidney. Lab Invest 45: 391–399

Johnson JR, Roberts PL, Stamm WE (1987) P fimbriae and other virulence factors in *Escherichia coli* urosepsis: association with patient's characteristics. J Infect Dis 156: 225–229

Jokinen M, Ehnholm C, Väisänen-Rhen V, Korhonen T, Pipkorn R, Kalkkinen N, Gahmberg CG (1985) Identification of the major human red cell sialoglycoprotein, glycophorin A^M, as the receptor for *Escherichia coli* IH11165 and characterization of the receptor site. Eur J Biochem 147: 47–52

Källenius G, Winberg J (1978) Bacterial adherence to periurethral cells in girls prone to urinary-tract infections. Lancet ii: 540–543

Källenius G, Möllby R, Winberg J (1980) In vitro adhesion of uropathogenic *Escherichia coli* to human periurethral cells. Infect Immun 28: 972–980

Källenius G, Möllby R, Svenson SB, Helin I, Hultberg H, Cederberg B, Winberg J (1981) Incidence of P-fimbriated *Escherichia coli* in urinary tract infections. Lancet ii: 1369–1371

Keith BR, Maurer L, Spears PA, Orndorff PE (1986) Receptor-binding function of type 1 pili affects bladder colonization of *Escherichia coli*. Infect Immun 53: 693–696

Korhonen TK, Edén S, Svanborg-Edén C (1980) Binding of purified *Escherichia coli* pili to human urinary tract epithelial cells. FEMS Microbiol Lett 7: 237–240

Korhonen TK, Leffler H, Svanborg-Edén C (1981) Binding specificity of piliated strains of *Escherichia coli* and *Salmonella typhimurium* to epithelial cells, *Saccharomyces cerevisiae* cells, and erythrocytes. Infect Immun 32: 796–804

Korhonen TK, Valtonen MV, Parkkinen J, Väisänen-Rhen V, Finne J, Ørskov F, Ørskov I, Svenson SB, Mäkelä PH (1985) Serotypes, hemolysin production, and receptor recognition of *Escherichia coli* strains associated with neonatal sepsis and meningitis. Infect Immun 48: 486–491

Korhonen TK, Parkkinen J, Hacker J, Finne J, Pere A, Rhen M, Holthöfer H (1986a) Binding of *Escherichia coli* S fimbriae to human kidney epithelium. Infect Immun 54: 322–327

Korhonen TK, Virkola R, Holthöfer H (1986b) Localiziation of binding sites for purified *Escherichia coli* P fimbriae in the human kidney. Infect Immun 54: 328–332

Korhonen TK, Virkola R, Väisänen-Rhen V, Holthöfer H (1986c) Binding of purified *Escherichia coli* O75X adhesin to frozen sections of human kidney. FEMS Microbiol Lett 35: 313–318

Levine S (1987) Choroid plexus: target for systemic disease and pathway to the brain (Editorial). Lab Invest 56: 231–233

McGee ZA (1985) Bacterial meningitis: current status and directions for the future. In: Sande MA, Smith AL, Root RK (eds) Bacterial meningitis. Churchil Livingstone, New York, pp 1–10

Moch T, Hoschutzky H, Hacker J, Kröncke K-D, Jann K (1987) Isolation and characterization of the α-sialyl-β-2,3-galactosyl-specific adhesin from fimbriated *Escherichia coli*. Proc Natl Acad Sci USA 84: 3462–3466

Neser J-R, Koellreutter B, Wuersch P (1986) Oligomannoside-type glycopeptides inhibiting adhesion of *Escherichia coli* strains mediated by type 1 pili: preparation of potent inhibitors from plant glycoproteins. Infect Immun 52: 428–436

Nowicki B, Rhen M, Väisänen-Rhen V, Pere A, Korhonen TK (1984) Immunofluorescence study of fimbrial phase variation in *Escherichia coli* KS71. J Bacteriol 160: 691–695

Nowicki B, Holthöfer H, Saraneva T, Rhen M, Väisänen-Rhen V, Korhonen TK (1986a) Location of adhesion sites for P-fimbriated and for O75X-positive *Escherichia coli* in the human kidney. Microb Pathogen 1: 169–180

Nowicki B,Vuopio-Varkila J, Viljanen P, Korhonen TK, Mäkelä PH (1986b) Fimbrial phase variation and systemic *E. coli* infection studied in the mouse peritonitis model. Microb Pathogen 1: 335–347

Nowicki B, Moulds J, Hull R, Hull S (1988) A hemagglutinin of *Escherichia coli* recognizes the Dr blood group antigen. Infect Immun 56: 1057–1060

O'Hanley P, Lark D, Falkow S, Schololnik G (1985) Molecular basis of *Escherichia coli* colonization of the upper urinary tract in BALB/c mice: Gal-Gal pili immunization prevents *Escherichia coli* pyelonephritis in the BALB/c mouse model of human pyelonephritis. J Clin Invest 75: 347–360

Ørskov I, Ørskov F (1983) Serology of *Escherichia coli* fimbriae. Prog Allergy 33: 80–105

Ørskov I, Ørskov F, Birch-Andersen A (1980) Comparison of *Escherichia coli* fimbrial antigen F7 with type 1 fimbriae. Infect Immun 27: 657–666

Ott M, Hacker J, Schmoll T, Jarchau T, Korhonen TK, Goebel W (1986) Analysis of the genetic determinants coding for the S-fimbrial adhesin (*sfa*) in different *Escherichia coli* strains causing meningitis or urinary tract infections. Infect Immun 54: 646–653

Parkkinen J, Finne J, Achtman M, Väisänen V, Korhonen TK (1983) *Escherichia coli* strains binding neuraminyl α2-3 galactosides. Biochem Biophys Res Commun 111: 456–461

Parkkinen J, Rogers GN, Korhonen T, Dahr W, Finne J (1986) Identification of the O-linked sialyloligosaccharides of glycophorin A as the erythrocyte receptors for S-fimbriated *Escherichia coli*. Infect Immun 54: 37–42

Parkkinen J, Korhonen TK, Pere A, Hacker J, Soinila S (1988a) Binding sites in the rat brain for *Escherichia coli* S fimbriae associated with neonatal meningitis. J Clin Invest 81: 860–865

Parkkinen J, Virkola R, Korhonen TK (1988b) Identification of factors in human urine that inhibit the binding of *Escherichia coli* adhesins. Infect Immun 56: 2623–2630

Pere A, Leinonen M, Väisänen-Rhen V, Rhen M, Korhonen TK (1985) Occurrence of type-1C fimbriae on *Escherichia coli* strains isolated from human extraintestinal infections. J Gen Microbiol 131: 1705–1711

Pere A, Väisänen-Rhen V, Rhen M, Tenhunen J, Korhonen TK (1986) Analysis of P fimbriae on *Escherichia coli* O2, O4, and O6 strains by immunoprecipitation. Infect Immun 51: 618–625

Pere A, Nowicki B, Saxén H, Siitonen A, Korhonen TK (1987) Expression of *Escherichia coli* P, type-1 and type-1C fimbriae in the urine of patients with acute urinary tract infection. J Infect Dis 156: 567–574

Rhen M, Klemm P, Korhonen TK (1986) Identification of two new hemagglutinins of *Escherichia coli*, *N*-acetyl-D-glucosamine-specific fimbriae and a blood group M-specific agglutinin, by cloning the corresponding genes in *Escherichia coli* K-12. J Bacteriol 168: 1234–1242

Roberts J, Hardaway K, Kaack B, Fussel EN, Baskin G (1984) Prevention of pyelonephritis by immunization with P-fimbriae. J Urol 131: 602–607

Runnels PL, Moon HW, Schneider RA (1980) Development of resistance with host age to adhesion of K99+ *Escherichia coli* to isolated intestinal epithelial cells. Infect Immun 28: 298–300

Scheld WM, Park T-S, Dacey RG, Winn HR, Jane JA, Sande MAJ (1979) Clearance of bacteria from cerebrospinal fluid to blood in experimental meningitis. Infect Immun 24: 102–105

Svanborg-Edén C, Hanson LÅ, Jodal U, Sohl-Åkerlund A (1976) Variable adherence to normal human urinary tract epithelial cells of *Escherichia coli* strains associated with various forms of urinary tract infections. Lancet ii: 490–492

Svanborg-Edén C, Eriksson B, Hanson LÅ (1977) Adhesion of *Escherichia coli* to human uroepithelial cells in vitro. Infect Immun 18: 767–774

Väisänen-Rhen V (1984) Fimbria-like hemagglutinin of *Escherichia coli* O75 strains. Infect Immun 46: 401–407

Väisänen-Rhen V, Korhonen TK, Finne J (1983) Novel cell-binding activity specific for *N*-acetyl-D-glucosamine in an *Escherichia coli* strain. FEBS Lett 159: 233–236

Väisänen-Rhen V, Elo J, Väisänen E, Siitonen A, Ørskov I, Ørskov F, Svenson SB, Mäkelä PH, Korhonen TK (1984) P-fimbriated clones among uropathogenic *Escherichia coli* strains. Infect Immun 43: 149–155

Westerlund B, Merenmies J, Rauvala H, Miettinen A, Järvinen A-K, Virkola R, Holthöfer H, Korhonen TK (1987) The O75X adhesin of uropathogenic *Escherichia coli*: receptor-active domains in the canine urinary tract and in-vitro interaction with laminin. Microb Pathogen 3: 117–127

Westerlund B, Kuusela P, Risteli J, Risteli L, Vartio T, Rauvala H, Virkola R, Korhonen TK (1989a) The O75X adhesin of uropathogenic *Escherichia coli* is a type IV collagen-binding protein. Mol Microbiol 3: 329–337

Westerlund B, Kuusela P, Vartio T, Van Die I, Korhonen TK (1989b) A novel lectin-independent interaction of P fimbriae of *Escherichia coli* with immobilized fibronectin. FEBS Lett 243: 199–204

Virkola R (1987) Binding characteristics of *Escherichia coli* type-1 fimbriae in the human kidney. FEMS Microbiol Lett 40: 257–262

Virkola R, Westerlund B, Holthöfer H, Parkkinen J, Kekomäki M, Korhonen TK (1988) Binding characteristics of *Escherichia coli* adhesins in human urinary bladder. Infect Immun 56: 2615–2622

Colonization Factor Antigens of Human Pathogens

D. J. EVANS, Jr. and D. G. EVANS

1 Introduction

The concept of microbial pathogenesis as a complex process divisible into discrete steps, each mediated by a specific type of virulence factor, is not new. However, within this concept lies the key to future success in prevention and control of diseases for which there is no practical vaccine or prophylactic method at this time. Every pathogen has a characteristic mode of pathogenesis, some more obvious than others and hence more susceptible to human intervention, diphtheria being a classic example. Others such as *Vibrio cholerae* are thought to be well understood and cholera therapy is technically well advanced; however, a cost-effective and practical vaccine rendering complete and long-term protection has yet to be achieved.

Dissecting the infectious process, the first specific event may be visualized as entry of the pathogen through a suitable host portal, such as breathing in a respiratory virus, drinking water contaminated with an enteric pathogen, or being bitten by a rabies-infected animal. Early recognition of this type of event as a prerequisite to illness gave rise to the principles of disinfection, sanitation, quarantine, insect vector control, pasteurization, food preservation, and sewage and

The Bacterial Enteropathogen Laboratory, Digestive Disease Section, Veterans Administration Medical Center, Houston, Texas, and Department of Medicine, Baylor College of Medicine, Houston, Texas, USA

Current Topics in Microbiology and Immunology, Vol. 151

water treatment which are routinely practiced today. Various combinations of these basic measures designed to prevent pathogen transmission have been remarkably effective in controlling once devastating pathogens and continue to be the most important means of disease prevention. Unfortunately, it is also true that on a global scale most of these basic preventative measures are not practiced, because of lack of either education or physical resources.

Clearly, if one cannot always prevent exposure to microbial pathogens, then one must seek means of intervention based on an understanding of events subsequent to such exposure. In this chapter, we concentrate on the importance of the second step in the infectious process, which is actually the first specific physical interaction between pathogen and host, i.e., attachment to a specific host receptor mediated by specialised macromolecules exposed on the surface of the pathogen. This key step in pathogenesis is currently the subject of intense investigation for diseases as diverse as HIV-1 (AIDS virus), malaria, and Asiatic cholera.

Consider the problem of designing vaccines. Most microbial pathogens, whether parasite, virus, or bacterium, possess unique species- and/or strain-specific surface antigens which can theoretically be employed as immunogens in a vaccine, and this approach has had its share of success. However, in reality many pathogens have so far defied all attempts at artificial immunization, some because they possess exquisitely complex and/or variable antigenic surface structure(s) and others by equally exquisite mechanisms for avoiding exposure to antibody attack. We feel that many of these problems will be resolved by targeting as immunogens the specific surface-associated antigens which function as adhesins, simply because preventing colonization by the pathogen is essentially equivalent to aborting the infection.

In this chapter we consider the role of adhesins in the infectious process, using the enterotoxigenic *Escherichia coli* (ETEC) colonization factor antigens (CFAs) as a "model" example. We submit some hopefully accurate historical perspectives on this subject as well as a brief overview of its current status, speculations on the role of CFAs in the evolution of ETEC as they exist today, and, finally, a few ideas about those aspects of ETEC CFAs which we feel will be the focus of future research.

2 Recognition of Adhesins as Important Virulence Factors: Historical Perspective

Generally, microbial pathogens possess on their surface specific molecules which are configured so as to recognize and interact with receptors on the surface of specific host cells, usually those of the epithelial cell lining of the respiratory, gastrointestinal, or genitourinary tract. We define colonization as adhesion at a specific site followed by multiplication at that site. Thus colonization factors represent the mechanism by which pathogens such as the ETEC recognize and affix themselves to a specific target tissue. Adhesion, even colonization, rarely constitutes the entire disease process, a fact which is largely responsible for the relative delay in appreciating the significance of colonization factors. Historically, the existence of colonization factors as molecular entities was less obvious than that of the functionally secondary virulence factors which mediate actual damage to the host and therefore are responsible for the overt symptoms of infection. Virulence factors of bacteria include factors facilitating invasion of tissue (invasins), cell death (cytotoxins, cytolysins), neurologic dysfunction (neurotoxins), and diarrhea (enterotoxins).

Interestingly, diseases such as pneumonia, meningitis, and even acute diarrhea may be caused by either viral or bacterial pathogens and even today the task of differentiating between these can be challenging. In practical terms, an intact infectious viral particle is itself an adhesin and replication (colonization) occurs intracellularly. Viruses do possess on their surface specific molecular entities which mediate adhesion to host receptors and these adhesins represent potentially important targets for designing antiviral defenses such as vaccines.

3 Multiplicity of *Escherichia coli* Fimbrial Adhesins and Correlations Between Fimbrial Type and Pathogenicity

Approximately a century ago it was recognized that the common enteric bacillus *Bacterium coli* (*Escherichia coli*) had the property of readily adhering to other cells, including plant cells, yeast cells, and animal erythrocytes. Later it was found that most, if not all, genera of gram-negative enteric bacteria have this property. It became convenient to classify such adherence as either mannose sensitive (MS) (i.e., readily blocked by mannose) or mannose resistant, since the most common type of adherence factor was mannose sensitive, and to employ erythrocytes to detect both MS- and MR-type adhesins. Hemagglutination (HA) tests have an additional attraction in that adhesins can be classified according to the type of erythrocyte producing a positive test. By consensus the current test for common fimbriae, which mediate MSHA, employs guinea pig erythrocytes. Common fimbriae are generally long, thin hair-like projections on the bacterial cell surface but as a group these are somewhat morphologically diverse. Common fimbriae are also quite diverse antigenically, even among individual species such as *E. coli*.

A large variety of functions have been ascribed to common fimbriae, or common pili, including a role in virulence. However, since the receptor for these fimbriae (mannose-containing glycoproteins) appears to be universally distributed, cell- or tissue-specific adherence is an unlikely function for common fimbriae. There is some experimental evidence, from work with animal models, that these fimbriae may play a secondary but important role in colonization of the lower urinary tract (ORNDORFF 1987). It is interesting that ETEC are capable of producing common fimbriae but this is not surprising since (a) production of these fimbriae is encoded by the chromosome and (b) ETEC certainly evolved from *E. coli* of the normal flora, which typically possess common fimbriae. There is no evidence for the role of common fimbriae in ETEC virulence although this possibility has been given serious consideration.

As discussed below, there is evidence that ETEC possess control mechanisms which dictate that either common fimbriae or CFA-type fimbriae will be produced in a particular environment (EVANS et al. 1977a, b). Serologic evidence has shown that CFA-type fimbriae are produced in vivo (DEETZ et al. 1979; STOLL et al. 1986). It is possible that common fimbriae may be one of the ancestral proteins from which evolved one or more of the tissue-specific fimbrial CFAs of the ETEC and/or other gram-negative pathogenic bacteria.

Pathogenic *E. coli* generally fall into one of two large groups, i.e., those serotypes commonly associated with intestinal infections and those associated with extra-intestinal infections (EVANS and EVANS 1983; ORSKOV et al. 1977). There are subdivisions within each major group, each sharing the ability to produce a particular set of virulence factors. Briefly, one hallmark of *E. coli* which cause extraintestinal infection is production of fimbrial colonization factors (PAP, or pyelonephritis-associated pili), the receptor of which is a digalactoside moiety (LEFFLER and SVANBORG-EDEN 1980; BOCK et al. 1985) plus one or more other virulence factors such as hemolysin and/or certain carbohydrate capsular antigens, particularly the K-1 antigen (EVANS et al. 1981). PAP-type fimbriae, like common fimbriae, are an antigenically heterogeneous group (PERE et al. 1986); however they do share a highly conserved region the function of which is receptor recognition. Correlations have been noted between possession of PAP receptors and suscepti-bility to urinary tract infection with this type of *E. coli*. However, not all *E. coli* associated with upper urinary tract infections produce PAP-type fimbriae. There are bioserotypes which possess functionally equivalent fimbriae that recognize different host receptors and also those which possess adhesins that are not fimbrial at all but are actually capsular in morphology (ORSKOV et al. 1985; RHEN et al. 1986; GOLDHAR et al. 1987).

The enteropathogenic group of *E. coli* can be subdivided into at least three subgroups on the basis of mechanism of pathogenesis; these are the ETEC, the *Shigella*-like enteroinvasive *E. coli*, or EIEC, serotypes, and the enteroadherent serotypes which include the classical EPEC serogroups. Some individual isolates of enteropathogenic *E. coli* possess unusual combinations of virulence factors and appear to belong to more than one subgroup, e.g., a classic EPEC serogroup hosting ETEC virulence plasmids (GUTH et al. 1985). These represent exceptional cases since it is well established that the three subgroups of enteropathogenic *E. coli* possess different specialized sites of attachment (ROBINS-BROWNE 1987). Only the ETEC preferentially colonize the small intestine. With the *Shigella*-like *E. coli* it appears that adherence to the intestinal cell may be the trigger which initiates invasion of the host cell by the bacteria.

4 ETEC Colonization Factor Antigens: Historical Perspective

Species-specific CFAs of the ETEC were not originally discovered by prospectively searching for fimbriae with unique adhesive properties (although today this is a well-accepted approach) but rather as surface antigen unique to diarrhea-associated isolates of *E. coli*. Upon discovery, both K88 (on swine-associated ETEC) and K99 (on calf-associated ETEC) were characterized as heat-labile capsular antigens and only later recognized as a surface layer possessing fimbrial structure (ORSKOV and ORSKOV 1966; ORSKOV et al. 1975; BURROWS et al. 1976). Subsequent studies on K88 and K99 revealed several properties which also helped in defining the CFAs of human-associated ETEC. K88 and K99 both mediate MRHA but are detectable with different species of erythrocytes and thus recognize different receptors, as has been confirmed by studies on their individual adherence properties. Also, the genes determining production of either K88 or K99 were found to be located on plasmids, which explains the close correlation between enterotoxin production (also plasmid mediated) and production of either K88 or K99.

Studies which detected piglets genetically resistant to K88-positive ETEC produced the first evidence that possession of a specific intestinal epithelial cell receptor is a determining factor in susceptibility to ETEC diarrhea (JONES and RUTTER 1972). That the tissue specificity and host specificity of the ETEC is a function of their CFAs has been repeatedly confirmed by epidemiologic observations and by in vivo and in vitro studies on the affinity of ETEC for intestinal epithelial cells of man and animals (BURROWS et al. 1976; SMITH and LINGOOD 1971; EVANS et al. 1978b; KNUTTON et al. 1984; CHENEY and BOEDEKER 1983). The observed pattern of tissue affinity is in accord with the natural host of the ETEC isolates; K88-positive ETEC do not adhere to human intestine and do not cause diarrhea in man.

Colonization factor antigen receptors are di- or trisaccharide sequences, often including sialic acid moieties, which occur on surface-associated sialoglycoproteins of the target cells. (BEACHEY 1981; BOCK et al. 1985, EVANS et al. 1988a). These same receptor sequences often occur on various cell types, including erythrocytes, but it is their occurrence on epithelial cells of the small intestine which is relevant to susceptibility to colonization with ETEC. One can speculate that host-specific receptor recognition is the result of an evolutionary process by which certain *E. coli* bioserotypes evolved the means to maximize their ability to colonize. In essence, potential host diversity appears to have been sacrificed in exchange for efficiency of colonization.

The existence of K88 and K99 led us to speculate that human-associated ETEC might possess plasmid-encoded fimbriae detectable as unique antigens having unique receptor specificity, detectable as MR-type hemagglutinins, and having a demonstrable correlation with enterotoxin production. We searched for, and found within cultures of ETEC isolated from cases of diarrhea, clones possessing a new antigenic type of fimbriae which satisfied the definition of a colonization factor and the production of which correlated with enterotoxin production and possession of a particular plasmid; this fimbrial antigen is CFA/I (EVANS et al. 1975). CFA/II also has these properties.

The major clue to the existence of CFA/II was that ETEC isolates belonging to the commonly isolated ETEC serotype O6:H16 never expressed CFA/I (EVANS and EVANS 1978). It is now known that O6:H16 is the most commonly isolated CFA/II-positive ETEC serotype. Both CFA/I and CFA/II have a specific affinity for attachment to human intestinal epithelial cells and both have been found, associated with different ETEC serotypes, among ETEC isolated from cases of acute diarrhea endemic in numerous countries. CFA/I- and CFA/II-positive ETEC account for a large proportion of ETEC isolated from cases of traveler's diarrhea and pediatric diarrhea and have been identified as the cause of hospital and community common-source outbreaks of acute diarrhea.

Receptors for both CFA/I and CFA/II are found on a major sialoglycoprotein of the bovine erythrocyte and MRHA of bovine erythrocytes is a reliable test for these CFAs since other *E. coli* rarely exhibit this property. Both CFA/I and CFA/II produce MRHA of human erythrocytes, but this is a fairly common property among non-ETEC mainly because PAP-type fimbriae also mediate MRHA of human erythrocytes. Erythrocytes from all human donors react with CFA/I since the receptor for CFA/I resides on the same sialoglycoconjugate as the ABO antigen determinants. The exact chemical structure of the CFA/I receptor has not yet been determined. The human erythrocyte receptor for CFA/II is not universal; in one study only 9% of 582 donors were positive for MRHA with CFA/II-positive ETEC. This was due in part to instability of the human red cell receptor for CFA/II and in part to the fact that possession of this receptor is related to racial/ethnic background, with 88% of black donors having red cells reactive with CFA/II (EVANS et al. 1988c).

5 Human-Associated ETEC: From CFA/I to Present

For many years after the discovery of the fimbrial CFAs, CFA/I and CFA/II (EVANS et al. 1975; EVANS and EVANS 1978) the relative significance of these CFAs, in terms of the overall ETEC diarrhea problem, was uncertain. The concept that possession of a CFA is a necessary prerequisite to virulence of ETEC for man could not be tested in a simple or straightforward manner with these ETEC since man is their only natural host. However, studies comparing CFA-positive strains and their spontaneous isogeneic CFA-negative derivatives (the same strain minus the CFA-encoding plasmid) in a susceptible animal model, the laboratory rabbit, did demonstrate that at realistically low doses only the CFA-positive strains caused diarrhea (EVANS et al. 1975). The same result was later observed with volunteers (EVANS et al. 1978). In retrospectively testing ETEC isolates from cases of diarrhea in Bangladesh and Mexico and in prospectively testing ETEC isolated from cases of travellers' diarrhea, a significant percentage of the strains were found to possess either CFA/I or CFA/II (EVANS et al. 1984a). Serum antibody responses to CFAs in natural cases of ETEC diarrhea and in volunteers, and also anti-CFA colostral antibody levels in mothers living in ETEC-endemic areas, were demonstrated, showing that both CFA/I and CFA/II are produced in vivo in the course of natural infection (EVANS et al. 1984a; STOLL et al. 1986).

In recent years more definitive evidence supporting the importance of CFAs as virulence factors of the ETEC has been obtained from prospective epidemiologic studies in countries with high rates of endemic acute diarrhea (BACK et al. 1980; GUTH et al. 1985; GOTHEFORS et al. 1985; THOMAS and ROWE 1982; ROWE et al. 1983).

Basically, these studies support our current concept of the role of CFAs in ETEC virulence: ETEC possessing a CFA are more likely to cause serious diarrhea than ETEC without a CFA (STOLL et al. 1986). Factors which increase host susceptibility such as poor nutritional status and previous or current infection with other enteric pathogens can lead to infection with ETEC strains lacking a full complement of virulence factors. Hence, both ETEC virulence and host susceptibility vary along a wide spectrum and both of these factors, plus others such as dose and manner of infection, determine the outcome of the host–pathogen encounter.

It must be emphasized that ETEC virulence is not an absolute property because of the multiplicity of ETEC virulence factors and the complex genetics of these factors. By definition, ETEC are *E. coli* which produce a heat-stable (ST) and/or a heat-labile (LT) enterotoxin. Virtually all CFA-encoding plasmids also encode for one or both types of ETEC enterotoxin; however, there are enterotoxin-encoding plasmids which do not encode for a CFA and both types of plasmid readily coexist in ETEC (SMITH et al. 1979; PENARANDA et al. 1983; EVANS and EVANS 1983; MURRAY et al. 1983; DANBARA et al. 1987). A complicating factor is that an ETEC strain may lose the plasmid encoding for a CFA and still retain a plasmid encoding only for enterotoxin, resulting in a CFA-minus ETEC clone. How often this occurs in vivo is an unanswered question but it has been demonstrated that the presence of antibody against a CFA can favor growth of CFA-minus derivatives which arise either by plasmid loss or mutation within the CFA operon. Also, for unknown reasons, CFA-encoding plasmids tend to be lost from ETEC isolates in vitro. This significantly increases the value of prospective studies in which *E. coli* isolates are tested as rapidly after isolation as possible for the property of CFA production, preferably in conjunction with serologic examination of the test population for anti-CFA responses (GOTHEFORS et al. 1985; STOLL et al. 1986).

For the reasons just stated, CFA-negative ETEC isolates are to be expected; however, ETEC isolates lacking both CFA/I and CFA/II could possess another, as yet unidentified CFA. Clearly, it is important to identify all of the epidemiologically important CFAs in order to obtain a valid assessment of the contribution of the individual CFAs to the overall ETEC diarrhea problem. Also, CFAs represent the ideal basis for an anti-ETEC vaccine and it must be assumed that an anti-CFA vaccine should include all of the most common CFAs in order to be successful. An important clue to the existence of CFAs other than CFA/I and CFA/II derives from the observation that each of these CFAs is produced by ETEC belonging to particular prevalent serotypes, for example CFA/I-positive O78:H11 and CFA/II-positive O6:H16 or O8:H9. Although the majority of the frequently isolated ETEC serotypes were found with either CFA/I or CFA/II, there were still other prevalent ETEC serotypes which did not possess an identifiable CFA. Thus CFA-negative isolated belonging to these serotypes are the most likely candidates for possession of "new" CFAs.

THOMAS et al. (1982) found a new antigenic type of fimbriae on ETEC isolates belonging to serotypes O25:H42, O115:H40, and O167:H5; this putative CFA was designated E8775. E8775 production is genetically linked to enterotoxin production and this CFA produces MRHA with human erythrocytes, a well-known property of CFA/I. Interestingly, 6 of 11 ETEC isolates belonging to serotype O115:H51 were found to produce CFA/II. The number of ETEC isolates possessing E8775 is minor compared to those with CFA/I or CFA/II (THOMAS and ROWE 1982). This is simply a reflection of the relative predominance and distribution of ETEC serotypes, which varies from time to time and from one location to another (BACK et al. 1980). DARFEUILLE et al. (1983) described as CFA/III a fimbrial antigen found on an ETEC strain belonging to serogroup O128. However, this putative CFA/III may in fact be a variant of CFA/I since we and at least three other laboratories have reported that isolates of enterotoxigenic O128 produce CFA/I (MURRAY et al. 1983; THOMAS and Rowe 1982; GOTHEFORS et al. 1985; GUTH et al. 1985).

HONDA et al. (1984) reported a new antigenic type of CFA on strains of ETEC. This CFA was described as highly hydrophobic, as are CFA/I, CFA/II, K88, and K99 (WADSTROM et al. 1980), and genetically linked to enterotoxin production. This pilus antigen, designated 260-1, did not produce MRHA with human or bovine erythrocytes but did mediate colonization in both infant rabbits and suckling mice. The relative prevalence of ETEC with 260-1 pili has yet to be tested and the possible association with particular ETEC serotypes also remains to be determined.

With the discovery of the E8775 fimbrial ETEC antigen, only a few frequently isolated ETEC serotypes (namely O148:H28 and O159:H4) were considered unaccounted for in terms of their CFAs. However, SEN et al. (1984) reported that eight of ten ETEC isolates belonging to serotype O148:H28 were positive for CFA/II. Recently, TACKET et al. (1987) examined isolates of ETEC belonging to serotype O159:H4 and negative for MRHA with human erythrocytes and found that six of ten such isolates, from diverse sources, possessed a new fimbrial antigen. This O159:H4-associated fimbriae may represent a new CFA since its production is plasmid encoded with ST and LT genes on the same plasmid. Definitive studies demonstrating a link between colonization activity and this putative CFA have not yet been reported.

KNUTTON et al. (1987) identified a unique fimbrial structure, consisting of very numerous curly fibrils, approximately 3 nm in diameter, on human enterocyte-adhesive ETEC strains belonging to serotype O148:H28. The O148:H28-associated fibrils are apparently nonhemagglutinating; the genetics and further antigenic characterization of this putative CFA have not yet been reported.

6 Role of ETEC Virulence Factors in the Transmission Dynamics of Acute Diarrhea and Speculations on the Evolution of ETEC

In numerous studies of ETEC diarrhea it has been observed that a small number of ETEC serotypes predominate against a background of much less frequently isolated serotypes (EVANS and EVANS 1983; ORSKOV et al. 1976; EVANS et al. 1984b). Serotypes which produce a known CFA are isolated in different regions around the world and are the same ones which predominate in specific locations (ROWE et al. 1983; DANBARA et al. 1987). Also, CFA-positive ETEC are responsible for the most serious cases of ETEC diarrhea. Serologic evidence, cited elsewhere, further confirms the importance of CFAs in ETEC virulence (STOLL et al. 1986). However, circulating, IgG, antibody against either CFA(s) or LT has proven to be a poor indicator of antigen exposure in the intestine even in cases of acute diarrhea. Specific intestinal secretory IgA responses are a much better measure of exposure to ETEC antigens but appropriate intestinal samples are difficult to obtain except under very convenient circumstances such as in volunteer studies (EVANS et al. 1978; STOLL et al. 1986).

Dependence of ETEC on possession of a CFA for successful colonization of their natural host has broad implications. As is true for many bacterial pathogens, those strains which possess a full complement of virulence factors (in this case, ST, LT, and a CFA) are the ones which infect and successfully colonize the largest number of individual hosts. This is essentially a function of the number of bacteria released from one host and thus available to infect other individuals. ETEC bioserotypes seem to be well adapted to survival in the small intestine but not in the colon which hosts *E. coli* of the normal flora. In fact bacterial enteropathogens in general appear to be at a disadvantage in the ability to take up long-term residence as part of the normal colonic flora. Survival of ETEC in a given population of hosts is well served by the infection–transmission–infection cycle which is driven by the combination of diarrheagenic toxin(s) plus highly efficient colonizing capabilities afforded by the CFAs. If one accepts the premise that with ETEC survival of the "species" is dependent upon virulence then it is not difficult to appreciate the competitive advantage derived from packaging the genetic information for both diarrheagenic toxin(s) and CFA(s) in the same plasmids or in different but coexisting plasmids. Thus the fact that the genes encoding for the ETEC virulence factors are located on extrachromosomal elements, rather than on the chromosome, may be directly related to their combined survival value.

Whether the genetic elements for ST, for LT, and for the individual CFAs evolved separately and later combined into the same plasmids or whether these elements evolved together on a particular type of plasmid is unknown. It is also uncertain why ETEC which produce particular H antigens frequently produce a particular type of CFA. Presuming that all flagellar antigens are functionally equivalent (and observing that H-minus ETEC derivatives are in no way impaired in their expression of CFA fimbriae), one must conclude that the observed CFA–H pairings are based on either indirectly related or coincidental selective factors, akin to the "clonal" hypothesis. On the other hand, there may be direct factors involved in this phenomenon such as interactions between the genetic mechanisms which control expression of these surface structures or even shared accessory elements such as anchor proteins.

If ETEC plasmids evolved from elements such as phage-like particles then it may be that particular O groups, particular O:H combinations, or particular bioserotypes were the preferred hosts for these different ancestral phage-like particles. Even more speculative is the possibility that the different CFAs could have evolved from attachment factors employed by the phage-like elements to infect their bacterial hosts. It is a fact that some bacteriocins, such as pyocins, structurally resemble bacteriophage particles and are thought to have evolved from bacteriophages (HOLLOWAY and KRISHNAPILLAI 1975). Also, not all transmissible plasmids encode for sex pili which are as large as F-type pili. Many of these sex pili (fimbriae) are in the same size range as CFA fimbriae and many are expressed in large numbers per cell, as are CFA fimbriae. LT may have evolved on the bacterial chromosome and later become incorporated into the extra-chromosomal element(s) encoding for a CFA, with ST being the last to join the system since the identity of ST as a transposon is still evident today. Recently perfected techniques for application to "genetic archeology" (YAMAMOTO et al. 1984; VINAL and DALLAS 1987; TAMIYA and YAGI 1985) may someday provide answers to these questions.

From the above, it can be seen that CFAs, as a functional class of macro-molecules, may be an excellent example of convergent evolution and that different CFAs may have evolved via different pathways as different *E. coli* bioserotypes adapted to their particular species of host. This could well account for the antigenic and morphologic diversity seen among the CFAs. Both CFA/II and E8775 consist of at least three different surface-associated components and these components can be expressed in various combinations, or even alone, by different ETEC strains. This is very unlike CFA/I, which has a single and consistent fimbrial morphology, is composed of only one type of subunit, and behaves like a single antigen. Most CFA/II-positive ETEC produce a fimbrial structure, which has strong MRHA activity with bovine erythrocytes, plus a very narrow (2–3 nm diameter) fibrillar structure. CFA/II fimbriae may be of one or another antigenic type, termed CS1 and CS2. Strains produce either CS1 or CS2 but not both (LEVINE et al. 1984; SMYTH 1984). The CS3 fibrillar antigen is either very weakly positive or negative for MRHA. Evidently CS3 functions to strengthen the attachment of CFA/II-positive ETEC to target cells since CS3-only strains, lacking fimbriae, are also associated with cases of diarrhea.

The three-component CFA/II system is essentially duplicated by the CFA E8775, which also consists of three components, termed the CS4, CS5, and CS6 antigens (THOMAS et al. 1985). CS4 and CS5 have fimbrial morphology whereas CS6 is difficult to visualize by electron microscopy but may also be a 3-nm-diameter curly fibrillar structure (KNUTTON et al. 1987). While the distribution of CS4, CS5, and CS6 among ETEC needs further clarification, the isolation of virulent CS5-only ETEC of serotype O115:H40 has already been documented and the ETEC serogroup O148:H28 (MANNING et al. 1987) may include a CS6-only variety (KNUTTON et al. 1987).

The following may be a reasonable hypothesis about the evolution of the ETEC as they exist today. There is evidence that both ST and LT originated in an ancestral microbe, probably as chromosomal genes, which had as its host one particular animal species, either man or swine. Both ETEC which cause acute diarrhea in man and those which are animal pathogens produce ST and/or LT. With minor variations ST from either type of host and LT from either type of host are basically the same although there are molecular subspecies of each which seem to have been host influenced. This indicates that both ST and LT orginated in one source and divergence has since taken place as the result of host factors such as differences in intestinal receptors in the case of ST or immunologic pressure in the case of LT. This process may have occurred in parallel with evolution of the CFA adhesins. Different bioserotypes produce different CFAs and associate with different animal species; since host specificity is a function of the CFAs, it follows that each CFA most likely evolved from a different ancestral protein system. Antigenically and morphologically the CFAs are more heterogeneous than are the known varieties of ST and LT. The particular ETEC "clonal" types prevalent today may simply be those bioserotypes (ex: O6:H16:CFA/II; CS1 + CS3) which harbored those plasmids in which a particular physical event occurred, that event being emplacement of an enterotoxin gene in the same plasmid as a CFA operon, or a physical combination of an ST + LT plasmid with a CFA-encoding plasmid. This evolutionary event may still be evident today in that O78:H11 ETEC usually carry a plasmid encoding for both ST and CFA/I but not LT, which coexists in a separate plasmid in this "clone." Further, O6:H16 ETEC carry a plasmid encoding for ST, LT, and CFA/II. One might predict that similar "clonal" evidence will be found in the case of E8775, K88, K99, and other ETEC CFAs. One might argue that size differences between particular types of plasmid carried by the "clones" isolated today are incompatible with the above hypothetical evolutionary events. However, it is only important that virulence factor operons be conserved.

7 Colonization Factor Antigens as the Basis for Anti-ETEC Vaccine(s)

The multiplicity of ETEC, CFAs and the heterogeneity of antigens within particular CFAs such as CFA/II have important consequences. For maximum effectiveness, a CFA-based vaccine should be designed to include all of the major antigens. However, it should be mentioned that although CFAs are seemingly numerous, these antigens are not nearly as numerous as the O groups of ETEC. Also, *E. coli* O antigens would be a poor choice upon which to base an anti-ETEC vaccine since many closely related O groups frequently colonize man as normal flora of the large intestine. The other obvious possibility is an antienterotoxin vaccine and this possibility is under active investigation (KLIPSTEIN et al. 1985; SVENNERHOLM et al. 1984). However, because of the mode of pathogenesis of the ETEC it is possible that those strains capable of heavily colonizing the epithelial surface of the small intestine may overwhelm an antienterotoxin defense. We agree with the generally accepted idea that incorporation of LT (or cholera B subunit) into an anti-ETEC or anticholera CFA or whole-cell vaccine would enhance efficacy (CLEMENTS et al. 1986).

The anti-CFA approach to vaccine development has several theoretically favorable aspects and some of these have already been demonstrated experimentally. The primary consideration is that ETEC do not attain significant, i.e., diarrheagenic, numbers in the small intestine without attachment to the mucosal surface. This has been demonstrated in the laboratory rabbit (EVANS et al. 1981) and also with volunteers in which anti-CFA immunization has prevented illness in the face of challenge with virulent ETEC (EVANS et al. 1984a). Studies with both animals and man support the premise that intestinal IgA is the immunoglobulin protective against ETEC colonization.

Development of an effective anti-ETEC vaccine is one of our major goals and although it seems clear that an oral vaccine based on CFAs will be the key to long-lasting protection, there is the task of maximizing delivery, via the oral route, of such antigens to the gut immune system. Although purified CFAs were effective in stimulating both intestinal anti-CFA IgA and protection in rabbits it has been difficult to reproduce this result with volunteers (EVANS et al. 1984b). We have concluded that CFAs administered in the form of CFA-positive bacterial cells are for some reason more effective in stimulating a protective antibody response. Whole killed ETEC as a vaccine has another attractive aspect, namely the easy adaptability of this approach for including cells containing all of the important CFAs in combination with other ETEC surface antigens such as plasmid-mediated outer membrane proteins which might contribute to a protective immune response.

We recently reported the results of a double-blind vaccination/challenge study in which two doses, $1 \times 10 \{10\}$ each, of colicin E2-killed ETEC strain H-10407, given orally 1 month apart, protected against challenge 8 weeks later with $3 \times 10 \{9\}$ living H-10407 (EVANS et al. 1988a). Eight of nine volunteers in the placebo group experienced diarrhea upon challenge, whereas only two of ten vaccinated individuals became ill. One of the two vaccinees who were not protected had failed to develop an intestinal IgA anti-CFA/I response and the other showed a weak antibody response and experienced only mild diarrhea. Thus, eight of ten vaccinees had a significant intestinal anti-CFA response and were protected.

The above results do not prove that anti-CFA/I IgA was the protective factor. In fact, more recent volunteer studies (unpublished data) showed that the colicin E2-killed H-10407 vaccine not only protected six of eight vaccinees against challenge with an O63:H-:CFA/I (ST + LT) strain but also protected six of eight vaccinees challenged with an O6:H16:CFA/II (ST + LT) ETEC strain. It is of interest that in recent studies we demonstrated that 19 of 22 vaccinees who had been administered two doses of the colicin E2-killed H-10407 vaccine did exhibit an intestinal anti-LT response. Previously, we had confirmed that colicin E2 (an endonuclease which destroys both the chromosomal and plasmid DNA) kills *E. coli* without damage to the integrity of the cells (KONISKY 1982; EVANS et al. 1988a). For example, lethal treatment with this colicin does not cause the release of LT from the bacterial cells; colicin E2-killed cells and control, untreated, cells released the same amount of LT upon treatment with polymyxin (unpublished results).

The data cited above show that the gut immune system is very effectively stimulated by both CFA/I and LT administered in the form of whole killed ETEC cells. Although the molecular events responsible for this result remain unknown, we believe that our original contention that colicin E2-killed cells would more naturally mimic actual infection than chemically or heat-killed cells has been justified, and that the prospects for development of an effective anti-ETEC vaccine appear to be good although the majority of this work lies in the future.

8 Evidence for Specific Mechanisms Controlling Expression of Colonization Factor Antigens by ETEC

Enterotoxigenic *E. coli* virulence plasmids encoding for a CFA represent an ideal experimental model for elucidating the molecular mechanisms which control expression of plasmid-encoded products in *E. coli*. It is clear that expression of a CFA is in large part dependent on molecular signals mediated either directly or indirectly by products of the bacterial chromosome. For example, expression of both CFA/I and CFA/II in vitro is regulated such that under certain growth conditions the cells express CFA fimbriae but not common fimbriae, and vice versa. Thus expression of CFA and common fimbriae are coordinated in a negative fashion. More direct evidence for such control was found when we isolated a mutant of strain H-10407 which is hyperproductive for CFA/I (unpublished results). This mutant, termed strain SB-3456, produces large amounts of CFA/I even under conditions which favor production of common fimbriae. Common fimbriae were not detectable on this mutant under any growth conditions although revertants of this mutant did show normal control over CFA/I production and did produce common fimbriae under appropriate growth conditions.

It is well known that broth culture favors common fimbriae production whereas agar culture favors CFA production. However, the effect of broth versus agar substrate can be overcome by manipulation of nutrients; for example growth on CFA agar results in CFA production whereas growth on peptone agar results in the production of common fimbriae (EVANS et al. 1977a, b). We have found that expression of CFA/I is regulated in part by the availability of iron in that CFA/I production in a defined medium is maximal when iron is decreased to a concentration which limits growth. Conversely, addition of iron in excess of that required for maximum cell yield, using the same basal medium, significantly decreased CFA/I production. These results seem reasonable since the human small intestine, where CFA/I production is relevant to virulence, is normally a low-iron environment. Individual carbon substrates, particularly acetate, were also found to affect expression of CFA/I. Replacement of glucose by acetate as the substrate in an excess-iron defined medium was found to abolish CFA/I production completely. Interestingly, the above-mentioned mutant (SB-3456) of strain H-10407 is unaffected by acetate. On an agar medium containing excess iron and glucose plus acetate as the major carbon sources strain H-10407 is negative for CFA/I but positive for common fimbriae; on this same medium the derivative strain SB-3456 is negative for common fimbriae and positive for CFA/I.

The above-cited results show that the control mechanism(s) which regulates the expression of common fimbriae, which is a chromosomal function, and that which regulates expression of CFA fimbriae, which is a plasmid-encoded function, are not independent. The location of the mutation responsible for the properties exhibited by the derivative strain SB-3456 (i.e., plasmid versus chromosome) and the exact nature of the gene product(s) or control gene functions involved are being investigated.

9 Summary and a View Toward the Future

In this chapter we have cited only a few of the many researchers who have contributed to the current state of knowledge about the ETEC and their significance as a major human health problem. Furthermore, there are volumes of data which could be cited concerning the economic cost of ETEC infection of domestic animals, CFAs of these animal-associated ETEC other than K88 and K99, and the important strides of progress which have been made in developing an effective vaccine approach to deal with these ETEC.

Our basic message is that the CFAs are the key to survival of the ETEC in any given population, be it man or animal, and we postulate that an anti-ETEC vaccine aimed at the CFAs, especially if combined with an antienterotoxin stimulus, will prove eventually to be very successful. In terms of economic feasibility, one must consider the current cost of ETEC diarrhea in morbidity and mortality in countries with a high endemicity of ETEC diarrhea, not to mention treatment costs, costs to travelers, and the effects of ETEC diarrhea on child development and increased susceptibility to the devastating effects of other pathogens. We also postulate that one major benefit which will be derived from studies on the ETEC CFAs will be elucidation of the CFA(s) of *Vibrio cholerae* and that this achievement will provide the final step in development of an oral vaccine remarkably effective against cholera.

Discovery of the new ETEC CFAs, or putative CFAs, cited here makes it imperative that epidemiologic studies on susceptible populations continue, preferably based on an organised surveillance approach rather than on short-term or retrospective studies. Certainly it would be timely, and hopefully economical, to institute newer, rapid identification techniques such as a battery of gene probes which could account for the known ETEC CFAs as well as the adhesive factors of the non-ETEC enteropathogenic *E. coli*.

Finally, it will be intersting to see the evolutionary history of the ETEC CFAs unfold as newer techniques such as computer-assisted restriction endonuclease analysis and protein/antigen analysis are put to the task. Here, we suggest that the range of *E. coli* fimbriae selected for examination be expanded to include all of the known types of sex pili, i.e., those fimbriae involved in DNA transfer between *E. coli* cells. These seemingly irrelevant pili may prove to be related, in an ancestral fashion, to the fimbrial/fibrillar CFAs. Also, elucidation of the molecular mechanics by which chromosomal and plasmid control mechanisms interact may lead to practical applications, even to completely new approaches to prophylaxis and treatment of diseases caused by pathogenic bacteria which are dependent on plasmid-encoded virulence factors.

References

Back E, Mollby R, Kaijser B, Stintzing G, Wadstrom T, Habte D (1980) Enterotoxigenic *Escherichia coli* and other gram-negative bacteria of infantile diarrhea: surface antigens, hemagglutinins, colonization factor antigens, and loss of enterotoxigenicity. J Infect Dis 142: 318–327

Beachey EH (1981) Bacterial adherence: adhesin-receptor interactions mediating the attachment of bacteria to mucosal surfaces. J Infact Dis 143: 325–345

Bock K, Breimer ME, Brignole A, Hansson GC, Karlsson K-A, Larson G, Leffler H, Samuelsson BE, Stromberg N, Eden CS, Thurin J (1985) Specificity of binding of a strain of uropathogenic *Escherichia coli* to Gal alpha 1-4-Gal-containing glycosphingolipids. J Biol Chem 260: 8545–8551

Burrows MR, Sellwood R, Gibbons RA (1976) Haemagglutinating and adhesive properties associated with the K99 antigen of bovine strains of *Escherichia coli*. J Gen Microbiol 96: 269–275

Cheney CP, Boedeker EC (1983) Adherence of an enterotoxigenic *Escherichia coli* strain, serotype O78:H11, to purified human intestinal brush borders. Infect Immun 39: 1280–1284

Clements JD, Sack DA, Harris JR, Chakraborty J, Khan MR, Stanton BF, Kay BA, Khan MU, Yunus M, Atkinson WA, Svennerholm A-M, Holmgren J (1986) Field trial of oral cholera vaccines in Bangladesh. Lancet 2: 124–129

Danbara H, Komase K, Kirii Y, Shinohara M, Arita H, Makino S, Yoshikawa M (1987) Analysis of the plasmids of *Escherichia coli* O148:H28 from travelers with diarrhea. Microb Pathogen 3: 269–278

Darfeuille A, Lafeuille B, Joly B, Cluzel R (1983) A new colonization factor antigen (CFA/III) produced by enteropathogenic *Escherichia coli* O128:B12. Ann Inst Pasteur Microbiol 134A: 53–64

Deetz TR, Evans DJ, Evans DG, DuPont HL (1979) Serologic responses to somatic O and colonization-factor antigens of enterotoxigenic *Escherichia coli* in travelers. J Infect Dis 140: 114–118

Evans DG, Evans DJ (1978) New surface-associated heat-labile colonization factor antigen (CFA/II) produced by enterotoxigenic *Escherichia coli* of serogroups O6 and O8. Infect Immun 21: 638–647

Evans DJ, Evans DG (1983) Classification of pathogenic *Escherichia coli* according to serotype and the production of antigens. Rev Infect Dis 5 (Suppl): 692–701

Evans DG, Silver RP, Evans DJ, Chase DG, Gorbach SL (1975) Plasmid-controlled colonization factor associated with virulence in *Escherichia coli* enterotoxigenic for humans. Infect Immun 12: 656–667

Evans DG, Evans DJ, DuPont HL (1977a) Virulence factors of enterotoxigenic *Escherichia coli*. J Infect Dis 136 (Suppl): 118–123

Evans DG, Evans DJ, Tjoa W (1977b) Hemagglutination of human group A erythrocytes by enterotoxigenic *Escherichia coli* isolated from adults with diarrhea: correlation with colonization factor. Infect Immun 18: 330–337

Evans DG, Satterwhite TK, Evans DJ, DuPont HL (1978) Differences in serological responses and excretion patterns of volunteers challenged with enterotoxigenic *Escherichia coli* with and without colonization factor antigen. Infect Immun 19: 883–888

Evans DJ, Evans DG, Hohne C, Noble MA, Haldane EV, Lior H, Young LS (1981) Hemolysin and K antigens in relation to serotype and hemagglutination type of *Escherichia coli* isolated from extraintestinal infections. J Clin Microbiol 13: 171–178

Evans DG, Cabada FJ, Evans DJ (1982) Correlation between intestinal immune response to colonization factor antigen/I (CFA/I) and acquired resistance to enterotoxigenic *Escherichia coli* diarrhea in an adult rabbit model. Eur J Clin Microbiol 1: 178–185

Evans DG, Graham DY, Evans DJ (1984a) Administration of purified colonization factor antigens (CFA/I, CFA/II) of enterotoxigenic *Escherichia coli* to volunteers. Response to challenge with virulent enterotoxigenic *Escherichia coli*. Gastroenterology 87: 934–940

Evans DG, Evans DJ, Sack DA, Clegg S (1984b) Entertoxigenic *Escherichia coli* pathogenic for man: biological and immunological aspects of fimbrial colonization factor antigens. In: Boedeker EC (ed) Attachment of organisms to the gut mucosa, vol I. CRC, Boca Raton, chap 8

Evans DJ, Evans DG, Opekun AR, Graham DY (1988a) Immunoprotective oral whole cell vaccine for enterotoxigenic *Escherichia coli* prepared by in situ destruction of chromosomal and plasmid DNA with colicin E2. FEMS Microbiol Immunol 47: 9–18

Evans DJ, Evans DJ, Moulds JJ, Graham DY (1988b) *N*-acetylneuraminyllactose-binding fibrillar hemagglutinin of *Campylobacter pylori*: a putative colonization factor antigen. Infect Immun 56: 2896–2906

Evans DJ, Evans DG, Diaz SR, Graham DY (1988c) Mannose-resistant hemagglutination of human erythrocytes by entertoxigenic *Escherichia coli* with colonization factor antigen II. J Clin Microbiol 26: 1626–1629

Goldhar J, Perry R, Golecki JR, Hochutzky H, Jann B, Jann K (1987) Non-fimbrial, mannose-resistant adhesins from uropathogenic *Escherichia coli* O83:K1:H4 and O14:K?:H11. Infect Immun 55: 1837–1842

Gothefors L, Ahren C, Stoll B, Barua DK, Orskov F, Salek MA, Svennerholm A-M (1985) Presence of colonization factor antigens on fresh isolates fecal *Escherichia coli*: a prospective study. J Infect Dis 152: 1128–1133

Guth BEC, Silva MLM, Scaletsky ICA, Toledo MRF, Trabulsi LR (1985) Enterotoxin production, presence of colonization factor antigen I, and adherence to HeLa cells by *Escherichia coli* O128 strains belonging to different O subgroups. Infect Immun 47: 338–340

Holloway BW, Krishnapillai V (1975) Bacteriophages and bacteriocins. In: Clarke PH, Richmond MH (eds) Genetics and biochemistry of *Pseudomonas* Willey, London, pp 99–132

Honda T, Arita M, Miwatani T (1984) Characterization of new hydrophobic pili of human enterotoxigenic *Escherichia coli*: a possible new colonization factor. Infect Immun 43: 959–965

Jones GW, Rutter JM (1972) Role of the K88 antigen in the pathogenesis of neonatal diarrhea caused by *Escherichia coli* in piglets. Infect Immun 6: 918–927

Klipstein FA, Engert RF, Houghten RA (1985) Mucosal antitoxin response in volunteers to immunization with a synthetic peptide of *Escherichia coli* heat-stable enterotoxin. Infect Immun 50: 328–332

Knutton S, Lloyd DR, Candy DCA, McNeish AS (1984) In vitro adhesion of enterotoxigenic *Escherichia coli* to human intestinal epithelial cells from mucosal biopsies. Infect Immun 44: 514–518

Knutton S, Lloyd DR, McNeish AS (1987) Identification of a new fimbrial structure in enterotoxigenic *Escherichia coli* (ETEC) serotype O148:H28 which adheres to human intestinal mucosa: a potentially new human ETEC colonization factor. Infect Immun 55: 86–92

Konisky J (1982) Colicins and other bacteriocins with established mode of action. Annu Rev Microbiol 36: 125–144

Leffler H, Svanborg-Eden C (1980) Chemical identification of a glycosphingolipid receptor for *Escherichia coli* attaching to human urinary tract epithelial cells and agglutinating human erythrocytes. FEMS Microbiol Lett 8: 127–134

Levine MM, Ristaino P, Marley G, Smythe C, Knutton S, Boedeker E, Black R, Young C, Clements ML, Cheney C, Patnaik R (1984) Coli surface antigens 1 and 3 of colonization factor antigen II-positive enterotoxigenic *Escherichia coli*: morphology, purification, and immune responses in humans. Infect Immun 44: 409–420

Manning PA, Higgins GD, Lumb R, Lanser JA (1987) Colonization factor antigens and a new fimbrial type, CFA/V, on O115:H40 and H-strains of enterotoxigenic *Escherichia coli* in Central Australia. J Infect Dis 156: 841–844

Murray BE, Evans DJ, Penaranda ME, Evans DG (1983) CFA/I-ST plasmids: comparison of enterotoxigenic *Escherichia coli* (ETEC) of serogroups O25, O63, O78, and O128 and mobilization from an R factor-containing epidemic ETEC isolate. J Bacteriol 153: 566–570

Orndorff PE (1987) Genetic study of piliation in *Escherichia coli*: implications for understanding microbe–host interactions at the molecular level. Pathol Immunopathol Res 6: 82–92

Orskov I, Orskov F (1966). Episome-carried surface antigen K88 of *Escherichia coli*. I. Transmission of the determinant of the K88 antigen and influence on the transfer of chromosomal markers. J Bacteriol 91: 69–75

Orskov I, Orskov F, Smith HW, Sojka WJ (1975) The establishment of K99, a thermolabile, transmissible, *Escherichia coli* K antigen, previously called "Kco", possessed by calf and lamb enterotoxigenic strains. Acta Pathol Microbiol Scand [B] 83: 31–36

Orskov F, Orskov I, Evans DJ, Sack RB, Sack FA, Wadstron T (1976) Special *Escherichia coli* serotypes among enterotoxigenic strains from diarrhoea in adults and children. Med Microiol Immunol (berl) 162: 73–80

Orskov I, Orskov F, Jan B, Jann K (1977) Serology, chemistry, and genetics of O and K antigen of *Escherichia coli*. Bacteriol Rev 41: 667–710

Orskov I, Birch-Anderson A, Duguid JP, Stenderup J, Orskov F (1985) An adhesive protein capsule of *Escherichia coli*. Infect Immun 47: 191–290

Penaranda ME, Evans DG, Murray BE, Evans DJ (1983) ST:LT:CFA/II plasmids in enterotoxigenic *Escherichia coli* belonging to serogroups O6, O8, O80, O85, and O139. J Bacteriol 154: 980–983

Pere A, Vaisanen-Rhen V, Rhen M, Tenhunen J, Korhonen TK (1986) Analysis of P fimbriae on *Escherichia coli* O2, O4, and O6 strains by immunoprecipitation. Infect Immun 51: 618–625

Rhen M, Klemm P, Korhonen TK (1986) Identification of two new hemagglutinins of *Escherichia coli*, N-acetyl-D-glucosamine-specific fimbriae and a blood group M-specific aglutinin, by cloning the corresponding genes in *Escherichia coli* K-12. J Bacteriol 168: 1234–1242

Robins-Browne RM (1987) Traditional enteropathogenic *Escherichia coli* of infantile diarrhea. Rev Infect Dis 9: 28–53

Rowe B, Gross R, Takeda Y (1983) Serotyping of enterotoxigenic *Escherichia coli* isolated from diarrhoeal travelers from various Asian countries. FEMS Microbiol Lett 20: 187–190

Sen D, Ganguly U, Saha MR, Bhattacharya SK, Datta P, Datta D, Mukherjee AK, Chakravarty R, Pal SC (1984) Studies on *Escherichia coli* as a cause of acute diarrhea in Calcutta. J Med Microbiol 17: 53–58

Smith HR, Cravioto A, Willshaw GA, McConnell MM, Scotland SM, Gross RJ, Rowe B (1979) A plasmid coding for the production of colonisation factor antigen I and heat-stable enterotoxin in strains of *Escherichia coli* of serogroup O78. FEMS Microbiol Lett 6: 255–260

Smith HW, Linggood MA (1971) Observations of the pathogenic properties of the K88, Hly and Ent plasmids of *Escherichia coli* with particular reference to porcine diarrhoea. J Med Microbiol 4: 467–485

Smythe CJ (1984) Serologically distinct fimbriae on enterotoxigenic *Escherichia coli* of serotype O6:K15:H16 or H-. FEMS Microbiol Lett 21: 51–57

Stoll BJ, Svennerholm A-M, Gothefors L, Baura D, Huda S, Holmgren J (1986) Local and systemic antibody responses to naturally acquired enterotoxigenic *Escherichia coli* diarrhea in an endemic area. J Infect Dis 153: 527–534

Svennerholm A-M, Gotherfors L, Sack DA, Bardhan PK, Holmgren J (1984) Local and systemic antibody responses and immunological memory in humans after immunization with cholera B subunit by different routes. Bull WHO 62: 909–918

Tacket CO, Maneval DR, Levine MM (1987) Purification, morphology, and genetics of a new fimbrial putative colonization factor of enterotoxigenic *Escherichia coli* O159:H4. Infect Immun 55: 1063–1069

Tamiya N, Yagi T (1985) Non-divergence theory of evolution: sequence comparison of some proteins from snakes and bacteria. J Biochem 98: 289–303

Thomas LV, Rowe B (1982) The occurrence of colonisation factors (CFA/I, CFA/II, and E8775) in enterotoxigenic *Escherichia coli* from various countries in South East Asia. Med Mcrobiol Immunol 171: 85–90

Thomas LV, Cravioto A, Scotland SM, Rowe B (1982) New fimbrial antigenic type (E8775) that may represent a colonization factor in enterotoxigenic *Escherichia coli* in humans. Infect Immun 35: 1119–1124

Thomas LV, McConnell MM, Rowe B, Field AM (1985) The possession of three novel coli surface antigens by enterotoxigenic *Escherichia coli* strains positive for the putative colonization factor PCF8775. J Gen Microbiol 131: 2319–2326

Vinal AC, Dallas WS (1987) Partition of heat-labile-enterotoxin genes between human and animal *Escherichia coli* isolates. Infect Immun 55: 1329–1331

Wadstrom T, Faris A, Freer J, Habte D, Hallberg D, Ljungh A (1980) Hydrophobic surface properties of enterotoxigenic *E. coli* (ETEC) with different colonization factors (CFA/I, CFA/II, K88 and K99) and attachment to intestinal epithelial cells. Scand J Infect Dis 24 (Suppl): 148–153

Yamamoto T, Nakazawa T, Miyata T, Kaji A, Yokota T (1984) Evolution and structure of two ADP-ribosylation enerotoxins, *Escherichia coli* heat-labile toxin and cholera toxin. FEBS Lett 169: 241–246

Colonization Factor Antigens of Enterotoxigenic *Escherichia coli* in Animals

H. W. MOON

1 Introduction

Colonization factor antigens probably play a critical role in the pathogenesis of most bacterial diseases. Such speculation is warranted because most bacterial infections arise from mucosal or cutaneous surfaces, and when such processes have been investigated the pathogens involved have been found to produce antigens that enable them to survive and multiply in the special environments at these surfaces. This implies that there are myriad colonization factor antigens that act at skin and mucosal surfaces and that have not yet been discovered. In some diseases, the bacterial pathogens (*Mycoplasma, Bordetella, Bacteroides, Campylobacter, Moraxella*, and *Clostridium*, as well as some strains of *Escherichia coli, Salmonella, Streptococcus*, and *Staphylococcus*) are characteristically confined to mucosae or skin and cause localized disease. In other diseases, the pathogens (*Pasteurella, Hemophilus*, and *Yersinia*, as well as some strains of *E. coli. Salmonella, Streptococcus*, and *Staphylococcus*) characteristically invade from mucosae to cause systemic disease. In both groups, subclinical carrier states, with bacteria in or on mucosae, maintain and disseminate the agent in the population. Clinical disease occurs less frequently than the carrier state, and results from opportunistic expansions in these localized bacterial populations (increased bacterial numbers leading to local disease or translocation and systemic infection).

U.S. Department of Agriculture, Agricultural Research Service, National Animal Disease Center, P.O. Box 70, Ames, Iowa, USA

Colonization depends on multiple host–bacterial interactions, and the pathogens have a constellation of attributes which act by different mechanisms to facilitate colonization. Adhesion to epithelial surfaces is one such attribute. Because adhesion is well recognized as critical to colonization by some pathogens, the antigens mediating adhesion are sometimes referred to as colonization factor antigens. This nomenclature is useful to the extent that it emphasizes the role of adhesion. It is misleading to the extent that it oversimplifies the colonization process by ignoring the probability that bacterial antigens that are not involved in adhesion are also necessary for colonization. The colonization factors of enterotoxigenic *E. coli* (ETEC) have been the focus of extensive multidisciplinary research during the last 20 years. As a result, they are comparatively well understood. This chapter will emphasize the adhesive interactions between the colonization factors of ETEC and the intestinal epithelium of their animal hosts. This will illustrate some of the principles involved in the colonization of mucosal surfaces in animals by bacterial pathogens.

2 Enterotoxigenic *Escherichia coli* of Calves and Pigs

Colonization of the small intestine by *E. coli* was recognized as a characteristic of a commonly occurring diarrheal disease in neonatal calves and pigs well before the discovery of *E. coli* enterotoxins or the fulfillment of Koch's postulates with ETEC. This abnormal colonization pattern was apparently the evidence that originally led to the hypothesis that the disease was caused by *E. coli*. The term colonization factor antigen (CFA) was originally applied to fimbriae (pili) of ETEC pathogenic for humans. The role of adhesive CFAs in ETEC infections of animals, and the molecular biology of the fimbrial CFAs, has been comprehensively reviewed (BRINTON et al. 1977; JONES and ISAACSON 1982; KORHONEN et al. 1985; GAASTRA and DE GRAAF 1982; ISAACSON 1985; KLEMM 1985; LEVINE et al. 1983; MOOI and DE GRAAF 1985; MOON et al. 1979b). This chapter presents the author's perspectives on the role of fimbriae and capsules in colonization of pig and calf small intestine by ETEC.

2.1 Fimbriae

Fimbrial CFA types known or suggested to facilitate adhesion and colonization by ETEC in the small intestine of pigs and calves are F1 (type 1), F4 (K88), F5 (K99), F6 (987P), F41, F42, and FY (Table 1).

The evidence that K88, K99, and 987P mediate adhesion in, and are required for colonization of, the small intestine by ETEC is convincing (reviewed, MOON et al. 1979b). For example, wild-type strains which bear one of these antigens adhere and colonize in the small intestine of the appropriate host, and also adhere to intestinal epithelial cells from such hosts in vitro. Isogenic derivatives of the wild-type strains

Table 1. Characteristics of known or putative fimbrial CFAs produced by ETEC isolated from animals

Antigen type	Colonization factor status	Morphology[a]	Hemag-glutinin	Location of genes	Host range of ETEC
F4 or K88	Known	Fine[b] Irregular[c]	Mannose-resistant	Plasmid	Pigs
F5 or K99	Known	Fine Irregular	Mannose-resistant	Plasmid	Calves, lambs, Pigs, mice
F6 or 987P	Known	Thick Straight	None	Chromosome	Pigs
F1 or type 1	Putative	Thick Straight	Mannose-sensitive	Chromosome	Animals, humans
F41	Putative	Fine Irregular	Mannose-resistant	Chromosome	Calves, lambs, Pigs, mice
F42	Putative	Thick Straight	Mannose-resistant	Unknown	Pigs
FY	Putative	Fine	Unknown	Unknown	Calves

[a] See Fig. 1;
[b] Diameter < 7 nm (Fig. 1A);
[c] Frequently kinked or curled (Fig. 1A).

which no longer bear these antigens (via mutation, loss of plasmid, or phase variation) lack the adhesive and colonizing attributes of the wild-type strains. The adhesive and colonizing attributes of K88 and K99 ETEC are transferred to suitable *E. coli* hosts along with K88 or K99 plamids. K88, K99, or 987P antibody inhibits adhesion and colonization in an antigen-specific fashion. The evidence that the other (putative) CFAs listed in Table 1 play a critical role in colonization of the small intestine by ETEC is less convincing than that for K88, K99, and 987P, and will be discussed in more detail.

2.1.1 F41 Fimbriae

F41 (Fig. 1A) was recognized more recently and has been less fully characterized than K88, K99, and 987P; in addition its role in colonization has been less well defined. Although F41 was originally referred to as anionic K99 (MORRIS et al. 1978, 1980), F41 is now recognized to be physically, antigenically, and genetically distinct from K99 (CHANTER 1982, 1983; ISAACSON 1985; MORRIS et al. 1982, 1983; MOSELEY et al. 1986; To 1984). It is also distinct from K88, although the chromosomal genetic determinant for production of F41 has some sequence homology with the plasmid-born determinant for K88 (MOSELEY et al. 1986). Most F41[+] ETEC belong to serogroups O9 and O101, and produce K99 as well as F41. Several of the strains originally designated as K99[+] and used in experiments to define the role of K99 in colonization (MOON et al. 1977; SMITH and LINGGOOD 1972) were subsequently shown to produce F41 in addition to K99 (MORRIS et al. 1980). Nevertheless, the conclusions about the role of K99 drawn from those experiments are valid because they are based, in part, on K99[+] strains which do not produce F41 and on the effects of curing or transferring the K99 plasmid (SMITH 1976). The fact that K99-facilitated colonization can occur independent of F41 makes it difficult to assess the role of F41 in colonization by K99[+] F41[+] ETEC. Most evidence in this regard is based on

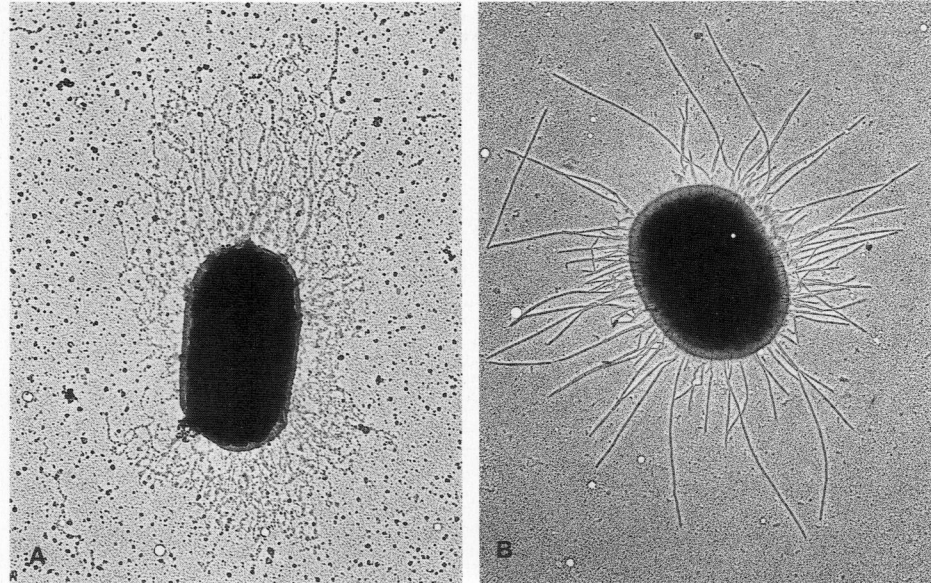

Fig. 1A, B. Electron micrographs of fimbriated *E. coli* (courtesy of Dr. S.M. To). The specimens were prepared for examination by rotary shadowing with platinum (magnification approximately × 28 000). **A** F41 fimbriae, to illustrate the fine irregular morphology characteristic of some fimbriae. **B** Type 1 fimbriae, to illustrate the thick straight morphology characteristic of some fimbriae

the use of phenotypic variant K99⁻ F41⁺ ETEC strains (MORRIS et al. 1983) which carry the gene for K99 (MOSELEY et al. 1986) but do not express the antigen. However, the possibility that these K99⁻ F41⁺ variants produce still another unrecognized fimbrial colonization factor cannot be excluded.

On the one hand, there is strong suggestive evidence that F41 can mediate colonization by adhesion. A K99⁻ F41⁺ ETEC strain has been shown to adhere to calf intestinal epithelial cells in vitro (MORRIS et al. 1982). In vitro adhesion is competitively inhibited by isolated F41 antigen and prevented by growing the bacteria at 18°C to inhibit expression of F41 (and other fimbriae). Both K99⁺ F41⁺ and K99⁻ F41⁺ ETEC adhere, colonize, and produce F41 antigen in the small intestine of pigs and calves during disease (MORRIS et al. 1982; RUNNELS et al. 1987). Vaccination with F41 antigen protects pigs against colonization with K99⁻ F41⁺ ETEC (RUNNELS et al. 1987). Although infant mice are not known to be natural hosts for ETEC infections, they are susceptible to colonization by adhesion and the resultant fatal diarrhea following oral inoculation with K99⁺ F41⁺ ETEC (DUCHET-SUCHAUX 1983; NAGY and USHE 1985). A K99⁻ F41⁺ variant ETEC strain was shown to be virulent in infant mice, while a K99⁻ F41⁻ derivative from it was not (BERTIN 1985).

On the other hand, the preponderance of evidence suggests that F41 is currently of minor importance or practical significance in naturally occurring disease. F41 is less prevalent than K88, K99, or 987P and is usually accompanied by K99 (MORRIS et al. 1980; MOSELEY et al. 1986). The few strains originally characterized as naturally occurring, virulent 3P⁻ or F41 only ETEC (AWAD-MASALMEH et al. 1982; MOON et al. 1980), were subsequently shown to carry the K99 gene, even though they do not express it in vitro or in vivo (MOSELEY et al. 1986). It seems likely that these strains may have expressed K99 at the time of their original isolation and lost that ability during growth in the laboratory. Thus, in naturally occurring disease, most ETEC that produce F41 apparently have redundant fimbrial adhesins (both F41 and K99) to facilitate colonization of the small intestine. In theory, if one of these redundant adhesins is blocked by antibody, then colonization and disease would still result based on adhesion mediated by the other. However, pregnant swine and cattle given purified fimbrial vaccines containing K99 but not F41, protected their suckling newborn pigs (MORGAN et al. 1978) and calves (ACRES et al. 1979) against challenge with ETEC strains 431 and B44, respectively. Both these strains are now known to produce F41 in addition to K99 (MORRIS et al. 1980; RUNNELS et al. 1987). Furthermore, monoclonal antibody against K99 can protect calves and pigs against challenge with the F41⁺ K99⁺ ETEC strains B44 and B41 (SADOWSKI et al. 1983; SHERMAN et al. 1983).

Thus, as a practical matter, K99 vaccines can be used to potect against strains that produce both adhesins. There are no known antigenic cross-reactions between F41 and K99. This suggests that vaccine-induced protection acts by mechanisms in addition to or other than, combining directly with the sites on fimbriae that mediate adhesion. Alternatively, there may be some unrecognized antigenic relatedness between F41 and K99, or K99 may be more important or critical to colonization than is F41.

The observation that vaccination with F41 protected pigs against challenge with a K99⁻ F41⁺ ETEC strain, but not against challenge with the K99⁺ F41⁺ ETEC strain 431 (RUNNELS et al. 1987), is consistent with the notion that K99 is more important than F41 for colonization by strain 431. However, we have recently found that variants of strains 431 that have lost the gene for K99 (MAINIL et al. 1987) still carry the gene for and produce F41 and are still virulent for newborn pigs (T.A. CASEY and H.W. MOON, unpublished material). Furthermore, F41 is apparently at least as important as K99 in the virulence of K99⁺ F41⁺ ETEC for infant mice and, in contrast to our experience with pigs, vaccination with F41 was more effective than vaccination with K99 in protecting infant mice against K99⁺ F41⁺ ETEC (DUCHET-SUCHAUX 1988). Obviously, more definitive data are needed on the role of F41 in colonization and on the relative importance of K99 and F41 in disease caused by ETEC that produce both antigens. Although F41 may currently be of minor importance, as suggested above, theoretically F41 has the potential to become highly important under the selection pressure of vaccines or other factors directed against K99.

2.1.2 Type 1 Fimbriae

The comparatively thick and straight (Fig. 1B), chromosomally encoded fimbriae which cause mannose-sensitive hemagglutination (MSHA) and are subject to phase variation are designated as type 1 (F1 or common pili). The question as to whether or not type 1 fimbriae are involved in the pathogenesis of ETEC infection is controversial (BRINTON et al. 1977, 1983; KARCH et al. 1987). The predominance of the evidence suggests that they are not. Most *E. coli* (pathogens and nonpathogens) produce type 1 fimbriae. In contrast to the other fimbriae listed in Table 1, type 1 is at least as prevalent among non-ETEC as it is among ETEC. Most animal ETEC probably have the capacity to produce type 1 in addition to one or more of the other fimbriae listed in Table 1. However, our results indicate that ETEC which have the capacity to produce type 1 fimbriae in vitro may not produce them in the small intestine during disease (To et al. 1984). The relatively low prevalence of type 1 fimbriae found in surveys of porcine ETEC also suggests that most isolates are in the type 1 negative phase when isolated from the intestine of diseased pigs (TRUSZCZYNSKI and OSEK 1987; WILSON and FRANCIS 1986).

However, type 1 fimbriae can mediate adhesion of type 1 positive phase *E. coli* to isolated porcine intestinal epithelial cells in vitro (ISAACSON et al. 1978). Furthermore, pigs inoculated orally with ETEC, in the type 1 positive phase, had bacteria bearing type 1 antigen adherent to their intestinal villi 9 h later (JAYAPPA et al. 1985). One possible explanation for the conflicting reports might be variation among type 1 fimbriae of ETEC (ADEGBOLA and OLD 1987; DE GRAAF and KLASSEN 1987; DUGUID 1985). It has been concluded (based substantially on unpublished data) that there are at least 12 different serotypes in the *E. coli* "type 1 pilus family" (JAYAPPA et al. 1983). The type 1$^+$ (MSHA) ETEC strain M8 colonizes pig small intestine, while a type 1$^-$ variant derived from it does not (NAKAZAWA et al. 1986). Strain M8 was subsequently found (E.A. DEAN and H.W. MOON, unpublished material) to produce 987P antigen in pig small intestine in vivo. Workers in another laboratory (BROES et al. 1988) found several porcine ETEC strains that caused MSHA and reacted with antisera against both type 1 and 987P fimbriae. It is not known if such strains produce two fimbrial antigens (type 1 and 987P) or variants of type 1 which react with polyclonal 987P antiserum. Nor is it known if impaired colonization by the variant of M8 is due to impaired expression of the type 1 antigen, the 987P antigen, or some other defect. Type 1 and 987P fimbrial antigens share some epitopes (ABRAHAM et al. 1986). However, cross-reactions are not detected in routine serologic tests using polyclonal antisera.

Type 1 and 987P fimbriae appear to be more closely related to each other than to the other fimbriae listed in Table 1. Both are chromosomally encoded and subject to phase variation. They are morphologically identical and have similar physical–chemical properties. They differ in amino acid composition, peptide subunit size, hexosamine content, receptor specificity, and hemagglutinating capability, as well as in antigenicity (BRINTON et al. 1977; DEAN and ISAACSON 1985; ISAACSON 1985; ISAACSON and RICHTER 1981). Epidemiologically, type 1 fimbriae are ubiquitous, while 987P is confined to porcine ETEC. 987P fimbriae apparently also differ from the type 1 fimbriae on most *E. coli* in the tendency of 987P$^+$ strains to enter, or be selected for, the fimbriate phase during growth in porcine small intestine. Perhaps 987P and type 1 fimbriae are closely related phylogenetically and intermediates exist, but have not been recognized to be intermediates.

2.1.3 *FY, F42, and Others*

Some *E. coli* from diarrheal calves in Europe (CONTREPOIS et al. 1982) and the United Kingdon (MORRIS et al. 1985) produce the fimbrial adhesin provisionally designated as FY or Att 25. FY commonly occurs along with K99 on calf ETEC in this region. The prevalence of FY in other regions of the world or among *E. coli* from other hosts is not known. In addition to ETEC, the enteropathogenic (nonentero-toxigenic, attaching and effacing) calf *E. coli* strain X114/83 has been shown to produce FY (HALL et al. 1988). Although the role (if any) of FY in colonization has not been defined, FY is suspected to facilitate colonization. There is evidence that when combined with K99 and F41 antibodies, FY antibody adds to the protection of calves against challenge with a $K99^+$ $F41^+$ FY^+ ETEC strain (CONTREPOIS and GIRARDEAU 1985).

Some porcine ETEC from pigs in Brazil produce fimbriae that are distinct from the others in Table 1 and are provisionally designated as F42. YANO et al (1986) demonstrated that F42 mediates mannose-resistant hemagglutination and adhesion to brush borders from porcine intestinal epithelial cells in vitro. They hypothesized that F42 may function as an adhesive colonization factor. However, data on the colonizing abilities of $F42^+$ ETEC and the role of F42 in the small intestine during ETEC infection are not yet available.

ETEC from calves and pigs produce fimbriae in addition to those listed in Table 1 (ANING and THOMLINSON 1983; CONTREPOIS et al. 1982; FAIRBROTHER et al. 1988; GIRARDEAU et al. 1988). One of them, designated CS31A, appears to be associated with bacteremia rather than intestinal colonization (GIRARDEAU et al. 1988). Furthermore, it seems likely that many such "new" or currently unrecognized fimbriae exist. Most of them can probably mediate adhesion in vitro, because adhesion occurs via both hydrophobic and receptor-mediated interactions. Some of the receptors or their analogues [i.e., mannose for type 1 and sialic acid for K99 (LINDAHL et al. 1987; MOURICOUT and JULIEN 1987; SMIT et al. 1984)] are widely distributed in animal cells. However, experience with K88 (below) demonstrates that, at least for some fimbriae, hydrophobic interactions or interactions with receptor analogues is not sufficient, and that colonization requires highly specific receptors. The experience with type 1 and 987P fimbriae (above) illustrates that, in addition to adhesive capability, in vivo expression of the fimbriae (or selection for the bacterial cells which bear them) is also a critical determinant as to which types of fimbriae facilitate colonization. Economic importance of a specific CFA is also dependent on its prevalence, and probably on whether or not it is redundant (usually accompanied by another CFA). Thus, as a practical matter, at present the diagnostic tests and vaccines based on K88, K99, and 987P are globally adequate for ETEC infections in calves and neonatal pigs. It seems reasonable to expect that other fimbriae will become practically important in time because of immunity, genetic change in livestock populations, or management practices which tend to select for alternative fimbriae or against those that predominate at present.

2.1.4 Host Specificity

The host specificity of ETEC is largely determined by fimbrial CFAs. Naturally occurring disease caused by $K88^+$ and 987^+ ETC is essentially confined to pigs. Humans are not susceptible to colonization by $K88^+$ ETEC, even following experimental challenge (DUPONT et al. 1971). Some pigs are genetically susceptible to colonization by $K88^+$ ETEC, and others are genetically resistant (GIBBONS et al.

1977). Genetic susceptibility depends on receptors for K88 on the brush borders of epithelial cells in the small intestine (SELLWOOD 1980). The occurrence of the receptors is controlled by an autosomal dominant gene(s). K88 antigens consist of a constant determinant designated as a, and variable determinants designated as b, c, and d (GUINÉE and JANSEN 1979). Adhesion of K88$^+$ ETEC to the K88-specific receptors on pig intestinal epithelial cell brush borders depends on the variable determinants. Some pigs are susceptible to adhesion by all three subtypes of K88 fimbriae (K88ab, K88ac, and K88ad), some are only susceptible to one or two subtypes, and some are resistant to all three subtypes of K88 (BIJLSMA et al. 1982, 1984; RAPACZ and HASLER-RAPACZ 1986). Thus, the pathogenicity of K88$^+$ ETEC in swine requires highly specific brush border receptors for K88. Furthermore, in vitro tests for these receptors correlate with the susceptibility of the pigs to disease, provided that the tests are conducted under the appropriate conditions (SELLWOOD 1980; SARMIENTO et al. 1988). Thus, as a practical matter, disease caused by K88$^+$ ETEC is narrowly restricted to those genotypes of swine that produce the appropriate subtypes of brush border receptors. The receptors are probably galactosyl-containing residues of glycoproteins (ANDERSON et al. 1980; SELLWOOD 1980).

There is reason to think that genetic intraspecies regulation of susceptibility to ETEC, based on the availability of receptors for fimbriae, extends beyond K88. SMITH and HALLS (1968) found that pigs from an inbred herd were susceptible to colonization by an O141:K85ac strain of ETEC, but that pigs from other herds were not. They suggested that the intestinal epithelium of the pigs in the inbred herd was genetically predisposed to adhesion by ETEC. The O141:K85ac ETEC strain used did not produce K88. However, the nature of the fimbriae that were produced by the strain (if any) was not determined.

In contrast to the specificity required for pathogenicity in swine, under some in vitro conditions K88 mediates adhesion to a variety of cells and cell products. These include erythrocytes from several species, buccal epithelial cells, intestinal epithelial cells from several species, intestinal mucus from mice, brush borders from mouse intestinal epithelial cells, and even from K88-resistant pigs (COX and HOUVENAGHEL 1987; JONES and RUTTER 1974; LAUX et al. 1986; PARRY and PORTER 1978; RUNNELS et al. 1980; SELLWOOD 1980; TZIPORI et al. (1984)). It is not known whether such interactions are mediated by the hydrophobicity of K88, by interactions with analogues to the specific receptor (which exists in the brush border of susceptible pigs), or by other mechanisms.

K88 mediated hemagglutination of porcine erythrocytes does not depend on the receptors that are required for adhesion to intestinal brush borders (COX and HOUVENAGHEL 1987). Thus, hemagglutination tests do not detect the genetically susceptible and resistant pigs. There is suggestive evidence that K88 can facilitate colonization in neonatal mice (DAVIDSON and HIRSH 1975; KÉTYI et al. 1978). The studies cited are based on clearance of K88$^+$ ETEC from the intestine but do not report directly on either adhesion or number of ETEC in the small intestine. Intestinal transit is comparatively sluggish in neonatal mice, and clearance of bacteria from their small intestine is markedly influenced by the ambient temperature (MOON et al. 1979a). These variables may have influenced the results in the reports cited. Thus, the relevance of the mouse model for K88 mediated adhesion and colonization is unknown. Apparently, horses are resistant to colonization by $K88^+$ ETEC, even though K88 does mediate adhesion to their small intestinal brush borders in vitro (TZIPORI et al. 1984). Perhaps this is because the brush borders do not contain sufficient (qualitatively or quantitatively) receptors to facilitate colonization in vivo.

In contrast to the narrow host species ranges characteristic of $K88^+$ and $987P^+$ ETEC, $K99^+$ and $K99^+$ $F41^+$ ETEC intensively colonize the small intestines of pigs, calves, lambs, and mice (BERTIN 1985; DUCHET-SUCHAUX 1983; MOON et al. 1977, 1979a; MORRIS et al. 1983; NAGY and USHE 1985; SMITH and HUGGINS 1978). However, the host specificity of $K99^+$ and $K99^+$ $F41^+$ ETEC is sharply age restricted (SMITH and HALLS 1967; reviewed, MOON and RUNNELS 1984; WILSON and FRANCIS 1986). Pigs commonly develop ETEC-induced diarrhea during the immediate neonatal period (< 1 week old) and after weaning ($\geqslant 3$ weeks old). $K88^+$ ETEC cause disease in both age groups, but $K99^+$ ETEC are confined to the immediate neonatal period. Calves and lambs infrequently develop ETEC-induced diarrhea after the immediate neonatal period. $K99^+$ ETEC, which rapidly colonize and cause fatal diarrhea following inoculation into neonatal calves and pigs, do not colonize or cause disease when given to 1-week-old calves or pigs. There are probably several factors that contribute to this innate age-dependent resistance of calves and pigs to $K99^+$ ETEC (reviewed by MOON and RUNNELS 1984). Age-dependent resistance is apparently due in part to decreased availability of epithelial cell receptors for K99 with age, in that $K99^+$ ETEC (now known to be $K99^+$ $F41^+$) adhere to epithelial cells from 1-day-old pigs, calves, and mice in greater numbers than to epithelial cells from older animals of the same species (RUNNELS et al. 1980).

Age restriction is also characteristic of $987P^+$ ETEC, in that they are also commonly associated with diarrhea in neonatal pigs, but infrequently associated with diarrhea in pigs after weaning (WILSON and FRANCIS 1986). Experimentally, a strain of $987P^+$ ETEC that adheres and colonizes intensively and causes fatal disease in neonatal pigs does not colonize or cause diarrhea in 3-week-old pigs (DEAN et al. 1987). In this case, development of resistance with age to $987P^+$ ETEC does not correlate with decreased availability of brush border receptors [probably galactose or fucose-containing residues of glycoproteins (DEAN and ISAACSON 1985)], in that $987P^+$ ETEC adhere equally well to brush borders isolated from intestinal epithelial cells from neonatal or 3-week-old pigs. In neonates, $987P^+$ bacteria appear to adhere directly to epithelium, leaving comparatively few bacteria free in the intestinal lumen (NAGY et al. 1977). In contrast, in 3-week-old pigs, $987P^+$ bacteria tend to be associated with the mucus and cellular debris in the intestinal lumen (DEAN et al. 1987). One of the hypotheses proposed in explanation was that 987P receptors remain bound to epithelial cells in the neonate, but are released from the epithelium and trapped by mucus in older pigs. Receptors in mucus would then hypothetically compete with those bound to the epithelial cells and protect the older pigs from colonization by $987P^+$ ETEC.

In addition to the host species specificity, specificity for some genotypes within a host species, and host age specificity, fimbriae also contribute to the site specificity of ETEC. For example, $K88^+$ ETEC colonize throughout the small intestine in neonatal pigs, while colonization by strains now recognized to be K99$^+$ or 987P$^+$ is restricted to the ileum (NAGY et al. 1976).

2.1.5 Vaccination

Fimbrial CFAs are protective antigens in vaccines against ETEC infections (RUTTER and JONES 1973; reviewed, BRINTON et al. 1977; DOUGAN and MORRISSEY 1984/85; LEVINE et al. 1983; MOON 1978; MOON and RUNNELS 1981). Vaccines based on fimbrial CFAs have come into routine commercial use to control the disease in neonatal calves and pigs (ACRES 1985; JAYAPPA et al. 1983; MOON et al. 1988; SIMONSON et al. 1983; SODERLIND et al. 1982). Although broad-based, well-controlled data on the efficacy of most of these commercial vaccines are not available, the available data as well as the anecdotal experience and prevailing opinion of veterinarians and producers indicate that they are effective. These vaccines depend on passive protection of suckling neonates via colostral antibody stimulated by vaccination of the dam during pregnancy. Colostral antibody in the intestinal lumen of the suckling neonate protects against colonization by ETEC bearing the fimbrial antigens represented in the vaccine. Intestinal antibody levels are replenished at each suckling during the colostral period (the first several days after birth). This short-lived protection is practically useful because of the age specificity of ETEC infection in these species. The disease is essentially confined to the colostral period in calves. Thus, even a single feeding of monoclonal K99 antibody of mouse origin, which is presumably only available (not absorbed, not digested) in the intestine for a few hours during the most susceptible period, can provide significant protection against the disease (SADOWSKI et al. 1983; SHERMAN et al. 1983). In swine, the disease also occurs during the later suckling and immediate postweaning periods. However, the greatest mortality due to ETEC infection in swine occurs during the colostral period, and can be controlled by the vaccines. The practicality of the existing vaccination programs is also enhanced by the fact that one antigen (K99) or three antigens (K88, K99, and 987P) account for more than 90% of the ETEC currently afflicting calves and neonatal pigs, respectively. Thus, vaccines of reasonably restricted valence provide broad protection.

The use of K88 vaccines was suggested to cause the emergence of K99$^+$ ETEC and the ad variant of K88 in European swine (GUINÉE and JANSEN 1979; SODERLIND et al. 1982). The efficacy of fimbrial CFA-based vaccines could conceivably be short-lived, because they may facilitate the emergence of ETEC bearing novel fimbrial CFAs or ETEC that do not depend on fimbrial CFAs for colonization. This does not seem likely to be a major practical limitation to the fimbrial vaccine approach for several reasons:

1. Most swine apparently acquire some fimbrial CFA antibody naturally, without developing clinical disease and without vaccination (ISAACSON et al. 1980; MOON et al. 1988; SIMONSEN et al. 1983). Presumably, CFAs and ETEC evolved under the constant selection pressure of such naturally acquired antibody. In spite of such general, long, and continued selection, a comparatively few CFA antigen types predominate globally at present. Furthermore, all pathogenic ETEC investigated to date require fimbrial CFA for colonization. Thus, the rate of antibody-induced

change must be quite slow, or the options to maintain pathogenicity as the result of such change must be quite limited. In swine, the vaccines tend to raise the level of antibody in the population, rather than to induce CFA antibody which would not otherwise have occurred. This "booster" effect of the vaccines will probably have comparatively little added selective or evolutionary effect on CFAs.

2. Even if vaccines do accelerate the emergence of new or alternative CFAs, this need not be a major limitation to the efficacy of fimbrial vaccines. Now that the role of fimbriae in colonization is recognized, and the technology to identify the relevant fimbrial CFAs and use them in vaccines has been developed, it should be rather straightforward to adjust the valence of the vaccines to fit the prevailing CFA types every few years.

3. The evidence that the ab, ac, and ad variants of K88 emerged as the result of selection pressure from K88 antibody (GUINÉE and JANSEN 1979) is not convincing If one assumes that ability to colonize small intestine is the major selective advantage influencing variation in CFAs, then variations in host receptors for K88ab, ac, and ad seem to be more powerful selective factors than antibody. Genetic resistance to K88 based on lack of receptors confers essentially complete protection from colonization and disease (GIBBONS et al. 1977; SARMIENTO et al. 1988; SMITH and HUGGINS 1978), while the protection conferred by K88 antibody (or antibody to other fimbrial CFAs) is partial. Receptor-mediated adhesion to swine brush borders has been shown to be specific for the b, c, and d variants of K88 (BILJSMA et al. 1982, 1984; RAPACZ and HASLER-RAPACZ 1986). On the other hand, the degree of variant specificity required for K88 antibody-mediated protection is not clear. Experiments on the inhibition of adhesion of K88+ ETEC to porcine epithelial cells by antibody in vitro have produced conflicting results. Experiments conducted by PARRY and PORTER (1978) indicated that antibody against the constant K88a determinant would inhibit adhesion of *E. coli* bearing either the ab or ac variants of K88. Those conducted by WILSON and HOHMANN (1974) indicated that inhibition depended on antibody specific for the b or c determinant of the K88+ test strain. Some commercial vaccines contain all three subtypes of K88 (ab, ac, and ad). It is not known if such trivalent K88 vaccines are more effective than those containing only one subtype of K88. Data on the effect of vaccinating dams with one subtype (such as K88ab) on the subsequent susceptibility of their suckling newborn pigs to challenge with another subtype (such as K88ac or ad) are not available. Monoclonal K88 antibody (raised in mice) protected pigs against challenge with K88+ ETEC, and antibodies directed against epitopes in either the constant K88a or the variable K88c determinants provided comparable levels of protection (SADOWSKI et al. 1984). These latter authors suggested that K88a antibody might protect by steric interference with the K88b, K88c, or K88d adhesion sites on the fimbriae.

4. The observation that the percentage of K99+ ETEC from diarrheal pigs in K88 vaccinated herds was higher than that in nonvaccinated herds (SODERLIND et al. 1982) does not demonstrate that K88 vaccination increased the prevalence of K99+ ETEC in the vaccinated population. The observation more likely reflects a reduction in the incidence of diarrhea in the total population, due to reduction in the incidence of K88+ ETEC infection. The idea that protection against ETEC of one

CFA type might somehow predispose to or facilitate colonization with ETEC of another CFA type is intuitively appealing, but is not substantiated by the evidence. It assumes competition between strains for receptors or nutrients, or antimicrobial effects of one strain on another, sufficient to impede colonization. However, each CFA type reacts with a different receptor, and simultaneous intensive colonization of the small intensive with two strains of ETEC belonging to two different CFA types occurs naturally (FRANCIS and WILSON 1985) and is readily reproduced experimentally (SMITH and HUGGINS 1978).

K88$^-$ and K99$^-$ variants of K88$^+$ and K99$^+$ ETEC do emerge during the course of ETEC infections (KÉTYI et al. 1978; LINGGOOD and PORTER 1981; MAINIL et al. 1987). The emergence of these variants is thought to be due to the selection pressure of antibody against the fimbriae. We found that such K99$^-$ variants from a K99$^+$ F41$^+$ parental strain still produced F41, intensively colonized pig small intestine, and caused disease (T.A. CASEY and H.W. MOON, unpublished data). Presumably, such K99$^-$ F41$^+$ variants are commonly emerging from K99$^+$ F41$^+$ ETEC strains in nature, and have been for some time. In spite of that, the prevalence of K99$^-$ F41$^+$ ETEC associated with naturally occurring disease remains low. Apparently, there is some as yet unrecognized selective advantage for the K99$^+$ F41$^+$ wild type. It would be interesting to determine whether the prevalence of K99$^-$ F41$^+$ ETEC would increase in the face of widespread use of K99 vaccines. However, that "experiment" is not likely to be done now, because the manufacturers are adding F41 to their vaccines. The inclusion of F41 in commercial vaccines appears to be based more on competitive marketing strategy than on the prevalence of K99$^-$ F41$^+$ ETEC or the need for F41 antigen to protect against K99$^+$ F41$^+$ ETEC. Vaccination with purified K99 protects against K99$^+$ F41$^+$ETEC in calves and pigs (ACRES et al. 1979; MORGAN et al. 1978). Conversely, vaccination with F41 protects against K99$^+$ F41$^+$ ETEC in mice (M. DUCHET-SUCHAUX, personal communication). In contrast, we have been unable to protect pigs against K99$^+$ F41$^+$ ETEC by vaccination with F41 (RUNNELS et al. 1987). Perhaps our experiments did not attain sufficiently high titers of F41 antibody to protect or, alternatively, there may be hierarchic relationships such that K99 antibody is somehow more effective than F41 antibody in protecting pigs against K99$^+$ F41$^+$ ETEC. There is increasing evidence that many (perhaps even most) ETEC will eventually be shown to produce more than a single fimbrial CFA (BRINTON et al. 1983; CONTREPOIS and GIRARDEAU 1985; LEVINE et al. 1984; MORRIS et al. 1980; SCHNEIDER and TO 1982; SVENNERHOLM et al. 1988). Thus, the question arises as to whether protection can be generally conferred by antibody against one of the several CFAs carried by a strain (as with K99 antibody for K99$^+$ F41$^+$ ETEC), or whether several or all of the CFAs carried must be targeted in order to protect. If the latter holds generally, and if myriad fimbrial CFAs do emerge, then the practical use of fimbrial vaccines could be short-lived. However, to date the breadth of protection conferred by K99 and the limited number of CFAs required to account for the prevailing ETEC suggest that this will not be a major limitation.

It is assumed that CFA antibody protects by reacting with the receptor combining sites on the fimbriae and, thus, blocking adhesion by the ETEC. However, additional and less highly specific mechanisms are apparently also involved. For example, the evidence for protection by K88a antibody cited above suggests that CFA antibody can protect even if the antibody is not directed against the receptor combining sites. Furthermore, the fact that K99 antibody protects against ETEC which produce both K99 and F41 suggests that CFA antibody is protective even if many fimbriae remain completely free of antibody. Thus, mechanisms such as steric hindrance, changes in surface charge, agglutination, binding of bacteria to CFA antibody in mucus rather than to receptors on the brush border, and opsonization may be involved in protection mediated by CFA antibody. Perhaps the efficacy of CFAs as protective antigens derives mostly from the facts that they are good antigens which are expressed on the surface of ETEC in the intestine during disease, and that a limited number of antigens represent the prevailing ETEC. That is, protection may not depend on directly blocking the receptor-combining sites of the fimbriae. If this is true, then any good antigen available on the surface of the bacteria in the intestine may protect as well as CFAs.

Although the existing vaccines can be used to protect neonates via passive lacteal immunity, they are not suitable for protection against postweaning ETEC infections in swine. This disease occurs after the pigs have been deprived of access to lacteal antibody from their dams. The existing vaccines are not suitable for active immunization of pigs before weaning, because they are injected parenterally and stimulate the systemic, rather than the mucosal immune system. The resulting IgG antibody in blood is unlikely to reach the intestinal lumen in quantities sufficient to protect. Furthermore, the fimbrial CFAs (other than K88) associated with the postweaning disease are not known (ANING and THOMLINSON 1983; BROES et al. 1988; LARSEN 1976; RIISING et al. 1975; SMITH and HALLS 1968; TRUSZCZYŃSKI and OSEK 1987). Oral vaccines stimulate the mucosal immune system, resulting in local production of secretory antibodies which are secreted into the intestinal lumen and provide active immunity. Live ETEC are effective oral fimbrial vaccines, but killed ETEC or isolated fimbriae are much less effective (DOUGAN and MORRISSEY 1984/85; EVANS et al. 1984; LEVINE et al. 1983; MOON 1981; SVENNERHOLM et al. 1988). Thus, once the fimbrial CFAs associated with postweaning diarrhea in swine are more completely defined, it should be possible to construct or isolate strains of *E. coli* that produce the fimbriae but not enterotoxin, and to use them in live oral vaccines to prevent postweaning ETEC infections in swine.

2.2 Capsules

Many of the ETEC which cause disease in calves, lambs, and pigs produce discrete capsular polysaccharide K antigens of the A (mucoid, heat-stable) variety. These capsules enhance the ability of ETEC to colonize the small intestine; thus, they are also CFAs. Acapsular mutants of some such strains can still colonize intensively enough to cause diarrhea, but some strains can do so only in the encapsulated form (HADAD and GYLES 1982; NAGY et al. 1976; SMITH and HUGGINS 1978; SOJKA et al. 1977). The mechanisms by which capsules enhance the virulence of ETEC are not known. They probably do not function directly as adhesins, because capsules actually impede the adhesion of fimbriated ETEC in vitro (RUNNELS and MOON

1984). It has been suggested that capsules act by stabilizing or maintaining the in
vivo adhesion initiated by fimbriae, or protect the bacteria from antibody, serum
factors, or phagocytes in the intestinal lumen (CHAN et al. 1982). Capsular antigens
can be protective in vaccines to provide passive immunity of sucklings against
ETEC infection (MOON and RUNNELS 1983). Some commercial vaccines include
both capsular and fimbrial CFAs. The potential value of capsular antigens in
vaccines is limited because many ETEC do not produce capsules and because there
are many different capsular antigen types among encapsulated ETEC (ØRSKOV and
ØRSKOV 1972, 1978).

3 Concluding Comments

Adhesion mediated by fimbriae is thought to be an essential step in most bacterial
infections that arise from mucosal surfaces in animals. This is best understood in
ETEC infections. A single CFA (K99) predominates globally among ETEC afflicting
calves, while three CFAs (K88, K99, and 987P) predominate globally among those
afflicting neonatal pigs. These predominant CFAs are routinely and effectively used
as the basis for diagnostic tests and vaccines (to stimulate passive lacteal immunity)
to control ETEC infections in suckling neonatal calves and pigs. In addition to these
currently predominant CFA types, most ETEC probably produce other fimbriae,
and there may be myriad fimbrial antigens among ETEC. There is good evidence
that some of these additional or currently less recognized or less prevalent fimbriae
(F41, F42, FY, type 1 variants, and others) can also mediate adhesion and facilitate
colonization in the intestine of calves and pigs. Thus, current control measures based
on the predominant CFA types could conceivably be short-lived, due to vaccine-
induced selection pressure. It is argued that this is unlikely to be a major limitation,
because such selection pressure has existed in nature for a long time, and yet only a
few CFA types predominate. There appear to be unrecognized factors that tend to
favor or select for K88, K99, and 987P. Intraspecies genetic regulation of K88
receptor occurrence indicates that K88 infections could be prevented by selective
breeding in swine. However, this has not come into practical use because of the
technical problems in identification of resistant breeding stock and because K88-
resistant pigs still have receptors for K99 and 987P.

    In contrast to the situation with the neonatal disease, the current understanding
of CFAs has not yet contributed significantly to the control of the ETEC infections
that occur in swine after weaning. Existing knowledge of CFAs and related
technology can probably be extended to allow control of the disease in these older
animals as well. Inhibition of CFA-facilitated colonization via active local
immunity, or by feeding analogues of fimbrial CFAs or analogues of their receptors,
and genetic selection for swine lacking receptors for CFAs are attractive approaches
for research toward that goal.

Acknowledgment. The author thanks Dr S.M. To for the electron micrographs
used in Fig. 1.

References

Abraham SN, Schifferli DM, Banks D, Beachey EH (1986) Immunological cross-reactivity between type 1 and 987P fimbriae of *Escherichia coli*. Abstracts, 86th annual meeting of the American Society of Microbiologists, p 84

Acres SD (1985) Enterotoxigenic *Escherichia coli* infections in newborn calves: a review. J Dairy Sci 68: 229–256

Acres SD, Isaacson RA, Babiuk LA, Kapitany RA (1979) Immunization of calves against enterotoxigenic colibacillosis by vaccinating dams with purified K99 antigen and whole cell bacterins. Infect Immun 25: 121–126

Adegbola RA, Old DC (1987) Antigenic relationships among type-1 fimbriae of enterobacteriaceae revealed by immunoelectronmicroscopy. J Med Microbiol 24: 21–28

Anderson MJ, Whitehead JS, Kim YS (1980) Interaction of *Escherichia coli* K88 antigen with porcine intestinal brush border membranes. Infect Immun 29: 897–901

Aning KG, Thomlinson JR (1983) Adhesion factor distinct from K88, K99, F41, 987P, CFAI and CFAII in porcine *Escherichia coli*. Vet Rec 112: 251

Awad-Masalmeh M, Moon HW, Runnels PL, Schneider RA (1982) Pilus production, hemagglutination, and adhesion by procine strains of enterotoxigenic *Escherichia coli* lacking K88, K99, and 987P antigens. Infect Immun 35: 305–313

Bertin A (1985) F41 antigen as a virulence factor in the infant mouse model of *Escherichia coli* diarrhoea. J Gen Microbiol 131: 3037–3045

Bijlsma IGW, De Nijs A, Van Der Meer C, Frik JF (1982) Different pig phenotypes affect adherence of *Escherichia coli* to jejunal brush borders by K88ab, K88ac, or K88ad antigen. Infect Immun 37: 891–894

Bijlsma IGW, De Nijs A, Frik JF (1984) Serological variants of *Escherichia coli* K88 antigen: differences in adhesion and haemagglutination. Rev Sci Tech Off Int Epiz 3: 871–879

Brinton CC, Fusco P, To AC-C, To SC-M (1977) The piliation phase syndrome and the uses of purified pili in disease control. Proc XIIIth US-Japan conference on cholera, Atlanta, Georgia, National Institute of Health, Sept 19–21, pp 33–70

Brinton CC, Fusco P, Wood S, Jayappa HG, Goodnow RA, Strayer RG (1983) A complete vaccine for neonatal swine colibacillosis and the prevalence of *Escherichia coli* pilli on swine isolates. Vet Med Small Anim Clin 78: 962–966

Broes A, Fairbrother JM, Lariviére S, Jacques M, Johnson WM (1988) Virulence properties of enterotoxigenic *Escherichia coli* O8:KX105 strains isolated from diarrheic piglets. Infect Immun 56: 241–246

Chan R, Acres SD, Costerton JW (1982) Use of specific antibody to demonstrate glycocalyx, K99 pili, and the spatial relationships of K99 $^+$ enterotoxigenic *Escherichia coli* in the ileum of colostrum-fed calves. Infect Immun 37: 1170–1180

Chanter N (1982) Structural and functional differences of the anionic and cationic antigens in K99 extracts of *Escherichia coli* B41. J Gen Microbiol 128: 1585–1589

Chanter N (1983) Partial purification and characterization of two non K99 mannose-resistant haemagglutinins of *Escherichia coli* B41. J Gen Microbiol 129: 235–243

Contrepois MG, Girardeau J-P (1985) Additive protective effects of colostral antipili antibodies in calves experimentally infected with enterotoxigenic *Escherichia coli*. Infect Immun 50: 947–949

Contrepois M, Girardeau JP, Gouet P, Desmattre P (1982) Vaccination contre L'antigene K99 et D'autres "adhesines" des *Escherichia coli* bovins. Proc XIIth World congress on diseases of cattle, Sept 7–10, Amsterdam, NL, pp 332–338

Cox E, Houvenaghel A (1987) In vitro adhesion of K88ab-, K88ac-, and K88ad-positive *Escherichia coli* to intestinal villi, to buccal cells and to erythrocytes of weaned piglets. Vet Microbiol 15: 201–207

Davidson JN, Hirsh DC (1975) Use of the K88 antigen for in vivo bacterial competition with porcine strains of enteropathogenic *Escherichia coli*. Infect Immun 12: 134–136

Dean EA, Isaacson RE (1985) Purification and characterization of a receptor for the 987P pilus of *Escherichia coli*. Infect Immun 47: 98–105

Dean EA, Whipp SC, Moon HW (1987) The role of epithelial receptors for 987P pili in age-specific intestinal colonization by enterotoxigenic *Escherichia coli*. In: Sack RB, Zinnaka Y (eds) Advances in research on cholera and related diarrheas. Proc US-Japan conference on cholera, Williamsburg, VA, Nov 9–12, 1987. KTK Scientific, Tokyo

De Graaf FK, Klaasen P (1987) Nucleotide sequence of the gene encoding the 987P fimbrial subunit of *Escherichia coli*. FEMS Microbiol Lett 42: 253–258

Dougan G, Morrissey P (1984/85) Molecular analysis of the virulence determinants of enterotoxigenic

Escherichia coli isolated from domestic animals: applications for vaccine development. Vet Microbiol 10: 241–257

Duchet-Suchaux M (1983) Infant mouse model of *E. coli* diarrhoea: clinical protection induced by vaccination of the mothers. Ann Rech Vet 14: 319–331

Duchet-Suchaux M (1988) Protective antigens against enterotoxigenic *Escherichia coli* O101:K99, F41 in the infant mouse diarrhea model. Infect Immun 56: 1364–1370

Duguid JP (1985) Antigens of type-1 fimbriae. In: Stewart-Tull DES, Davis M (eds) Immunology of the bacterial cell envelope. Wiley, New York, chap 11, pp 301–318

DuPont HL, Formal SB, Hornick RB, Snyder MJ, Libonati JP, Sheahan DG, LaBrec EH, Kalas JP (1971) Pathogenesis of *Eschericha coli* diarrhea. New Engl J Med 285: 3–11

Evans DG, Graham DY, Evans DJ, Opekun A (1984) Administration of purified colonization factor antigens (CFA/I, CFA/II) of enterotoxigenic *Escherichia coli* to volunteers. Gastroenterology 87: 934–940

Fairbrother JM, Lariviere S, Johnson WM (1988) Prevalence of fimbrial antigens and enterotoxins in nonclassical serogroups of *Escherichia coli* isolated from newborn pigs with diarrhea. Am J Vet Res 49: 1325–1328

Francis DH, Wilson RA (1985) Concurrent infection of pigs with enterotoxigenic *Escherichia coli* of different serogroups. J Clin Microbiol 22: 457–458

Gaastra W, De Graaf FK (1982) Host-specific fimbrial adhesins of noninvasive enterotoxigenic *Escherichia coli* strains. Microbiol Rev 46: 129–161

Gibbons RA, Sellwood R, Burrows M, Hunter PA (1977) Inheritance of resistance to neonatal *E. coli* diarrhoea in the pig: examination of the genetic system. Theor Appl Genet 51: 65–70

Girardeau JP, der Vertanian M, Ollier JL, Contrepois M (1988) CS31A, a new K88-related fimbrial antigen on bovine enterotoxigenic and septicemic *Escherichia coli* strains. Infect Immun 56: 2180–2188

Guinée PAM, Jansen WH (1979) Behavior of *Escherichia coli* K antigens K88ab, K88ac, and K88ad in immunoelectrophoresis, double diffusion, and hemagglutination. Infect Immun 23: 700–705

Hadad JJ, Gyles CL (1982) The role of K antigens of enteropathogenic *Escherichia coli* in colonization of the small intestine of calves. Can J Comp Med 46: 21–26

Hall GA, Chanter N, Bland AP (1988) Comparison in gnotobiotic pigs of lesions caused by verotoxigenic and non-verotoxigenic *Escherichia coli*. Vet Pathol 25: 205–210

Isaacson RE (1985) Pili of enterotoxigenic *Escherichia coli* from pigs and calves. Adv Exp Med Biol 185: 83–99

Isaacson RE, Richter P (1981) *Escherichia coli* 987P pilus: purification and partial characterization. J Bacteriol 146: 784–789

Isaacson RE, Fusco PC, Brinton CC, Moon HW (1978) In vitro adhesion of *Escherichia coli* to porcine small intestinal epithelian cells: pili as adhesive factors. Infect Immun 21: 392–397

Isaacson RE, Dean EA, Morgan RL, Moon HW (1980) Immunization of suckling pigs against enterotoxigenic *Escherichia coli*-induced diarrheal disease by vacinating dams with purified K99 or 987P pili: antibody production in response to vaccination. Infect Immun 29: 824–826

Jayappa HG, Strayer JG, Goodnow RA, Fusco PC, Wood SW, Haneline P, Cho H-J, Brinton CC (1983) Experimental infection and field trial evaluations of a multiple-pilus phase-cloned bacterin for the simultaneous control of neonatal colibacillosis and mastitis in swine. Proceedings 4th Int symp on neonatal diarrhea, Vet Infect Disease Organization, University of Saskatchewan, Canada, Oct 3–5, pp 518–535

Jayappa HG, Goodnow RA, Geary SJ (1985) Role of *Escherichia coli* type 1 pilus in colonization of porcine ileum and its protective nature as a vaccine antigen in controlling colibacillosis. Infect Immun 48: 350–354

Jones GW, Isaacson RE (1982) Proteinaceous bacterial adhesins and their receptors. CRC Crit Rev Microbiol 10(3): 229–260

Jones GW, Rutter JM (1974) The association of K88 antigen with haemagglutinating activity in porcine strains of *Escherichia coli*. J Gen Microbiol 84: 135–144

Karch H, Heesemann J, Laufs R, Kroll H-P, Kaper JB, Levine MM (1987) Serological response to type 1-like somatic fimbriae in diarrheal infection due to classical enteropathogenic *Escherichia coli*. Microb Pathogen 2: 425–434

Kétyi I, Kuch B, Vertényi A (1978) Stability of *Escherichia coli* adhesive factors K88 and K99 in mice. Acta Microbiol Acad Sci Hung 25: 77–86

Klemm P (1985) Fimbrial adhesins of *Escherichia coli*. Rev Infect Dis 7: 321–340

Korhonen TK, Rhen M, Vaisanen-Rhen V, Pere A (1985) Antigenic and functional properties of enterobacterial fimbriae. In: Stewart-Tull DES, Davies M (eds) Immunology of the bacterial cell envelope. Wiley, New York, pp 319–354

Larsen JL (1976) Differences between enteropathogenic *Escherichia coli* strains isolated from neo-natal *E. coli* diarrhoea (N.C.D.) and post weaning diarrhoea (P.W.D) in pigs. Nord Vet Med 28: 417–429

Laux DC, McSweegan EF, Williams TJ, Wadolkowski EA, Cohen PS (1986) Identification and characterization of mouse small intestine mucosal receptors for *Escherichia coli* K-12(K88ab). Infect Immun 52: 18–25

Levine MM, Kaper JB, Black RE, Clements ML (1983) New knowledge on pathogenesis of bacterial enteric infections as applied to vaccine development. Microbiol Rev 47: 510–550

Levine MM, Ristaino P, Marley G, Smyth C, Knutton S, Boedeker E, Black R, Young C, Clements ML, Cheney C, Patnaik R (1984) Coli surface antigens 1 and 3 of colonization factor antigen II-positive enterotoxigenic *Escherichia coli*: morphology, purification, and immune responses in humans. Infect Immun 44: 409–420

Lindahl M, Brossmer R, Wadstrom T (1987) Carbohydrate receptor specificity of K99 fimbriae of enterotoxigenic *Escherichia coli*. Glycoconjugate 4: 51–58

Linggood MA, Porter P (1981) The antibody-mediated elimination of adhesion determinant from enteropathogenic strains of *Escherichia coli*. In: Berkeley RCW, Lynch JM, Melling J, Rutter PR, Vincent B (eds) Microbial adhesion to surfaces. Horwood, Chichester, chap 24, pp 441–453

Mainil JG, Sadowski PL, Tarsio M, Moon HW (1987) In vivo emergence on enterotoxigenic *Escherichia coli* variants lacking genes for K99 fimbriae and heat-stable enterotoxin. Infect Immun 55: 3111–3116

Mooi FR, De Graaf FK (1985) Molecular biology of fimbriae of enterotoxigenic *Escherichia coli*. In: Goebel W (ed) Genetic approaches to microbiol pathogenicity. Springer, Berlin Heidelberg New York, pp 119–138 (Current topics in microbiology and immunology, vol 118)

Moon HW (1978) Pili as protective antigens in vaccines for the control of enterotoxigenic *Escherichia coli* infections. Proceedings, 2nd Intl symp on neonatal diarrhea, Vet Infect Disease Organization, University of Saskatchewan, Canada, Oct 3–5, pp 393–410

Moon HW (1981) Protection against enteric colibacillosis in pigs suckling orally vaccinated dams: evidence for pili as protective antigens. Am J Vet Res 42: 173–177

Moon HW, Runnels PL (1981) Prospects for development of a vaccine against diarrhea caused by *Escherichia coli*. In: Tholme T, Holmgren J, Merson MH, Mollby R (eds) Acute enteric infections in children. New prospects for treatment and prevention. Elsevier North-Holland Biomed, Amsterdam, pp 477–491

Moon HW, Runnels PL (1983) Trials with somatic (O) and capsular (K) polysaccharide antigens of enterotoxigenic *Escherichia coli* as protective antigens in vaccines for swine. Proceedings, 4th Int symp on neonatal diarrhea, Vet Infect Disease Organization, University of Saskatchewan, Canada, Oct 3–5, pp 558–569

Moon HW, Runnels PL (1984) The K99 adherence system in cattle. In: Boedeker EC (ed) Attachment of organisms to the gut mucosa, vol I. CRC, Boca Raton, pp 31–37

Moon HW, Nagy B, Isaacson RE, Ørskov I (1977) Occurrence of K99 antigen on *Escherichia coli* isolated from pigs and colonization of pig ileum by K99 + enterotoxigenic *E. coli* from calves and pigs. Infect Immun 15: 614–620

Moon HW, Fung PY, Isaacson RE, Booth GD (1979a) Effects of age, ambient temperature, and heat-stable *Escherichia coli* enterotoxin on intestinal transit in infant mice. Infect Immun 25: 127–132

Moon HW, Isaacson RE, Pohlenz J (1979b) Mechanisms of association of enteropathogenic *Escherichia coli* with intestinal epithelium. Am J Clin Nutr 32: 119–127

Moon HW, Kohler EM, Schneider RA, Whipp SC (1980) Prevalence of pilus antigens, enterotoxin types, and enteropathogenicity among K88-negative enterotoxigenic *Escherichia coli* from neonatal pigs. Infect Immun 27: 222–230

Moon HW, Rogers DG, Rose R (1988) Effects of a live oral pilus vaccine on duration of lacteal immunity to enterotoxigenic *Escherichia coli* in swine. Am J Vet Res 49(12): 2068–2077

Morgan RL, Isaacson RE, Moon HW, Brinton CC, To C-C (1978) Immunization of suckling pigs against enterotoxigenic *Escherichia coli*-induced diarrheal disease by vaccinating dams with purified 987 or K99 pili: protection correlates with pilus homology of vaccine and challenge. Infect Immun 22: 771–777

Morris JA, Stevens AE, Sojka WJ (1978) Anionic and cationic components of the K99 surface antigen from *Escherichia coli* B41. J Gen Microbiol 107: 173–175

Morris JA, Thorns CJ, Sojka WJ (1980) Evidence for two adhesive antigens on the K99 reference strain *Escherichia coli* B41. J Gen Microbiol 118: 107–113

Morris JA, Thorns CJ, Scott AC, Sojka WJ, Wells GA (1982) Adhesion in vitro and in vivo associated with an adhesive antigen (F41) produced by a K99 mutant of the reference strain *Escherichia coli* B41. Infect Immun 36: 1146–1153

Morris JA, Thorns CJ, Wells GAH, Scott AC, Sojka WJ (1983) The production of F41 fimbriae by piglet strains of enterotoxigenic *Escherichia coli* that lack K88, K99 and 987P fimbriae. J Gen Microbiol 129: 2753–2759

Morris JA, Sojka WJ, Ready RA (1985) Serological comparison of the *Escherichia coli* prototype strains for the F(Y) and AH 25 adhesins implicated in neonatal diarrhoea in calves. Res. Vet Sci 38: 246–247

Moseley SL, Dougan G, Schneider RA, Moon HW (1986) Cloning of chromosomal DNA encoding the F41 adhesin of enterotoxigenic *Escherichia coli* and genetic homology between adhesins F41 and K88. J Bacteriol 167: 799–804

Mouricout MA, Julien RA (1987) Pilus-mediated binding of bovine enterotoxigenic *Escherichia coli* to calf small intestinal mucins. Infect Immun 55: 1216–1223

Nagy B, Ushe TC (1985) Colonization of infant mice by K99 + *Escherichia coli* and lack of colonization by *Streptococcus faecium* M74. Acta Vet Hung 33: 143–147

Nagy B, Moon HW, Isaacson RE (1976) Colonization of porcine small intestine by *Escherichia coli*: ileal colonization and adhesion by pig enteropathogens that lack K88 antigen and by some acapsular mutants. Infect Immun 13: 1214–1220

Nagy B, Moon HW, Isaacson RE (1977) Colonization of porcine intestine by enterotoxigenic *Escherichia coli*: selection of piliated forms in vivo, adhesion of piliated forms to epithelial cells in vitro, and incidence of a pilus antigen among porcine enteropathogenic *E. coli*. Infect Immun 16: 344–352

Nakazawa M, Haritani M, Sugimoto C, Kashiwazaki M (1986) Colonization of enterotoxigenic *Escherichia coli* exhibiting mannose-sensitive hemagglutination to the small intestine of piglets. Microbiol Immunol 30: 485–489

Ørskov F, Ørskov I (1972) Immunoelectrophoretic patterns of extracts from *Escherichia coli* O antigen test strains O1 to O157 examinations in homologous OK sera. Acta Pathol Microbiol Scand [B] 80: 905–910

Ørskov F, Ørskov I (1978) Significance of surface antigens in relation to enterotoxigenicity of *E. coli*. Proceedings, cholera and related diarrheas, 43rd Nobel symposium, Stockholm, pp 134–141

Parry SH, Porter P (1978) Immunological aspects of cell membrane adhesion demonstrated by porcine enteropathogenic *Escherichia coli*. Immunology 34: 41–49

Rapacz J, Hasler-Rapacz J (1986) Polymorphism and inheritance of swine small intestinal receptors mediating adhesion of three serological variants of *Escherichia coli*-producing K88 pilus antigen. Anim Genet 17: 305–321

Riising H-J, Svendsen J, Larsen JL (1975) Occurrence of K88-negative *Escherichia coli* serotypes in pigs with post weaning diarrhoea. Acta Pathol Microbiol Scand [B] 83: 63–64

Runnels PL, Moon HW (1984) Capsule reduces adherence of enterotoxigenic *Escherichia coli* to isolated intestinal epithelial cells of pigs. Infect Immun 45: 737–740

Runnels PL, Moon HW, Schneider RA (1980) Development of resistance with host age to adhesion of K99 + *Escherichia coli* to isolated intestinal epithelial cells. Infect Immun 28: 298–300

Runnels PL, Moseley SL, Moon HW (1987) F41 pili as protective antigens of enterotoxigenic *Escherichia coli* that produce F41, K99, or both pilus antigens. Infect Immun 55: 555–558

Rutter JM, Jones GW (1973) Protection against enteric disease caused by *Escherichia coli*–a model for vaccination with a virulence determinant? Nature 242: 531–532

Sadowski PL, Acres SD, Sherman DM (1983) Monoclonal antibody for the protection of neonatal pigs and calves from toxic diarrhea. In: Hollaender A, Laskin AI, Rogers P (eds) Basic biology of new developments in biotechnology. Plenum, New York, pp 93–99 (Basic Life Sciences, vol 25)

Sadowski PL, Hermanson V, Hoffsis G (1984) Protection of pigs from fatal colibacillosis through the use of "a" and "c" specific monoclonal antibodies to the K88 pilus. Proceedings, 65th Annual Meeting of the Conference of Research Workers in Animal Disease, Chicago, Illinois, p 52

Sarmiento JI, Casey TA, Moon HW (1988) Postweaning diarrhea in swine: experimental model of enterotoxigenic *Escherichia coli* infection. Am J Vet Res 49(7): 1154–1158

Schneider RA, To SCM (1982) Enterotoxigenic *Escherichia coli* strains that express K88 and 987P pilus antigens. Infect Immun 36: 417–418

Sellwood R (1980) The interaction of the K88 antigen with porcine intestinal epithelial cell brush borders. Biochim Biophys Acta 632: 326–335

Sherman DM, Acres SD, Sadowski PL, Springer JA, Bray B, Raybould TJG, Muscoplat CC (1983) Protection of calves against fatal enteric colibacillosis by orally administered *Escherichia coli* K99-specific monoclonal antibody. Infect Immun 42: 653–658

Simonson RR, Isaacson RE, Jacob CR, Newman KZ (1983) The class specific antibody in passive protection against *Escherichia coli* diarrhea in pigs using a subunit genetically engineered pili vaccine. Proceedings, 4th Int symp on neonatal diarrhea, Vet Infect Disease Organization, University of Saskatchewan, Canada, Oct 3–5, pp 548–557

Smit H, Gaastra W, Kamerling JP, Vliegenthart JFG, De Graaf FK (1984) Isolation and structural characterization of the equine erythrocyte receptor for enterotoxigenic *Escherichia coli* K99 fimbrial adhesin. Infect Immun 46: 578–584

Smith HW (1976) Neonatal *Escherichia coli* infections in domestic mammals: transmissibility of pathogenic characteristics. Ciba Found Symp 42: 45–72

Smith HW, Halls S (1967) Observations by the ligated intestinal segment and oral inoculation methods on *Escherichia coli* infections in pigs, calves, lambs and rabbits. J Pathol Bacteriol 93: 499–529

Smith HW, Halls S (1968) The production of oedema disease and diarrhoea in weaned pigs by the oral administration of *Escherichia coli*: factors that influence the course of the experimental disease. J Med Microbiol 1: 45–59

Smith HW, Huggins MB (1978) The influence of plasmid-determined and other characteristics of enteropathogenic *Escherichia coli* on their ability to proliferate in the alimentary tracts of piglets, calves and lambs. J Med Microbiol 11: 471–492

Smith HW, Linggood MA (1972) Further observations on *Escherichia coli* enterotoxins with particular regard to those produced by atypical piglet strains and by calf and lamb strains: the transmissible nature of these enterotoxins and of a K antigen possessed by calf and lamb strains. J Med Microbiol 5: 243–250

Soderlind O, Olsson E, Smyth CJ, Mollby R (1982) Effect of parenteral vaccination of dams on intestinal *Escherichia coli* in piglets with diarrhea. Infect Immun 36: 900–906

Sojka WJ, Wray C, Morris JA (1977) Passive protection of lambs against experimental enteric colibacillosis by colostral transfer of antibodies from K99-vaccinated ewes. J Med Microbiol 11: 493–499

Svennerholm A-M, Vidal YL, Holmgren J, McConnel MM, Rowe B (1988) Role of PCF 8775 antigen and its coli surface subcomponents for colonization, disease, and protective immunogenicity of enterotoxigenic *Escherichia coli* in rabbits. Infect Immun 56: 523–528

To SCM (1984) F41 antigen among porcine enterotoxigenic *Escherichia coli* strains lacking K88, K99, and 987P pili. Infect Immun 43: 549–554

To SC-M, Moon HW, Runnels PL (1984) Type 1 pili (F1) of porcine enterotoxigenic *Escherichia coli*: vaccine trial and tests for production in the small intestine during disease. Infect Immun 43: 1–5

Truszczyński M, Osek J (1987) Occurrence of mannose resistant hemagglutinins in *Escherichia coli* strains isolated from porcine colibacillosis, Comp Immun Microbiol Infect Dis 10: 117–124

Tzipori S, Withers M, Hayes J (1984) Attachment of *E. coli*-bearing K88 antigen to equine brush-border membranes. Vet Microbiol 9: 561–570

Wilson MR, Hohmann AW (1974) Immunity to *Escherichia coli* in pigs: adhesion of enteropathogenic *Escherichia coli* to isolated intestinal epithelial cells. Infect Immun 10: 776–782

Wilson RA, Francis DH (1986) Fimbriae and enterotoxins associated with *Escherichia coli* serogroups isolated from pigs with colibacillosis. Am J Vet Res 47: 213–217

Yano T, Leite DDS, de Carmargo IJB, de Castro AFP (1986) A probable new adhesive factor (F42) produced by enterotoxigenic *Escherichia coli* isolated from pigs. Microbiol Immunol 30: 495–508

Inhibition of Bacterial Attachment: Examples from the Urinary and Respiratory Tracts

C. Svanborg Edén, B. Andersson, G. Aniansson, R. Lindstedt, P. de Man, A. Nielsen, H. Leffler, and A. Wold

1 Introduction

Bacterial attachment to components of mucosal surfaces promotes bacterial persistence in vivo and thus guides the colonization of mucosal sites (BEACHY 1981). Attachment is also the first in a series of pathogenetic events leading to disease. Bacterial binding enhances the tissue-damaging effect of exo- and endotoxins, and the subsequent tissue invasion by whole bacteria (FRETER 1969; SMITH and LINGGOOD 1971; JONES and RUTTER 1972; SVABORG EDÉN et al. 1982c) (Fig. 1).

Inhibition of attachment in vivo can modify bacterial colonization or interrupt infection (FRETER 1969; JONES and RUTTER 1972; WILLIAMS and GIBBONS 1972; SVANBORG EDÉN and SVENNERHOLM 1978). The detailed information on mechanisms

Department of Clinical Immunology, University of Göteborg, Guldhedsgatan 10, S-413 46 Göteborg, Sweden

Current Topics in Microbiology and Immunology, Vol. 151
© Springer-Verlag Berlin·Heidelberg 1990

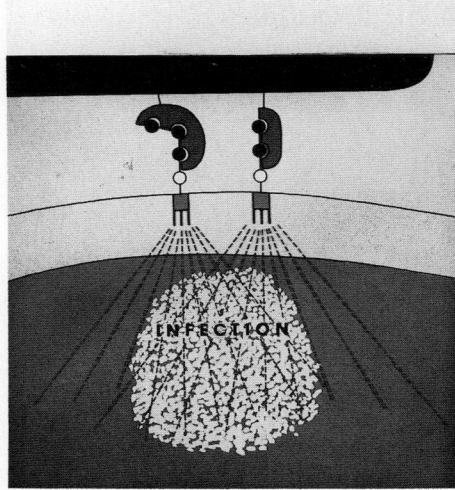

Fig. 1. Attachment of bacteria to target cells is mediated via bacterial adhesins specifically recognizing target cell receptors

of attachment which has been generated over the past decade should be used to examine the protective potential of adhesion inhibition in humans. This chapter describes approaches to inhibition of bacterial adhesion based on in vitro studies, animal experiments, and clinical studies and discusses the feasibility of applying this knowledge to prophylaxis and treatment of infections in man.

2 Approaches to Adhesion Inhibition

Attachment may be inhibited by substances that block the interaction between the binding site of the bacterial adhesin and the target cell receptor (Fig. 1). These include: the soluble form of the receptor-binding site (Fig. 2b), antibodies against the receptor-binding site of the adhesin (Fig. 2a), and antireceptor antibodies that occupy the receptor site or soluble adhesins. The synthesis of bacterial adhesins is influenced by, for example, antibiotics. Antibacterial agents thus may be used not only for their antibacterial effect but also to interfere with bacterial attachment. Compounds that alter the glycosylation of the host cells can change the receptor phenotype.

2.1 Blocking of the Adhesin with Soluble Receptor Oligosaccharides

The ability of soluble receptor oligosaccharides to inhibit attachment to the receptor-bearing target cell in vitro has been demonstrated in several systems.

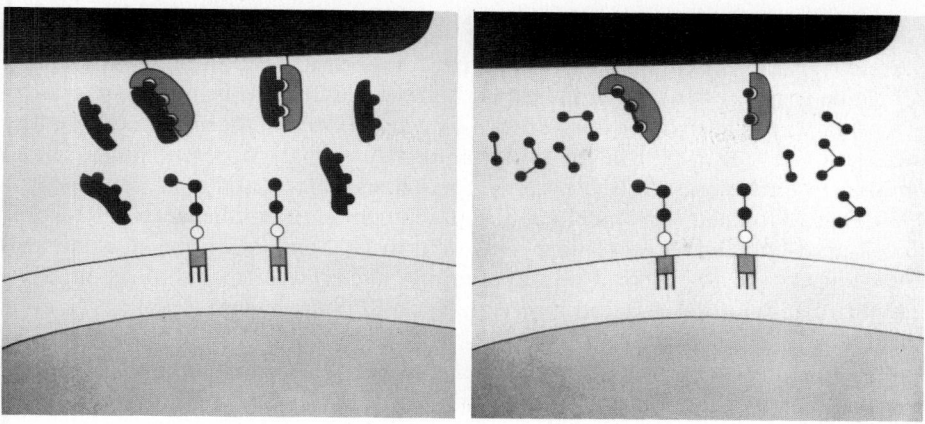

Fig. 2. Adhesion inhibition by blocking of the bacterial adhesin with **a** antibodies against the receptor-binding site and **b** soluble receptors which competitively block adhesin binding to the receptor

2.1.1 Mannose-Specific Binding

It has been recognized for several decades that the simple sugar mannose, or derivatives thereof, can inhibit bacterial binding to different target cells (COLLIER et al. 1955). The ability of bacteria to induce agglutination of erythrocytes from guinea pig or other species which is inhibited or reversed by mannose, provided the basis for the classification of adhesins on Enterobacteriaceae as mannose sensitive (MS) or mannose resistant (MR) (for review see DUGUID and OLD 1980).

Bacteria with mannose-reversible binding reactions are variably inhibited by different mannose-containing oligosaccharide structures (FIRON et al. 1982, 1983; SHARON and OFEK 1986; NEESER et al. 1986; SVANBORG EDÉN et al. 1983b). This variation in receptor specificity within the group with mannose-containing oligosaccharide structures, has also been shown by heterogeneity in the spectrum of target cells to which bacteria with type 1 fimbriae bind. The nature of the cell-bound receptors has recently been the focus of investigation. The trisaccharide Manα1-3[Man]α1-6Man occurs on asparagine-linked oligosaccharide chains of cell surface glycoproteins. The branched core is substituted in various ways, either with mannose residues alone or with N-acetylglucosamine, galactose, and sialic acid. The result is complex, high-mannose or hybrid types of oligosaccharide chains. So far, the oligosaccharide sequences of cell-bound glycoprotein receptors have not been identified. Glycoproteins from human polymorphonuclear leukocytes (PMNLs) and the BHK cell line have been shown to bind Escherichia coli fimbriae. Studies on mutants from the BHK cell line, defective in processing of the N-linked oligosaccharide chains, indicate that the complex type of N-linked chains are less effective as receptors than the oligomannose or hybrid types (GIAMPAPA et al. 1988; RODRIQUEZ-ORTEGA et al. 1987).

2.1.1.1 Receptor Activity of Secreted Glycoproteins

Secreted glycoproteins interact with type 1 fimbriae. The Tamm-Horsfall gly-coprotein is synthesized in the kidney and is secreted into the urine. It contains N-glycosidic chains both of the tetra-antennary complex and of the high-mannose type (SERAFINI-CESSI et al. 1984a, b). The Tamm-Horsfall glycoprotein binds type 1 fimbriae in a mannose-sensitive manner (SVANBORG EDÉN et al. 1982a; ØRSKOV et al. 1980; DULAWA et al. 1988) and blocks the attachment to uroepithelial cells. Recently, the Tamm-Horsfall glycoprotein was shown to opsonize for phagocytosis (SILVERBLATT et al. 1986) and to inhibit bacterial hemagglutination. The inhibitory activity was reduced by treatment with mannosidase (PARKKINEN et al. 1988).

Immunoglobulin molecules of all classes, as well as secretory component, possess N-linked oligosaccharide chains (KORNFELD and KORNFELD 1976). We recently demonstrated that secretory IgA, especially of the IgA2 subclass, functions as a receptor for mannose-specific bacterial lectins (SVANBORG EDÉN et al. 1987; WOLD et al. 1988a). This was first shown using myeloma proteins lacking antibody specificity against the *E. coli* strains used. Whereas IgG and IgM myeloma proteins were inactive, proteins of the IgA isotype agglutinated the bacteria with MS adhesins. The specificity for mannose was shown by inhibition with methyl-α-D-mannoside and by the lack of reactivity with bacteria without MS adhesins. Half of the polymeric IgA2 myelomas displayed strong reactivity and half of the IgA1 myelomas moderate activity. This was explained by truncation of the oligosac-charide with exposure of mannose in a terminal position.

Receptor activity for MS adhesins was also shown for total secretory IgA isolated from human milk and secretory IgA2, which is the predominant im-munoglobulin in the large intestine. Secretory IgA as well as IgA2 myeloma inhibited the attachment of mannose-binding *E. coli* strains to human colonic epithelial cells. Attachment mediated by adhesins specific for Galα1-4Galβ-containing receptors was unaffected. By providing oligosaccharide binding sites for mannose-specific bacterial lectins, secreted immunoglobulins may promote bacter-ial colonization, when bound to the mucus layer, and protect the host against bacterial attachment directly to the tissues—all without any need for specificity of the antigen-combining site against each bacterial antigen. Indeed, the myeloma IgA2 proteins which inhibited attachment of type 1 fimbriated *E. coli* to intestinal epithelial cells also protected mice against experimental urinary tract infection (UTI).

2.1.1.2 Consequences of Mannose-Specific Binding

The human large intestine is an important ecological niche for Enterobacteriaceae, of which, for example, *E. coli*, *Klebsiella*, and Enterobacter express MS adhesins. Human colonic epithelial cells carry mannose-containing receptors (DUGUID and OLD 1980) in both membrane-bound and surface-associated form (WOLD et al. 1988b), which may contribute to bacterial persistence in the large intestine.

The urinary tract is a second niche for the gram-negative species which frequently express mannose-specific adhesins (HAGBERG et al. 1981). The mannose-specific attachment to urinary tract epithelial cells is species dependent (HAGBERG et al. 1983a). Mannose-containing receptors mediate attachment to rat cells but less so to mouse cells. MS attachment has been shown to contribute to bacterial persistence in kidneys and bladders of rats and mice (ARONSON et al. 1979; HAGBERG et al. 1983b; FADER and DAVIS 1980; IWAHI et al. 1983), including after experimental infection with *E. coli* genetically manipulated to express the MS adhesins (HAGBERG et al. 1983b; IWAHI et al. 1983). Bacteria with MS adhesins bind variably, but mostly poorly, to human uroepithelial cells (SVANBORG EDÉN and HANSSON 1978).

Attempts have been made to deduce the role of MS binding for human UTI from the frequency of expression of mannose-specific adhesins in clinical isolates of different origins. A majority of clinical isolates contain chromosomal DNA homologous to *pil*, the chromosomal DNA region of the *E. coli* strain SH1 required for the expression of fimbriae and MS adhesins (HULL et al. 1981). MS adhesins are expressed by 95% of laboratory-grown acute cystitis isolates compared to about 60% of fecal strains (HAGBERG et al. 1981). It has, however, been claimed that all *E. coli* can be induced to express MS adhesins given the right culture conditions (C. BRINTON, personal communication). The frequency estimates based on phenotype are therefore difficult to interpret.

Mannose-containing receptors also mediate the binding of bacteria to, for example, PMNLs, macrophages, and lymphocytes (BAR-SHAVIT et al. 1977; BLUMENSTOCK and JANN 1972; SVANBORG EDÉN et al. 1984). Binding of MS adhesins to human PMNLs triggers a chemoluminescence signal. The degree of phagocytosis and killing depends, however, on the capsule and lipopolysaccharide of the bacteria. Phagocytosis is likely to reduce the survival of mannose-specific bacteria in compartments where PMNLs are active (HAGBERG et al. 1983a, b; SVANBORG EDÉN et al. 1987).

2.1.1.3 Protection

It is probably not a coincidence that the glycoproteins which are secreted in the large intestine and the urinary tract express oligosaccharide sequences which expose mannose and enable them to interact with enterobacterial adhesins. Mucosal membranes other than the large intestine and the urinary tract mainly produce IgA1, which showed poor reactivity with the mannose-specific bacterial adhesins (WOLD et al. 1988a). The reason for the association of type 1 fimbriae with colonization rather than disease may be that secreted glycoproteins provide continuous protection of the tissues as well as superficial mucosal binding sites. This hypothesis may be examined by analysis of the colonic flora and UTI rate in individuals lacking IgA2 or Tamm-Horsfall glycoprotein, respectively.

The protective potential of soluble methyl-α-D-mannoside has been analyzed in experimental UTI models (ARONSON et al. 1979; FADER and DAVIS 1980; SVANBORG EDÉN et al. 1982c). Using the diuresing mouse infection model, a reduction in bacterial excretion was demonstrated. Similar results were obtained in a rat UTI model. With a different end point, i.e., the number of bacteria in kidneys and bladders of mice 24 h after infection, no protective effect of mannose could be documented (SVANBORG EDÉN et al. 1982c).

2.1.2 Binding to the Globoseries of Glycolipids

The globoseries of glycolipids were the first cell-bound receptor molecules for attaching bacteria to be identified, and the specificity has been characterized extensively (LEFFLER and SVANBORG EDÉN 1986). Their identification as members of the globoseries is based on the presence of the Galα1-4Galβ disaccharide, which confers a bent conformation to the oligosaccharide chain (BOCK et al. 1985). The function of the globoseries of glycolipids as receptors has been derived from several lines of evidence:

Bacteria bind selectively to cells carrying the globoseries of glycolipids, e.g., to erythrocytes from individuals of blood group P1 and P2, but not of blood group P̄ with a genetic defect in the synthesis of these glycolipids (KÄLLENIUS et al. 1980; LEFFLER and SVANBORG EDÉN 1980). Purified glycolipids not only confer receptor activity either to cells or to inert surfaces (LEFFLER and SVANBORG EDÉN 1980, 1981: BOCK et al. 1985; DE MAN et al. 1987) but also inhibit bacterial binding to receptor-bearing target cells (LEFFLER and SVANBORG EDÉN 1980).

The fact that Galα1-4Galβ was the only structural unit shared by the components with receptor activity led to the proposal of Galα1-4Galβ as the recognition site. This was subsequently confirmed using synthetic disaccharides (DE MAN et al. 1987; LEFFLER and SVANBORG EDÉN 1986). The Galα1-4Galβ disaccharide did not, however, provide the optimal receptor conformation. Most Galα1-4Galβ-recognizing E. coli reacted more strongly with globotetraosylceramide, and globotetraose was about tenfold more active than Galα1-4Galβ as an inhibitor of attachment (ENERBÄCK et al. 1987; LEFFLER and SVANBORG EDÉN 1980) (see Table 1).

Recently, it has become apparent that E. coli may recognize components of the

Table 1. Adhesion inhibition by soluble receptor molecules

Structure	Concentration (mM) required for 50% inhibition, compared to the saline control
GalNAcβ1-3Galα1-4Galβ1-4Gelβ-O-Et	0.22
Galα1-4Galβ1-4Glcβ-O-Et	0.52
Galα1-4Galβ1-4GlcNAcβ-O-Et	0.71
Galα1-4Galβ-O-Et	0.88
Galα1-4Galβ-O- (OH ... OH)	0.32
Galα1-4Galβ	2.13
2Deoxy-Galα1-4Galβ-O-Et	2.05
GalAα1-4GalA	2.10
Galα1-4Galβ-O-$(CH_2)_2$—S—$C_6H_5NH_2$	>3.5
Glcα1-4Galβ-O-Et	>3.5
GalNacβ1-3Galα-O-Me	>3.5
Galβ1-4Galcβ-O-Me	>3.5

E. coli 2914 were preincubated for 30 min at 37°C with soluble saccharides at concentrations varying from 3.5 to 0.1 mg/ml. Urinary tract epithelial cells were subsequently added, and adherence testing performed as described. The number of attached bacteria was counted by interference contrast microscopy, and inhibition given in % of the saline control

globoseries other than Galα1-4Galβ. The second copy of the *pap* gene cluster in J96 encodes adhesins that reacts with sheep erythrocytes (LUND et al. 1988; HULL and HULL, in press) and binds to globotetraosylceramide and to the Frossman antigen on thin-layer chromatogram plates. Bacterial isolates from the urinary tract of dogs bound specifically to the globo-A structure, but not to Galα1-4Galβ or other disaccharide components of this structure (SENIOR et al. 1988; LINDSTEDT et al. 1989). Their specificity was indeed the same as that of the second copy of *pap* homologous DNA from *E. coli* J96 (SENIOR et al. 1988; LINDSTEDT et al. 1989). This did not, however, recognize GalNAcα1-3GalNAcβ as initially suggested (LUND et al. 1988), but rather the GalNAc terminal together with Galα1-4Galβ disaccharide in a binding site possibly provided by the approximation of the terminal and internal saccharides in the bent globo-A and α Forssman structures (LINDSTEDT et al. 1989). It remains to be shown whether these specificities confer functions on the bacteria different from those of the Galα1-4Galβ-recognizing strains.

These observations have considerable importance for the selection of receptor analogues for inhibition of attachment in vivo. Inhibition of binding to the globoseries of glycolipids is likely to be best achieved using more complete oligosaccharides or combinations of Galα1-4Galβ and other saccharide sequences (Table 1).

2.1.2.1 Functional Consequences

In contrast to the mannose-specific adhesins, which are widely distributed, the ability to bind to the globoseries of glycolipids is restricted to certain *E. coli* clones. These are identified by certain O:K:H serotypes, and are typical of isolates with the ability to colonize the large intestine and infect extraintestinal sites (for review see SVANBORG EDÉN and DE MAN 1987).

We recently obtained evidence for the presence of a receptor in the human large intestine mediating the attachment of Galα1-4Galβ-specific *E. coli* (WOLD et al. 1988b). The intestinal cell line HT-29 as well as surgical specimens of human colon specifically bound *E. coli* strains transformed with *pap*, i.e., the chromosomal DNA fragment encoding pili and adhesins with specificity for Galα1-4Galβ-containing receptors. The bacteria bound primarily to a material loosely associated with the surface of the intestinal cells, which is in contrast to the attachment to uroepithelial cells. Therefore, the Galα1-4Galβ-containing receptor may be associated with a mucin-like glycoprotein rather than with glycolipids in the large intestine. This may explain the ability of Galα1-4Galβ-binding clones to become quantitatively dominating and resident, e.g., to colonize the large intestine for extended periods of time. If this can be confirmed it is plausible that the evolutionary advantage provided by the recognition of the globoseries of glycolipids is related to colonization of the intestine rather than or in addition to the spread to the urinary tract and other sites.

The globoseries of glycolipids occur in epithelial and nonepithelial components of the urinary tract (LEFFLER and SVANBORG EDÉN 1980; BREIMER et al. 1985). Binding to these receptors explains the attachment to uroepithelial cells of uropathogenic *E. coli*. The specific attachment has several consequences. Attachment increases bacterial persistence in kidneys and bladders (HAGBERG et al. 1983b). This has been most clearly demonstrated in experimental UTI models using *E. coli* strains genetically manipulated to differ in adhesins. The advantage for attaching strains over nonattaching mutants or transformants was about tenfold. In man, the role of adherence for persistence is more difficult to evaluate. Persisting bacteriuria occurs mainly in the asymptomatic state, where the frequency of attaching *E. coli* strains is low (20%) (SVANBORG EDÉN et al. 1976).

The injection of attaching *E. coli* into the mouse urinary tract provokes inflammation. Adhesins and the lipid A moiety of endotoxin have a special role. Isolated lipid A coaggregated with adhesins specific for the globoseries of glycolipids induce inflammation synergistically. Since the signal was triggered by injection into the urinary tract of the isolated bacterial components, this implies that the epithelial cells when stimulated can mediate signals to the systemic inflammatory responses. Bacterial metabolism or invasion was not required for this to occur (LINDER et al. 1988). The crucial role of endotoxin for inflammation was also shown by the lack of inflammation in lipopolysaccharide nonresponder mice, and their consequent inability to clear the infection (HAGBERG et al. 1985; SHAHIN et al. 1987).

The role of Galα1-4Galβ-specific binding in the induction of inflammation has been confirmed in patients with UTI; those infected with *E. coli* with Galα1-4Galβ-specific adhesins had higher inflammation than those infected with other strains (MÅRILD et al. 1988; DE MAN et al. 1988). Since acute pyelonephritis is diagnosed by inflammatory signs, this may explain the association between attachment, binding to the globoseries of glycolipid receptors, and severity of disease in man, which was the initial observation leading to the elucidation of this problem complex (SVANBORG EDÉN et al. 1976).

From the two examples of colonic colonization and induction of inflammation it appears that the nature of the carrier may determine the functional consequences of adherence (SVANBORG et al. 1989). The Galα1-4Galβ-containing receptor material on colonic epithelial cells was shown to be loosely associated with the cells, giving a different adherence pattern than that to the uroepithelial cells, where the receptor is anchored in the membrane via the ceramide portion of the glycolipid receptor. In contrast to the membrane-bound form of the receptor, which may mediate inflammation, the secreted form may provide protection against tissue contact and facilitate colonization.

2.1.2.2 Protection

From the above it is clear that inhibition of binding to the globoseries of glycolipids or their analogues has a protective potential. Peroral administration of receptor bound to a solid phase might decrease bacterial persistence in the large intestine. A protective effect against UTI may be secondary to a more rapid turnover of the uropathogenic *E. coli* in the large intestine and a secondarily reduced satistical chance of spread to extraintestinal sites.

The protection of the urinary tract by globotetraose and Galα1-4Galβ-O-Et has been tested in mouse and primate UTI models, respectively (SVANBORG EDÉN et al.

1982c; ROBERTS et al. 1984). In mice, the infection, measured as bacterial persistence, was reduced but not abolished. In primates there was a transient decrease in the degree of infection, as measured by the leukocyte response. Since adherence in the urinary tract has a dual function in promoting both bacterial persistence and clearance through inflammation it is difficult to predict the protective potential of receptor analogues. In view of the clear-cut association of binding with clinical disease, and the well-defined receptor specificity, the UTI complex should be used as a model system to address these questions in man.

2.1.3 Lactoseries of Glycolipids

Attaching strains of *Streptococcus pneumoniae* interact with the lactoseries of glycolipids (ANDERSSON et al. 1983). The receptor function has been demonstrated in three ways: (a) Coating of epithelial cells with lactotetraosylceramide and neolacto-tetraosylceramide increased the attachment to those cells. (b) Soluble, natural or synthetic oligosaccharides inhibited attachment of pneumococci to, for example, epithelial cells, the most effective inhibitors being lactotetraose and neolacto-tetraose. (c) The oligosaccharides functioned as receptors when coupled to a solid phase. Pneumococci agglutinated latex beads covalently coupled with synthetic GlcNAcβ1-3Gal and Galβ1-3GlcNAc. These disaccharides provided the minimal receptor requirement.

2.1.3.1 Natural Occurrence of Receptor Saccharides in Human Milk

The lactoseries oligosaccharides occur in human milk. Consistent with this (ANDERSSON et al. 1986), the oligosaccharide fraction of milk was shown to inhibit the attachment of pneumococci to respiratory tract epithelial cells. The inhibitory activity of the oligosaccharides was higher than that of specific antipneumococcal antibodies, which also occur in milk. Inhibition of *Haemophilus influenzea* and *S. pneumoniae* was also associated with the k-casein component (ANIANSSON et al., in press).

The ability to attach to nasopharyngeal epithelial cells characterizes the majority of pneumococcal and *H. influenzae* isolates from that site (ANDERSSON et al. 1981, 1988). Also the majority of strains causing acute otitis media are adhering. Attachment thus appears to promote bacterial persistence in the nasopharynx, and secondary to that the spread of the isolates present in the nasopharynx to the middle ear when host resistance is reduced, e.g., by a viral infection. Accordingly, reduction of the nasopharyngeal colonization might secondarily reduce the frequence of acute otitis media.

Although the structure of the nasopharyngeal epithelial cell receptors for *S. pneumoniae* and *H. influenzae* has not been characterized, the finding of highly active inhibitors of attachment in milk provided the basis for an assessment of antiadherence in vivo. We have analyzed the ability of receptor analogues in milk to protect against bacterial colonization in the respiratory tract (ANIANSSON et al. 1987). Human milk is an excellent vehicle for antiadhesive substances, since it is

Table 2. Adhesion inhibition in vivo. Preliminary evidence for protection by antiadhesive components in human milk ($n = 271$)

Age of child (months)[a]	Feeding pattern		Colonization frequency (%)			Acute otitis media (%)	Px
			Hi[b]	Pnc[c]			
0–3	BM		4	10		1	<0.001
	Mixed		2	13		9	NS
	Other		8	8		4	
4–6	BM		6	19		8	
	Mixed		3	25		7	NS
	Other		4	22		12	
7–12	BM		0	11		0	NS
	Mixed	0.001	9	29		14	<0.001
	Other		8	20		19	
Total	BM		4	11	0.001	2	<0.01
	Mixed		4	23	0.001	8	<0.001
	Other		7	21		20	

[a] Data from 271 children;
[b] *H. influenzae*;
[c] Pneumococci

administered onto the site of bacterial colonization, e.g., the nasopharynx, several times daily.

The acquisition of the nasopharyngeal flora was followed prospectively from birth with cultures at 2, 6, and 10 months of age. Milk samples were collected and episodes of otitis media were recorded. The protection afforded by the milk was calculated as the difference in bacterial colonization and otitis media of breast-fed versus non-breast-fed babies and correlated to the content of oligosaccharide inhibitors of the individual milks. Some preliminary results are shown in Table 2. Breast-fed babies receiving milk with high antiadhesive activity had reduced nasopharyngeal colonization and frequency of otitis compared to non-breast-fed controls or children receiving milk with low inhibitory activity (not shown).

2.2 Adhesion Inhibition by Antibacterial Antibodies

Adhesion may be inhibited either directly by antibodies to the adhesin or indirectly by antibodies to other bacterial surface components, which may inhibit adherence by, for example, agglutination or steric hindrance (FRETER 1969; JONES and RUTTER 1972; WILLIAMS and GIBBONS 1972; SVANBORG EDÉN and SVENNERHOLM 1978; SVANBORG EDÉN et al. 1982a). For instance, IgG antibodies against the O antigen of *E. coli* lipopolysaccharides agglutinate bacteria and thereby reduce the number of bacteria available for attachment. Agglutinating antibodies may thus contribute to the immune-mediated protection of mucosal surfaces. F(ab) fragments have no such effect. Antibodies to bacterial surface antigens adjacent to the adhesin may sterically hinder access to the receptor. This effect is, in many systems, difficult to separate from the effect of adhesin-specific antibodies.

Antiadhesin antibodies appear to be a simple and straightforward means of blocking bacterial adhesion which has been successfully tested in veterinary praxis (see other contributors, this volume). This concept has, however, several major problems.

2.2.1 Different Antigenicity of Fimbriae and Adhesins

The adhesins of, for example, E. coli are associated with pili/fimbriae but are distinct from the fimbrial subunit (LUND et al. 1987; MOCH et al. 1987; HOSCHÜTZKY et al. 1989). The immune response against whole fimbriated bacteria or against isolated fimbriae is dominated by antibodies directed towards the fimbrial subunit (fimA of type 1 fimbriae, papA of $Gal\alpha1$-$4Gal\beta$-binding adhesins). Although polyclonal antifimbrial antisera were reported by several investigators to block adherence (ISAACSON 1977; KORHONEN et al. 1980), monoclonal antibodies to the structural protein did not. Monoclonal antiadhesin antibodies inhibited attachment. Antibodies from patients with acute pyelonephritis, although active against the fimbrial subunit, failed to block adhesion (DE REE et al. 1987). To achieve antiadhesive antibodies, the adhesin should thus be used to induce an immune response.

2.2.2 Antigenic Diversity

Several bacterial surface components vary their antigenic properties while retaining a common function. The E. coli lipopolysaccharides share the toxicity afforded by lipid A but vary the structure and antigenicity of the O polysaccharide (ØRSKOV et al. 1977). In analogy, fimbriae which share the receptor specificity vary in antigenicity. Fimbriae with adhesins specific for the $Gal\alpha1$-$4Gal\beta$-containing receptors express several fimbrial (F) antigens, including $F7_1$, $F7_2$, and F8–F14. The F antigens are conserved among strains of the same O:K:H serotype (ØRSKOV and ØRSKOV 1983). The extent of antigenic variation for the adhesins remains to be defined.

2.2.3 Antigenic Repertoire

The antigenic properties vary between isolates infecting different patient populations. Both the frequency of attaching strains and the receptor specificity of the adhesins and the serotypes of the fimbrial antigens are variable. The selection of antigens for the induction of antiadhesive antibodies thus needs to be based on information about the strains infecting the population for which the antiadhesive treatment will be targeted. Most of the work on bacterial vaccines has assumed that the selection of vaccine antigens should be based on the most virulent strains. In contrast, patients with recurrent bacterial infections often appear to be reinfected with bacteria different from those causing the primary infection and with reduced virulence. For example, patients with recurrent pyelonephritis, reflux, and renal scarring had about 30% attaching *E. coli*, compared with about 80% in those without reflux (LOMBERG et al. 1984). When such patient populations are the target of vaccination programs, antiadhesive immunity may not be the optimal choice (SVANBORG EDÉN et al. 1982a, b).

2.2.4 Induction of Antiadhesive Mucosal Immunity

The concept of adhesion inhibition presumes that the antiadhesive antibodies exert their action at mucosal surfaces, by preventing contact between bacteria and, for instance, epithelial cells. The dominating mucosal immunoglobulin class, secretory IgA (SIgA), is assembled by selective coupling of dimeric IgA to secretory component during transport through the epithelial cells into the lumen (KRAEHENBUHL et al. 1987). The mucosal immune responses require local antigenic stimulation. The plasma cells are recruited from the site of mucosal priming, e.g., in the Peyer's patches or the respiratory tract, to sites of antigenic stimulation and are induced to produce specific antibodies. Once induced, mucosal immune responses are short-lived (for references, see HANSON and SVANBORG EDÉN 1988).

The induction of stable antiadhesive immunity requires administration of intact adhesins and/or fimbrial antigens to the mucosal immune system. Soluble protein antigens administered to the surface mostly are poor immunogens. For this reason alternative modes of administration are being tested, including the use of live bacteria as antigen presentors. For example, a hybrid between *E.coli* and *Salmonella*, which colonizes the Peyer's patches, may be genetically manipulated to produce adhesins and act as a vehicle delivering these antigens to antigen-presenting cells in the mucosal immune system (KATZ et al. 1987).

An attempt to induce local antiadhesive immunity in the urinary tract of humans was recently made by deliberate installation into the urinary tract of bacteria which had been genetically manipulated in vitro. We selected a wild-type *E. coli* strain which had been carried in the urinary tract of one child for more than 3 years without giving rise to symptoms. The strain, which was highly sensitive

Table 3. Antibody response of patients receiving genetically manipulated *E. coli* in the urinary tract

Patient	Days after colonization	Urinary antibodies, ELISA titer[a]	
		IgA	IgM
EKJ	2	0.59	1.1
LEA	5	0.20	0.26
EJ	30	0.78	0.3
BL	4	0.53	0.26
JE	4	0.13	0.19

[a]Optical density at 405 nm after incubation for 100 min with substrate for alkaline phosphatase conjugated to antihuman IgA or IgM. A heat extract from the transformant strain 506 MR, containing lipopolysaccharide, fimbriae, and other heat-stable proteins, was used as antigen. Patients were colonized with *E. coli* 506 MR, introduced via a catheter into the urinary bladder

to antibacterial agents and had one small plasmid, was transformed with the chromosomal DNA fragments *pap* or *pil* encoding Galα1-4Galβ-specific and mannose-specific adhesins respectively. The transformants and the parent strain were introduced into the urinary bladders of patients by catherterization. The five patients who consented to participate all had neurogenic bladder disorders and recurrent pyelonephritis. The immune response to the introduced strains was measured by ELISA using as antigens heat extracts of whole bacteria containing fimbriae and lipopolysaccharide (Table 3). No serum antibody response was recorded. In contrast, a significant urinary antibody response was observed in the three patients who were stably colonized with the *E. coli* strain (EKJ, EJ, BC).

2.2.5 *Protection by Antiadhesive Antibodies*

Experimental UTI has been used as a model for the study of antiadhesive immunity. The advantages of the model include the fact that adherence is important both for bacterial persistence and for the induction of disease. In addition, the urinary tract is normally sterile and infections are caused by a single strain, which makes it easy to evaluate the antigens inducing or acting as targets for immunity and to detect protection. Protection against homologous bacterial challenge in experimental UTI models has been achieved using different vaccine antigens: whole bacteria, lipopolysaccharides, capsular polysaccharides, and fimbriae (KAIJSER et al. 1978; ROBERTS et al. 1981; SILVERBLATT 1974; for review see SVANBORG EDÉN et al. 1988a). The fimbrial preparations used have been contaminated with lipopolysaccharide, and in spite of extensive controls using either heterogeneous challenge or parallel induction of immunity to the O antigen, it has been difficult to evaluate the contribution of antifimbrial immunity per se. In addition, the protection has not been defined in terms of mucosal antibodies and antiadherence in vivo. The protective power of mucosal antibodies in the urinary tract thus remains to be defined. In view of the complications involved in the induction of antiadhesive immunity and since the receptor structure recognized by the adhesins is shared by antigenically different strains, the use of receptor molecules for competitive inhibition of binding appears as a more straightforward strategy for antiadherence.

2.3 Receptor Blockade with Antireceptor Antibodies

A large proportion of naturally occurring antibodies are directed against carbohydrate antigens. Such antibodies may reognize cell-bound glycoconjugates, e.g., in epithelial and nonepithelial tissues. Binding to structures which are receptors for bacterial adhesins may competitively inhibit bacterial binding to the receptor-bearing cell. The extent to which such antibodies modify bacterial attachment has not been studied.

The antibodies to the globoseries of glycolipids have been studied in relation to the P blood group system (WATKINS 1978). Individuals of blood group P2 produce antibodies to the P1 antigen. Antibodies to the P blood group antigen occur in both P1 and P2 individuals. Neither of these antibodies has specificity for the Galα1-

4Galβ disaccharide. Still, binding of antibodies to an adjacent determinant of the oligosaccharide might influence the access of bacterial adhesins to the Galα1-4Galβ disaccharide.

Another source of antireceptor antibodies is anti-idiotypes. Antibodies against the binding site of the bacterial adhesin resemble the combining site on the receptor. A subset of the antiantibodies may, in turn, recognize the receptor. In this way, an immune response triggered by the bacterial adhesin may lead to receptor blockade.

Monoclonal antibodies to Galα1-4Galβ have been used to map the distribution of receptors in the urinary tract (O'HANLEY et al. 1985; LARK et al. 1984). Their protective potential has, however, not been investigated. The binding of antibodies to host tissues involves the risk of triggering autoimmune events.

2.4 Phenotypic Changes of Bacteria or Target Cells

Antibacterial agents influence the expression of adhesins. This has been shown not only for inhibitors of protein synthesis, e.g., chloramphenicol, but also for streptomycin and penicillins. The contribution of adhesion inhibition compared to bacteriostatic or bactericidal effects of antibacterial agents in vivo has not been defined. It is possible that antiadhesive effects contribute to the positive effects of low-dose prophylaxis against UTIs (SANDBERG et al. 1979; for review see VOSBECK et al. 1982).

The receptor expression of the target cell may be altered by pharmacologic means. It is well established that agents affecting the glycosylation, such as tunicamycin, modify the cell surface if added to cells in culture. Commonly used drugs which affect glycosylation of cell surface glycoconjugates may influence the receptor expression in man. These effects have not been analyzed in relation to bacterial adherence.

3 Concluding Remarks

Adhesion inhibition is an attractive concept for prophylaxis and protection against infection. By interrupting the initial contact between bacteria and mucosal surfaces, the subsequent events in the disease process are aborted. In principle, inhibition of adherence may be achieved in all the ways described in this review. The blocking of binding by receptor analogues is easily achieved in vitro. The suggested relationship between respiratory tract infections and the antiadhesive effect of milk components, may provide the evidence required for larger scale testing of antiadhesive substances as a complement to the currently used antimicrobial agents. Antiadhesive immunity is a simple and attractive concept. In view of the complications involved in the induction of antiadhesive immunity, and since the receptor structures recognized by the adhesins are shared regardless of antigenicity, the use of receptor molecules appears to be the most straightforward approach.

References

Andersson B, Eriksson B, Falsen E, Fogh A, Hanson LÅ, Nylén O, Peterson H, Svanborg Edén C (1981) Adhesion of *Streptococcus pneumoniae* to human pharyngeal epithelial cells in vitro. Differences in adhesive capacity among strains isolated from subjects with otitis media, specticemia, or meningitis or from healthy carriers. Infect Immun 32: 311–317

Andersson B, Dahmén J, Frejd T, Leffler H, Magnusson G, Noori G, Svanborg Edén C (1983) Identification of a disaccharide unit of a glycoconjugate receptor for pneumococci attaching to human pharyngeal epithelial cells. J Exp Med 158: 559–570

Andersson B, Porras O, Hanson LÅ, Lagergård T, Svanborg-Edén C (1986) Inhibition of attachment of *Streptococcus pneumoniae* and *Heamophilus influenzae* by human milk and receptor oligosaccharides. J Infect Dis 153: 232–237

Andersson B, Gray BM, Dillon HC, Bahrmand A, Svanborg Edén C (1988) Role of adherence of *S. pneumoniae* in acute otitis media. Pediatr Inf Dis 7: 476–480

Aniansson G, Andersson B, Alm B, Larsson P, Nylén O, Pettersson H, Ringér P, Svanborg Edén C (1987) Protection by breastfeeding against bacterial colonization of the nasopharynx: a pilot study.

Aniansson G, Andersson B, Lindstedt R, Svanborg edén C (in press) Receptor activity of human casein for *S. pneumoniae* and *H. influenzae*

Aronson M, Medalia O, Schori L, Mirelman D, Sharon N, Ofek I (1979) Prevention of colonization of the urinary tract of mice with *Escherichia coli* by blocking of bacterial adherence with methyl α-D-mannopyranoside. J Infect Dis 139: 329–332

Bar-Shavit Z, Ofek I, Goldman R, Mirelman D, Sharon N (1977) Mannose residues on phagocytes as receptors for the attachment of *Escherichia coli* and Salmonella typhi. Biochem Biophys Res Commun 78: 455–460

Beachey EH (1981) Adhesin-receptor interactions mediating the attachment of bacteria to mucosal surfaces. J Infect Dis 143: 325–345

Blumenstock B, Jann K (1982) Adhesion of piliated *Escherichia coli* strains to phagocytes: difference between with mannose-sensitive pili and those with mannose-resistant pili. Infect Immun 35: 264–269

Bock K, Brignole A, Breimer Me et al. (1985) Specificity of binding of a strain of uropathogenic *Escherichia coli* to Galα1-4Gal containing glycosphingolipids. J Biol Chem 260: 8545–8551

Breimer M, Hansson GC, Leffler H (1985) The specific glycosphingolipid composition of human ureteral epithelial cells. J Biochem 98: 1169–1180

Collier WA, de Miranda G, von Loeuwenhoek AJ (1955) Microbiol Serol 21: 135–140

de Mann P, Cedergren B, Enerbäck S et al. (1987) Receptor-specific agglutination tests for detection of bacteria that bind the gnoboseries glycolipids. J Clin Microbiol 25: 401–407

de Mann P, Jodal U, Lincoln K, Svanborg Edén C (1988) Bacterial attachment and inflammation in the urinary tract. J Infect Dis 158: 29–35

de Ree IM, Schwillens P, van den Bosch JF (1987) Monoclonal antibodies raised against Pap fimbriae recognize minor component(s) involved in receptor binding. Microb Pathogen 2: 113–121

Duguid JP, Old DC (1980) Adhesive properties of Enterobacteriaceae. In: Beachey EH (ed) Bacterial adherence. Champman and Hall, New York, pp 187–215

Dulawa J, Jann K, Thomsen M, Rambausek M, Ritz E (1988) Tamm Horsfall glycoprotein interfers with bacterial adherence to human kidney cells. Eur J Clin Invest 18: 87–91

Enerbäck S, Larsson AC, Jodal U et al. (1987) Binding to galactoseα1-4galactoseβ-containing receptors as potential diagnostic tool in urinary tract infection. J Clin Microbiol 25: 407–411

Fader RC and Davis C (1980) Effect of piliation on *Klebsiella pneumaniae* infection in rat bladders. Infect Immun 30: 554–561

Firon N, Ofek I, Sharon N (1982) Interaction f mannose-containing oligosaccharides with the fimbrial lectin of *Escherichia coli*. Biochem Biophys Res Commun 105: 1426–1432

Firon N, Ofek I, Sharon N (1983) Carbohydrate specificity of the surface lectins of *Escherichia coli*, *Klebsiella pneumoniae* and *Salmonella typhimurium*. Carbohydr Res 120: 235–249

Freter R (1969) Studies of the mechanism of action of intestinal antibody in experimental cholera. Texas Rep Biol Med 27 (Suppl 1): 299–316

Giampapa CS, Abraham SN, Chiang TM, Beachy EH (1988) Isolation and characterization of a receptor for type 1 fimbriae of *Escherichia coli* from guinea-pig erythrocytes. J Biol Chem 263: 5362–5367

Hagberg L, Jodal U, Korhonen TK et al. (1981) Adhesion hemagglutination and virulence of *Escherichia coli* causing urinary tract infections. Infect Immun 31: 564–574

Hagberg L, Engberg I, Freter R et al. (1983a) Ascending unobstzucted urinary tract infection in mice caused by pyelonephritogenic *Escherichia coli* of human origin. Infect Immun 40: 273–283

Hagberg L, Hull R, Hull S et al. (1983b) Contribution of adhesion to bacterial persistence in the mouse urinary tract. Infect Immun 40: 265–272

Hagberg L, Briles D, Svanborg Edén C (1985) Evidence for separate genetic defects in C2H/HeJ and C3HeB/FeJ micm, that affect the susceptibility to gram-negative infection. J Immunol 134: 4118–4122

Hanson LÅ, Svanborg Edén C (eds) (1988) In: Mucosal immunology. Nobel symposium no 68. Monogr Allergy

Hoschützky H, Lottspeich F, Jann K (1989) Iso,ation and characterization of the α-Galactosyl-1,4-β-Galactosyl specific adhesin (Padhesin) from fimbriated *Escherichia coli*. Infect Immun 57: 76–81

Hull S, Hull R (in press) Linkage and duplication of copies of genes encoding P fimbriae and hemolysin in the chromosome of a uropathogenic *E. coli* isolate. In: Svanborg Edén C, Kass EH (eds) Host parasite interaction in the urinary tract. Chicago University Press, Chicago

Hull RA, Gill RE, Hsu P et al. (1981) Construction and expression of recombinant plasmids encoding type I or D-mannose resistant pili from a urinary tract infection *Escherichia coli* isolate. Infect Immun 33: 933–938

Isaacson RE (1977) K99 surface antigen on *Escherichia coli*: Purification and partial characterization. Infect Immun 15: 272–279

Iwahi T, Abe Y, Nakao H, Imada A, Tsuchiya K (1983) Role of type 1 fimbriae in the pathogenesis of ascending urinary tract infection induced by *Escherichia coli* in mice. Infect Immun 39: 1307–1315

Jones GW, Rutter JM (1972) Role of the K88 antigen in the pathogenesis of neonatal diarrhoea caused by *Escherichia coli* in piglets. Infect Immun 6: 918–927

Kaijser H, Larsson P, Olling S (1978) Protection against ascending *Escherichia coli* pyelonephritis in rats and significance of loca immunity. Infect Immun 20: 78–81

Källenuius G, Möllby R, Svensson SB et al. (1980) The p^k antigen as receptor for the haemagglutination of pyelonephritic *E. coli*. FEMS Microbiol Lett 7: 297–302

Katz J, Michalek SM, Curtiss R, Harmon C, Richardsson G, Mestecky J (1987) Novel oral vaccines: the effectiveness of cloned gene products on inducing secretory immune responses. In: McGhee JR, Mestecky J, Ogra Pl, Bienenstock J (eds) Recent advances in mucosal immunology. Plenum, New York, pp 1741–1747

Korhonen TK, Nurmiaho EL, Ranta H, Svanborg Edén C (1980) New method for isolation of immunologically pure pili from *Escherichia coli*. Infect Immun 27: 569–575

Kornfeld R, Kornfeld S (1976) Comparative aspects of glycoprotein structure. Annu Rev Biochem 45: 217–233

Kraehenbuhl J-P, Schaerer E, Weltzin K, Solari R (1987) Receptor-mediated transepthelial transport of secretory antibodies and engineering of mucosal antibodies. In; Mc Ghee Jr, Mestecky J, Ogra Pl, Bienenstock J (eds) Recent advances in mucosal immunology. Plenum, New York, pp 1053–1069

Lark DL, O'Hanley P, Weinstein WM, Schoolnik G (1984) Human uroepithelia: the distribution and density of receptor analogues for pyelonephritic *E. coli* pili. Clin Res 32: 379

Leffler H, Svanborg Edén C (1980) Chemical identification of a glycosphingolipid receptor for *E. coli* attaching to human urinary tract epithelial cells and agglutinating human erythrocytes. FEMS Microbiol Lett 8: 127–134

Leffler H, Svanborg Edén C (1981) Glycolipid receptors for uropathogenic *Escherichia coli* on human erythrocytes and uroepithelial cells. Infect Immun 34: 920–929

Leffler H, Svanborg Edén C (1986) Glycolipids as receptors for *Escherichia coli* lectins or adhesins. In: Mirelman D (ed) Microbial lectins. Wiley, New York pp 84–96

Linder H, Engberg I, Mattsby-Baltzer I, Jann K, Svanborg-Edén C (1988) Induction of inflammation by *Escherichia coli* on the mucosal levels: requirement for adherence and endotoxin. Infect Immun 56: 1309–1313

Lindstedt R, Falk P, Larsson G, Svanborg Edén C (1989) Comparative binding of ONAP and *prs* to the globoseries of glycolipids. Carbohydr Res (submitted for publication)

Lomberg H, Hellström M, jodal U, Leffler H, Lincoln K, Svanborg Edén C (1984) Virulence associated traits in *E. coli* causing first and recurrent episodes of urinary tract infections in children with or without vesicoureteric reflux. J Infect Dis 150: 561–569

Lund B, Lindberg F, Marklund B-I, Normark S (1987) The PapG protein is the α-D-galactopyranosol-(1-4)-β-D-galactopyranose-binding adhesin of uropathogenic *Escherichia coli*. Proc Natl Acad Sci USA 84: 5898–5902

Lund B, Marklund B-I, Strömberg N, Lindberg F, Karlsson K-A, Normark S (1988) Uropathogenic *E. coli* express srologically identical pili with different receptor binding specificities. Mol Microbiol 2: 255–263

Mårlid S, Wettergren B, Hellström M, Jodal U, Lincoln K, Orskov I, Orskov F, Svanborg Edén C (1988)

Bacterial virulence and inflammatory response in infants ith febrile urinary tract infection or screening bacteriuria. J Pediatr 112: 398–

Moch T, Hoschutzky H, Hacker J, Kröncke K-D, Jann K (1987) Isolation and characterization of the α-sialyl-β-2,3-galactosyl-specific adhesin from fimbriated *Escherichia coli*. Proc Natl Acad Sci USA 84: 3462–3466

NesserJ-R, Koellreutter B, Wuersch P (1986) Oligomannoside-type glycopeptides inhibiting adhesion of *E. coli* strains mediated by type 1 pili: preparation of potent inhibitors of plant glycoproteins. Infect Immun 52: 428–436

O'Hanley P, Lark D, Falkow S, Schoolnik G (1985) Molecular basis of *Escherichia coli* colonization fo the upper urinary tract in BALB/c mice. J Clin Invest 75: 347–360

Ørskov I, Ørskov F, Jann B, Jann K (1977) Serology, chemistry and genetics of O and K antigens of *Escherichia coli*. Bacteriol Rev 41: 667–710

Ørskov I, Ferencz A, Ørskov F (1980) Tamm-Horsfall protein or uromucoid is the normal urinary slime that traps type 1 fimbriated *Escherichia coli*. Lancet 1: 887

Ørskov I, Ørskov F (1983) Serology of *Escherichia coli* fimbriae. Prog Allergy 33: 80–105

Parkkinen J, Virkola R, Korkonen TK (1988) Identification of inhibitors in human urine for binding of *E. coli* adhesins. Infect Immun (in press)

Porras O, Dillon H, Gray B, Svanborg Edén C (1987) Lack of correlation of in vitro adherence of *Haemophilus influenzae* to epithelial cells with frequent occurrence of otitis media. Pediatr Infect Dis J 6: 41–45

Roberts JA, Domingue GJ, Martin LN, Kim JCS, Rangan SRS (1981) Immunology of pyelonephritis in the primate model. II. Effect of Immunosuppression. Invest Uro 19: 148–153

Roberts JA, Kaack B, Källenius G Et al. (1984) Receptors for pyelonephritogenic *Escherichia coli* in primates. Invest Urol 131: 163–168

Rodriguez-Ortega M, Ofek I, Sharon N (1987) Infect Immun 35: 968–973

Sandberg T, Stenquvist K, Svanborg Edén C (1979) Effects of subminimal inhibitory concentrations of ampicillin, chloramphenicol and nitrolmantoin on the attachment of *E. coli* to human uroepithelial cells in vitro. Rev Infect Dis 1: 838–844

Senior D, Baker N, Cedergren B, Falk P, Larsson G, Lindstedt R, Svanborg Edén C (1988) Globo A — A new receptor specificity for attaching *E. coli*. FEBS Lett 237: 123–127

Serafini-Cessi F, Malagolini N, Dall'Olio F (1984a) A tetraantennary glycopeptide from human Tamm-Horsfall glycoprotein inhibits agglutination of desialylated erythrocytes induced by leucoagglutinin. Biosci Rep 4: 973–978

Serafini-Cessi F, Dall'Olio F, Malagoni N (1984b) High-mannose oligosackharides from human Tamm-Horsfall glycoprotein. Biosci Rep 4: 269–274

Shahin R, Engberg I, Hagberg L, Svanborg Edén C (1987) Neutrophil recruitment and bacterial clearance correlated with non-responsiveness in local gram-negative infection. J Immunol 10: 3475–3480

Sharon N, Ofek I (1986) Mannose specific bacterial surface lectins. In: Miselman D (ed) Microbial lectins and agglutinins. Wiley, New York, pp 56–79

Silverblatt FJ (1974) Host-parasite interaction in the rat renal pelvis. A possible role for pili in the pathogenesis of pyelonephritis. J Exp Med 140: 1696–1711

Silverblatt F, Cohen L (1979) Anti-pili antibody affords protection against experimental, ascending pyelonephritis. J Clin Invest 64: 333–336

Smith HW, Lingood MA (1971) Observations on the pathogenic properties of the K88, HLY and Ent plasmids of *Escherichia coli* with particular reference to porcine diarrhoea. J Med Microbiol 4: 467–485

Svanborg Edén C, Hansson HA (1978) *Escherichia coli* pili as possible mediators of attachment to human urinary tract epithelial cells. Infect Immun 21: 229–237

Svanborg Edén C, de Man P (1987) Bacterial virulence in urinary tract infection. In:Andriole V (ed) Urinary tract infections. Infectious disease clinics of North America, vol 1. Karger, Base, pp 731–750

Svanborg Edén C, Svennerholm A-M (1978) Secretory immunoglobulin A and G antibodies prevent adhesion of *Escherichia coli* to human urinary tract epithelial cells. Infect Immun 22: 790–797

Svanborg Edén C, Hanson LÅ, Jodal U, Lindberg U, Sojl Åkerlund A (1976) Variable adherence to normal urinary tract epithelial cells of *Escherichia coli* strains associated with various forms of urinary tract infection. Lancet II: 490–492

Svanborg-Edén C, Hanson LÅ, Jodal U, Leffler H, Mårild S, Korhonen T, Brinton C, Jann B, Jann K, Silverblatt F (1982b) Receptor analogs and antipili antibodies as inhibitors of attachment of uropathogenic *Escherichia coli*. In: sell K, Hanson LÅ, Strobel W (eds) Recent advances in mucosal immunity. Raven, New York, pp 355–369

Svanborg Edén C, Fasth A, Hagberg L, Hanson LÅ, Korhonen TK, Leffler H (1982b) Host interaction with *Escherichia coli* in the urinary tract. In: Robbins J, Hill J, Sadoff J (eds) Bacterial vaccines. IV. Thieme, Stuttgart, pp 113–131

Svanborg Edén C, Freter R, Hagberg L et al. (1982c) Inhibition of experimental ascending urinary tract infection by receptor analogue. Nature 298: 560–562

Svanborg Edén C, Andersson B, Hagberg L, Hanson LÅ, Leffler H, Magnusson G, Noori G, Dahmén J, Söderström T (1983a) Receptor analogues and antipili antibodies as inhibitors of attachment in vivo and in vitro. Ann NY Acad Sci 409: 580–595

Svanborg Edén C, Hull R, Falkow S, Leffler H (1983b) Target cell specificity of wild type *E. coli* and mutants and clones with genetically defined adhesions. Prog Food Nutr Sci Res 7; 75–89

Svanborg Edén C, Bjursten LM, Hull R et al. (1984) Influence of adhesins on the interaction of *Escherichia coli* with human phagocytes. Infect Immun 44: 407–413

Svanborg Edén C, Andersson B, Aniansson G, Leffler H, Lomberg H, Mesteckley J, Wold AE (1987) Glycoconjugate receptors for bacteria attaching to mucosal sites. In: Mestecley J, McGhee JR, Bienenstock J, Ogra PJ (eds) Recent advances in mucosal immunology. B. New York, Plenum, New York, pp 937–939

Svanborg Edén C, de Man P, Linder H, Lomberg H (1988a) Natural versus induced resistance to urinary tract infection. In: Savanborg Edén C, Kass EA (eds) Host parasite interaction in the urinary tract. Chicago University Press, Chicago

Svanborg Edén C, Linder H, de Man P (1988b) Consequences of specific attachment of *E. coli* to Galα1-4Galβ-containing receptors. In: Schrinner E, Richmond MH, Seibert G, Schwartz U (eds) Surface structures of microorganisms and their interaction with the mammalian host. NCH Verlagsgesellschaft, pp 185–196 (workshop conferences, Hoechst, vol 18)

Svanborg Edén C, Hull S, Leffler H, Norgren S, Plos K, Wold A (1989) The large intestine as a reservoir for *Escherichia coli* causing extraintestinal infections. In: Midtredt T (ed) B. Gustavsson symposium (in press)

Vosbeck K, Mett H, Huber U, Bohn J, Petiquant m (1982) Effects of low concentrations of antibiotics on *E. coli* adhesion. Antimicrob Agents. Chemother 21: 864–869

Watkins WM (1978) Gnetics and biochemistry of some human blood groups. Proc R Soc Lond [Biol] 202: 31–53

Williams RC, Gibbons RJ (1972) Inhibition of bacterial adherence by secretory immunoglobulin A: a mechanism of antigenic disposal. Science 177: 697–699

Wold AE, Mestecky J, Svanborg Edén S (1988a) Agglutination of *Escherichia coli* by secretory IgA: a result of interaction between bacterial mannose-specific adhesins and immunoglobulin carbohydrate Monogr Allergy 24: 307–309

Wold AE, Thorssen M, Hull S, Svanborg Edén C (1988b) Attachment of *Escherichia coli* to mannose- or Galα1-4Galβ-containing receptors in human colonic epithelial cells. Infect Immun (in press)

Synthetic Peptides: Prospects for a Pili (Fimbriae)-Based Synthetic Vaccine

M. A. Schmidt

1 Introduction

Pathogenic bacteria owe their capacity to cause disease to the combined effects of their repertoires of virulence factors. These factors, in interaction with the various defense mechanisms of the host, are involved in triggering a succession of events leading from colonization and infection of local epithelia to the onset of disease which in most cases afflicts the host as a whole. In mucosal infections adherence of the pathogens to host epithelial cells is regarded as the essential first step. Adhesion is mediated by surface-associated proteinaceous *adhesins* which recognize and bind specific carbohydrate structures on the epithelial cell surface (GIBBONS 1977; BEACHEY 1981; FRETER 1981). Most of the adhesin systems described so far have been found to be associated with filamentous protein structures, fibrillar or fimbrial, though a few examples of afimbrial adhesins have been reported. The possibility of interfering with the adherence of pathogenic microorganisms to epithelial cell receptors might open up new ways of combating infectious diseases at a very early stage of the infectious process.

In this chapter some general aspects of the use of synthetic peptides for the induction of protective immunity will be discussed, as well as more recent

Zentrum für Molekulare Biologie Heidelberg (ZMBH), Im Neuenheimer Feld 282, D-6900 Heidelberg, FRG developments in the application of synthetic peptides corresponding in sequence to selected segments of structural proteins of fimbriae (pili) of *Neisseria gonorrhoeae* and P-specific strains of uropathogenic *Escherichia coli.*

2 Purified Whole Fimbriae as Vaccines

2.1 Fimbriae as Vaccines in Enterotoxigenic *Escherichia coli*-Induced Diarrheal Diseases

As fimbriae are important virulence factors and prominent antigens on the bacterial surface, thus readily interacting with the host defense system, and as these structures are furthermore relatively easy to purify, several attempts have been made to utilize whole isolated fimbriae as vaccines for the induction of protective immunity (KLEMM 1985). Especially in veterinary medicine this approach has been quite successful as there are relatively few fimbrial serotypes associated with enterotoxigenic *E. coli* (ETEC) strains responsible for diarrheal diseases of domestic animals. Fimbriae characterized so far include K88, K99, and 987P, which are also designated F4, F5, and F6 antigens, respectively (ISAACSON 1977; ØRSKOV and ØRSKOV 1983) and F41 (DE GRAAF and ROORDA 1982; MORRIS et al. 1982).

In 1973, RUTTER and JONES immunized pregnant sows with purified K88 fimbriae and achieved protection (RUTTER and JONES 1973). The same strategy was applied with similar success to the immunization of dams with purified K99 fimbriae, thus providing passive protection for the offspring of calves, lambs, and piglets due to anti-K99 antibodies taken up with the mother's colostrum (ACRES et al. 1979; MORGAN et al. 1978; SOJKA et al. 1978).

Piglets could also be protected against homologous challenge by immunization of pregnant swine with purified 987P or F41 antigens (MORGAN et al. 1978; NAGY et al. 1978; ISAACSON et al. 1980; RUNNELS et al. 1987). There is some evidence that the best protection can be achieved by a combination of K88, K99, and 987P antigens in the preparation used for active immunization (CONTREPOIS and GIRARDEAU 1985). For a number of years commercial vaccines based on K88, K99, and 987P fimbriae have been available which have been shown to be effective in the protection of newborn farm animals.

Because the results in the veterinary field have been very promising, pilot studies were undertaken to evaluate the use of fimbriae for the prevention of diarrheal diseases in humans. The best characterized fimbrial antigens found in ETEC strains isolated from patients with diarrheal diseases are the colonization factor antigens CFA/I and CFA/II (EVANS et al. 1978, 1979, 1982; EVANS and EVANS 1978) and the PCF8775 antigen (THOMAS et al. 1982, 1985, 1987). Whereas CFA/I seems to be a homogeneous antigen with a fimbrial structure, CFA/II and PCF8775 have been shown to be antigenically hetrogeneous. CFA/II consists of at least three *E. coli* surface antigens: CS1, CS2, and CS3 (CRAVIOTO et al. 1982). The corresponding antigens for PCF8775 are designated CS4, CS5, and CS6 (THOMAS et al. 1985).

Experiments in animal models have confirmed that oral immunization with purified fimbrial antigens gives significant levels of protection (DE LA CABADA et al. 1981; SVENNERHOLM et al. 1988). These findings prompted volunteer studies to assess the immune response (LEVINE et al. 1984; SACK 1980) as well as the degree of protection achieved by oral administration of purified colonization factor antigens in humans (EVANS et al. 1984; LEVINE et al. 1983; KAPER and LEVINE 1988). Oral immunization alone seemed not to give sufficient levels of protection. Only after subcutaneous priming did oral immunization protect against challenge with virulent ETEC strains (EVANS et al. 1984). As dominant surface antigens, fimbriae interact with the defense mechanisms of the host and are thus prone to selective immunologic pressure which should lead to the selection of new antigenic variations. Surprisingly, in intestinal infections caused by ETEC strains, relatively few serologically distinct adhesion factors have been described (GAASTRA et al. 1982; KLEMM and MIKKELSEN 1982; GROSS and ROWE 1985).

2.2 Fimbriae as Vaccines in Urinary Tract Infections

Most of the uropathogenic *Escherichia coli* strains isolated from pyelonephritis patients express P-specific fimbriae which recognize an α-(1-4)-linked digalactoside as minimal receptor. P fimbriae have been evaluated for their capacity to induce protective immunity in animal models of experimental pyelonephritis involving mice (O'HANLEY et al. 1985) and primates (ROBERTS et al. 1982, 1984b; KAACK et al. 1988), which have been shown to express the same glycolipid receptors on uroepithelia as humans (O'HANLEY et al. 1985; ROBERTS et al. 1984a). In these experiments, however, immunization and challenge have been performed with the corresponding wild-type strains. Though P fimbriae vary greatly immunologically due to the heterogeneity of the immunodominant structural fimbrillin protein (ABE et al. 1987), sequencing of the N-terminal regions of fimbrillins of several uropathogenic *E. coli* isolates indicated a conserved sequence pattern (KLEMM 1985; NORMARK et al. 1986) (see also Table 2). It turned out, however, that these regions are immunologically recessive in native fimbriae (SCHMIDT et al. 1984). Furthermore, more recently it has been shown that the adhesive properties of P, S, and type 1 fimbriae are conferred by minor constituents of the pilus filament (LINDBERG et al. 1986; LUND et al. 1987; MOCH et al. 1987; MIMION et al. 1986) by the major fimbrial subunit.

For P fimbriae three minor proteins designated PapE, PapF, and PapG have been shown to form an adhesive complex located at the tip adhesive complex (LINDBERG et al. 1987). PapG was identified as the digalactoside-binding protein (LUND et al. 1987; HOSCHÜTZKY et al. 1989). Investigation of the antigenic properties of the PapE, PapF, and PapG proteins using absorbed polyclonal antisera as well as comparison of the respective amino acid sequences of the F13, F11, and F7$_2$ fimbrial serotypes revealed that there is substantial antigenic as well as amino acid sequence heterogeneity between the three proteins. Furthermore, even the receptor-binding PapG proteins exhibited marked differences in their amino acid sequences, with only a few segments conserved (LUND et al. 1988). These findings are surprising as one might have expected to find proteins with identical functions to be more conserved as they seem not to be exposed to the selective pressure of the immune system to a large extent. This may also dampen current hopes of employing the receptor-binding PapG proteins as the basis for the development of a subunit vaccine directed against the actual adhesin complex, though a cross-reactive mono-clonal antibody has been described (HOSCHÜTZKY et al. 1989).

3 Synthetic Peptides and Antigen Presentation

The investigation of the antigenic make-up of proteins has been the focus of intense research efforts for several decades. Nearly 30 years ago evidence had already been accumulated (JERNE 1960; SELA 1969) suggesting that antiprotein antibodies specifically recognize rather short, distinct segments of the protein sequence and that antibodies against these determinants can be induced by short synthetic peptides corresponding in sequence to these domains (LERNER 1982). With the advent of DNA sequencing and the availability of deduced amino acid sequences of a fast growing number of proteins, interest in synthetic peptide technology increased even more. Synthetic peptides serve as "immunologic tools," i.e., to study the antigenicity of proteins, the interaction of the immune system, and the possible expression of proteins from unidentified reading frames. Furthermore, synthetic peptides are investigated regarding their potential use as synthetic vaccines (ROWLANDS 1986).

With regard to the antigenicity of proteins the term "immunogenic epitope" is used to describe a particular domain in the native protein that is recognized by the immune system and gives rise to antibodies able to bind synthetic peptides corresponding to this domain. "Antigenic epitope" refers to domains that are recognized in the native protein by antibodies engendered by synthetic peptides corresponding to that particular region of the protein.

Antibodies induced by native proteins can recognize two types of determinant: either sequential (also termed "continuous" or "primary") or conformational (also termed "discontinuous," "assembled," or "topographic"). While sequential epitopes are made up of linear amino acid sequences, conformational determinants are introduced by the juxtaposition of two or more segments which are separated on the linear sequence due to the three-dimensional folding of the amino acid chain. Both types of determinant occupy discrete regions of the protein surface and are thus subject to conformational changes (VAN REGENMORTEL 1986, 1987). Though experimental evidence suggests that most of the determinants establishing the overall antigenicity of a given protein are made up of conformational epitopes (BARLOW et al. 1986), synthetic peptides have been applied successfully for the elucidation of linear antigenic structures of several proteins (ATASSI 1984; NIMAN et al. 1983). More recently, a technique has been developed to synthesize peptides designed to mimic conformational epitopes in the form of "mimotopes" (GEYSEN et al. 1985, 1986). A mimotope is a sequence of amino acids which mimics the ability of a conformational determinant to bind a specific monoclonal antibody without necessarily bearing any resemblance to the sequence of the epitope (STEWARD and HOWARD 1987). An additional feature of the application of synthetic peptides is the possibility of engendering cross-reactive antibodies against areas of the protein found to be inaccessible to the immune system when the native protein is used as the immunogen (WILSON et al. 1984). This could have important implications for the development of synthetic vaccines, as discussed below.

A basic problem concerning the general application of synthetic peptides as immunogens still remains, i.e., how to select segments of a protein sequence for synthesis which are likely to encompass antigenic determinants and might be used to trigger protective immunity. Despite intensive efforts by numerous groups, only a few minimal properties concerning the nature of antigenic sites could be agreed on. For a region likely to be recognized as an antigenic determinant the prospective antigenic sites should be localized on the surface of the protein and should be accessible to the components of the immune system. There is a continuing debate on whether these regions present defined conformations (NOVOTNY et al. 1987) or represent areas of relatively high flexibility and mobility of the amino acid chain (TAINER et al. 1984, 1985). The latter argument might offer an explanation for the surprisingly frequent induction of cross-reactive antibodies by synthetic peptides (NIMAN et al. 1983; WILSON et al. 1985; JEMMERSON 1987), as very few short peptides (less than 20 amino acids) adopt a defined conformation in aqueous solutions.

Thus, for the prediction of possible antigenic determinants for the induction of cross-reactive antibodies there is no particular "best" method available. Rather, prediction is based on a combination of several available methods, or algorithms (Table 1; STEWARD and HOWARD 1987), supported if possible by additional experimental information.

The activation of B cells to synthesize and secrete specific antibodies is in most cases coincident with and dependent on the activation of antigen-specific helper T cells (T_h) to release B cell growth and differentiation factors. Unlike the B cell receptor (antibody), which directly recognizes native proteins free in solution, T_h

Table 1. Available methods for the prediction and detection of antigenic determinants (according to STEWARD and HOWARD 1987)

Hydrophilicity plots
Acrophilicity plots
Prediction of secondary structure (β-turns, β-sheets, α-helices)
Protrusion index plots
Segmental or atomic mobility analysis
Solvent accessibility
Knowledge-based modeling of amino acid sequences
Analysis of immunologic properties of purified fragments
Reactivity with monoclonal antibodies
Affinity of antibodies to native protein for peptides
Analysis of substitutions, deletions in variant proteins
Synthesis of overlapping peptides and mimotopes

cells usually cannot recognize foreign antigens in their native state. Rather, the antigen must be processed and presented on the surface of antigen presenting cells (APCs) in association with the Ia molecule, a class II major histocompatibility complex (MHC) antigen (MALE et al. 1987; BROWN et al. 1988; JANEWAY 1988; WATTS and DAVIDSON 1988). Processing in most cases involves antigen uptake, denaturation, and proteolysis in endosomes (UNANUE 1984; ALLEN 1987; PIERCE and MARGOLIASH 1988; CRESSWELL 1987). To enlist T cell help most short peptides have therefore to be conjugated to a larger carrier protein which serves as a source for T cell epitopes (ADA 1988).

The elucidation of the molecular basis for T cell recognition, a major goal in molecular immunology, has very recently seen a dramatic development. As a result of the description of the first crystallographic structure of an MHC glycoprotein (HLA-A2) (BJORKMAN et al. 1987a, b), more detailed investigation of possible contact sites as well as the molecular modeling of other class I and class II MHC antigens will be possible (PARHAM 1988; BROWN et al. 1988). Sequence analysis of T cell epitopes as well as the characterization of antigen-Ia complexes (BUUS et al. 1986; SETTLE et al. 1987; LAMB et al. 1987; ALLN et al. 1987; BABBITT et al. 1985) led to the proposal of a simple sequential motif for the prediction of T cell determinants (ROTHBARD et al. 1988; ROTHBARD and TAYLOR 1988) as well as to the suggestion of amphipathic α-helices as well-suited T cell epitopes (DELISI and BERZOFSKY 1985; BERZOFSKY 1988).

Thus, the exciting possibility of combining B cell epitopes recognized in the particular antigen with known or prospective T cell determinants for the respective MHC antigens in one synthetic peptide might open up new avenues for the development of highly efficient synthetic vaccines. Initial studies investigating this possible application for the malaria circumsporozoite protein (GOOD et al. 1987) as well as for the gp120 envelope protein of HIV (CEASE et al. 1987) have been reported.

4 Selection, Synthesis, and Conjugation of Fimbriae-Based Peptides as Immunogens

Synthetic peptides have so far been applied for the detection and identification of linear B cell epitopes in fimbriae of urinary tract infective *E. coli* and of *N. gonorrhoeae*. The straightforward use of synthetic peptides to identify antigenic or immunogenic epitopes is applicable only to such determinants which are encoded by linear amino acid sequences. In most proteins continuous epitopes represent only a fraction of the overall antigenicity. Nevertheless, antibodies induced by synthetic peptides can be successfully used as reagents for the identification of antigenic determinants and also as sequence-specific probes for structure–function analysis if they cross-react with high enough affinity with the native protein. The induction of such antibodies will only occur frequently if (in solution) the peptide adopts or can be induced to adopt a conformation that is identical or at least very similar to the conformation of the corresponding segment in the parent protein (ARNON 1986a, b, VAN REGENMORTEL 1987; SHINNIK et al. 1983). Of the basic secondary structures commonly found in proteins [α-helices, β-pleated sheets, and β (reverse)-turns], β-turns have the highest probability of being mimicked in a short isolated synthetic peptide of about 15 amino acid residues. Strands of a β-pleated sheet require interaction with the adjacent strand to be stabilized (CHOTIA et al. 1977). Helices can form in an isolated peptide but the amino acid chain must be long enough for sufficient intrahelical interaction to occur. Further stabilization of helices is often the result of additional intramolecular interactions with neighboring protein regions (CHOTIA et al. 1977). The forces involved in the formation of β-turns, however, are thought to be inherent in the linear amino acid sequence of the oligopeptide chain and are in particular dependent on the nature of the four residues actually forming the reverse turn (RICHARDSON 1981; ROSE et al. 1985; LESZCZYNSKI and ROSE 1986). Turns have been found in immunologically interesting regions of proteins (ROSE et al. 1985). Furthermore, short peptides have been demonstrated to be able to adopt β-turns as preferred conformations in aqueous solutions (DYSON et al. 1985). For the fimbriae-based synthetic peptides these considerations lead to the selection of regions of the protein sequence predicted to contain preferably hydrophilic reverse turns as prime targets for peptide synthesis.

For the prediction of regions with a high probability for hydrophilic β-turns, a variety of algorithms are available for the prediction of secondary structures as well as hydrophilic and acrophilic domains (e.g., ROBSON and SUZUKI 1976; CHOU and FASMAN 1978; HOPP and WOODS 1981; KYTE and DOOLITTLE 1982).

The choice of peptides corresponding to *N. gonorrhoeae* or *E. coli* fimbriae sequences followed the arguments discussed above and was biased with regard to secondary structure and hydrophilicity predictions according to the algorithms of CHOU and FASMAN (1978) and HOPP and WOODS (1981). Regions were selected that were predicted to contain hydrophilic reverse turns at either end of the peptide sequence. The methods employed to conjugate oligopeptides to carrier protein molecules have a critical influence on the ability of the synthetic peptide to adopt the desired conformation (BRIAND et al. 1985; DYRBERG and OLDSTONE 1986) and thus also on the specificity of the antipeptide antibodies induced. During synthesis of selected fimbrial sequences a natural or additional cystein residue was therefore placed at the N- or C-terminal end distal to the predicted reverse turn, so that the peptide could be coupled in a unique orientation via heterobifunctional cross-linking agents to the respective carrier protein (SCHMIDT et al. 1984; DYRBERG and OLDSTONE 1986).

Peptides are synthesized by standard Merrifield solid phase techniques (BARANY and MERRIFIELD 1980; STEWARD and YOUNG 1984). Each peptide is conjugated after reduction to thyroglobulin and bovine serum albumin (BSA) using *m*-malein-imidobenzoyl *N*-hydroxysuccinimide ester (MBS) and succinimidyl 4-(*N*-mileinimido-methyl) cyclohexane-1-carboxylate (SMCC), respectively, as hetero-bifunctional cross-linking agents. Peptide specific antibodies are induced with the thyroglobulin conjugates while the resulting antisera are tested with the corresponding BSA conjugates.

5 *Neisseria gonorrhoeae* Fimbriae-Based Synthetic Peptides as Prospective Vaccines

Among the sexually transmitted diseases gonorrhea has reached pandemic proportions in many parts of the world, reawakening intensive research efforts regarding various aspects of the gonococcus (GOTSCHLICH 1984). Gonococcal fimbriae have been demonstrated to be mediators of gonococcal adherence to epithelial cells (BRITIGAN et al. 1985). Fimbriae have therefore been proposed as essential constituents of any acellular gonorrhea vaccine (ROBBINS 1978). However, the serologic diversity of gonococcal fimbriae remains a problem, though it has been suggested that fimbriae from antigenically heterologous strains possess a common conserved determinant which is immunorecessive in the native protein (BRINTON et al. 1978; SCHOOLNIK et al. 1983). Thus, considerable efforts have been directed towards the characterization of fimbrial structure, receptor binding domains (SCHOOLNIK et al. 1984), and the possibility of employing fimbriae, subfragments, and corresponding synthetic peptides for the induction of protective immunity.

The complete 159 amino acid fimbrillin sequence from strain MS11 has been determined. Subfragments generated by CNBr and tryptic cleavage were analyzed for the presence of functional regions. TC-2 (31-111), a fragment overlapping with the CNBr-2 fragments (9-92), was found to bind to human endocervical cells but not to buccal or HeLa cells (SCHOOLNIK et al. 1983, 1984). Thus, these fragments were suggested to contain the receptor binding domain of the gonococcal fimbriae. Furthermore, sequence comparison with the corresponding fragments of heterologous strains (R10 and 2686) indicated that these regions are highly conserved among gonococcal fimbriae. However, as suspected, these regions were shown to be immunorecessive in the native fimbriae. When fragment CNBr-2 of the fimbrillin was used as an immunogen, the common fimbriae determinant became immunodominant. The resulting antibodies cross-reacted with fimbriae from heterologous strains (SCHOOLNIK et al. 1983), thus raising the possibility of circumventing the serologic diversity among gonococcal fimbriae.

Based on these results, synthetic peptides corresponding to conserved as well as variable regions of the MS11 fimbrillin sequence were used to map the antigenic structure of the fimbrillin more precisely as well as to identify linear regions of the protein that would elicit cross-reactive antisera possibly inhibiting the binding of gonococci to eukaryotic cells (ROTHBARD et al. 1984). Results of these studies supported the earlier finding that type-specific and conserved cross-reactive determinants are separated in different regions of the fimbrillin sequence.

The selection and synthesis of peptides followed the lines discussed in the two preceding paragraphs. Eight peptides were synthesized corresponding to amino acid sequences 21–35, 41–50, 48–60, 69–84, 95–107, 107–121, 121–134, and 135–151. Antisera to whole MS11 or R10 fimbriae bound only the regions corresponding to residues 48–60, thus indicating this region to contain the common immunorecessive epitope identified earlier in the TC-2 and CNBr-2 subfragments. The binding was significantly less than that of homologous antiserum to the strain-specific immuodominant epitopes in the cystein loop (R121–134 and R135–151). Less than

15% of the immunogenicity of the whole fimbriae was found to reside in the common determinant (R48–60). The results of these studies in combination with similar experiments employing monoclonal antibodies as probes (EDWARDS et al. 1984) supported the clustering of type-specific and conserved epitopes in different regions of the fimbrillin sequence. However, when the central CNBr-2 fragment (residues 8–92) was used as immunogen a much stronger cross-reactive response was generated. Antisera raised against the CNBr-2 fragment did not recognize R48–60 but bound to R69–84 and R41–50. This indicates that the immunogenicity of this region is determined by interactions in the whole intact fimbriae which are absent in the isolated fragments. Furthermore, it shows that peptides can elicit antibody populations that are different from those obtained using the intact protein as immunogen.

The antipeptide antisera were assessed for cross-reactivity with intact fimbriae of homologous and heterologous strains as well as for their ability to inhibit the binding of whole bacteria to human endometrial carcinoma cells (ROTHBARD et al. 1985). As shown in Fig. 1, the antipeptide antisera varied in their ability to recognize the homologous MS11 as well as the heterologous R10 fimbriae. Especially peptide R69–84 was effective in evoking a cross-reactive response. In Western blotting all antipeptide antisera cross-reacted with MS11 and most also recognized fimbriae from strain R10. Anti-R69–84 antiserum additionally recognized fimbriae from the heterologous strains F62, 1896, and 2686. Antisera raised against residues 21–35, 41–50, 48–60, and 69–84 immunoprecipitated the TC-2 fragment of MS11 fimbriae containing the proposed receptor-binding domain. Thus, these antisera were screened for their ability to block adherence of strain F62 to the human endometrial carcinoma cell line ENCA-4. Only antisera to residues 41–50 and 69–84 corresponding to the proposed conserved receptor-binding domain inhibited adherence (about 90%). Cross-reactive antibodies directed against the region containing type-specific epitopes (R135–151) were not able to reduce the binding of gonococci to ENCA-4 endometrial cells.

The peptides eliciting cross-reactive antibodies have been proposed as candidate immunogens for the prevention of gonorrhea (ROTHBARD et al. 1985) under the assumption that they would evoke the formation of antiadhesive antibodies. Success of any synthetic vaccine is clearly linked to the conservation of the respective antigenic epitope in the in vivo situation. As no animal model exists for gonorrhea, this question was addressed in a model of experimental gonorrhea in male volunteers (SWANSON et al. 1987). All reisolates from an infection with a defined strain MS11 expressed pili distinct from the strain used for infection and thus corroborated earlier results found with other MS11 variants and clinical isolates (HAGBLOM et al. 1985). Furthermore, amino acid changes have also always been found in the 69–84 oligopeptide region, suggesting that this region is not conserved in the in vivo situation and that therefore the peptide 69–84 will not be effective as a gonorrhea vaccine (SWANSON et al. 1987). Thus, the mechanism of chromosomal rearrangement involving recombination of minicassettes (SEGAL et al. 1985) gives rise to novel pili whose antigenic differences stem from amino acid changes in multiple regions so that the possible existence of a truly conserved domain needs to be reinvestigated.

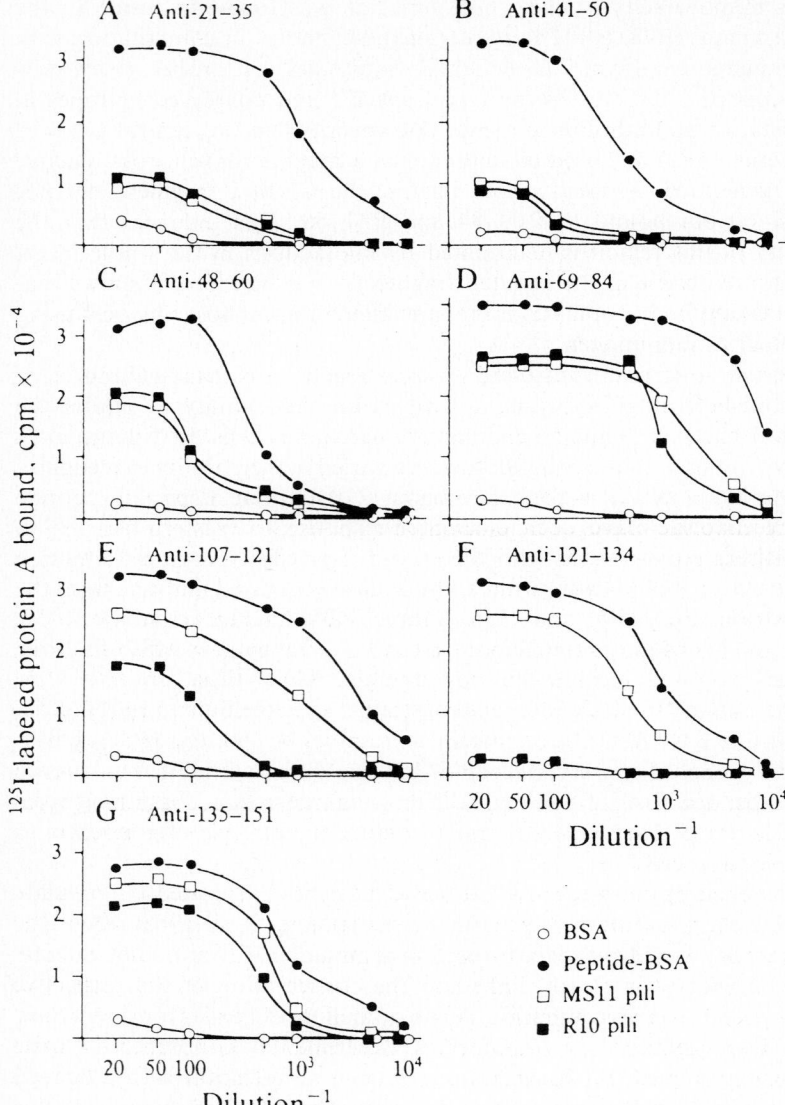

Fig. 1. Recognition of either homologous MS11 or heterologous R10 pili (fimbriae) by antipeptide sera in solid phase binding assays. Antisera elicited by peptide–thyroglobulin conjugates corresponding to residues 21–35 (*A*), 41–50 (*B*), 48–60 (*C*), 69–84 (*D*), 107–121 (*E*), 121–134 (*F*), and 135–151 (*G*) of MS11 pili were incubated with homologous peptide–BSA conjugate (●), isolated intact MS11 (□) and R10 (■) pili, or BSA (○). (ROTHBARD et al. 1985)

6 Synthetic Peptides Based on P-Specific Fimbriae of Uropathogenic *Escherichia coli* as Vaccines

Urinary tract infections caused by uropathogenic strains of *E. coli* represent one of the most common types of infectious disease, afflicting about 20% of all women at least once (ANDRIOLE 1987). About 80%–90% of all pyelonephritis isolates express P-specific fimbriae recognizing the α-Gal-(1-4)-β-Gal digalactoside as receptor. Though P-specific fimbriae from heterologous uropathogenic *E. coli* strains are structurally and functionally closely related and have also been shown to be protective immunogens for homologous infections in animal models (O'HANLEY et al. 1985; ROBERTS et al. 1982, 1984b; KAACK et al. 1988), their pronounced serologic diversity has so far impeded the development of a fimbriae-based vaccine. A way of circumventing this obstacle might be the development of synthetic peptides corresponding to possibly conserved sequences of the immunodominant fimbrillin.

N-terminal amino acid sequences of a number of fimbrillin proteins associated with P-specific fimbriae have been elucidated (Table 2), indicating conserved segments among heterologous strains. These regions, however, are immunorecessive when native isolated fimbriae are used as immunogens (SCHMIDT et al. 1984). Synthetic peptides corresponding to selected regions of the HU 849 fimbrillin (J96) sequence have been applied to map linear antigenic and immunogenic determinants (SCHMIDT et al. 1984). Three specific immunogenic epitopes (R25–38, R38–50, R48–61) were identified which jointly constitute the N-terminal cystein loop of the fimbrillin. Two prominent antigenic determinants were localized on peptides corresponding to R5–12 and R93–104 whereas R65–75 and R119–131 represented two minor antigenic determinants. However, only antisera raised against thyroglobulin conjugates of R5–12 and R93–104 cross-reacted with fimbriae from heterologous *E. coli* strains. Anti-R5–12 antiserum recognized about 85% of all P-specific fimbriae investigated so far by Western blotting (Fig. 2). Thus, the synthetic peptide R5–12 corresponds to a highly conserved antigenic determinant which was furthermore shown to be immunorecessive in the native fimbriae (SCHMIDT et al. 1984) and should thus not be prone to antigenic variation due to the selective pressure of the host's immune system.

Peptides corresponding to antigenic epitopes were assayed for their protective effect when applied as immunogens in the model of experimental pyelonephritis in mice. (O'HANLEY et al. 1985). Groups of Balb/c mice were immunized and boostered with peptide–thyroglobulin conjugates in Freund's adjuvant and subsequently challenged by intra-bladder inoculations of 10^8 colony-forming units (CFU) of the wild-type pyelonephritogenic *E. coli* strain J96. Two days later colonization of urine and kidneys with J96 bacteria was assessed as a measure of infection. It was shown that immunization with the synthetic peptides R5–12 and R65–75 protected against homologous infection just as efficiently as immunization with homologous isolated fimbriae (SCHMIDT et al. 1988). Thus peptides R5–12 and R65–75 correspond to "protective" epitopes (Fig. 3). Furthermore, as R5–12 encompasses a highly conserved cross-reactive epitope, these results suggest that this region might also be

Table 2. Comparison of N-terminal amino acid sequences[a] of UTI *E. coli* pilins and (afimbrial) adhesins

Bacteria	Protein	Isolate	N-terminal amino acid sequence
UTI *E. coli*	*Afimbrial*		
	AFA-I	KS52	N F T S S G T N G K V D L T I T E E C R V T V E
	AFA-V	MIR2194	A N Q G V V N S K G T V I D A T C G I D P D
	Fimbrial		
	F7₁	C1212	A A T I P Q G Q G E V A F K G T V V D A P
	F7₂	AD110	A P T I P Q G Q G K V T F N G T V V D A P C G
	F9	3669	A P S Q G S G Q V N F K G T V I D A P C G I E T Q S A N Q T I D F
	F11		A P T I P Q G Q G K V T F N G T V V D A P C S I S Q K S A D Q S I
	F12	C1979	A P T I P E G Q G K V T F N G T V V
	F13	J96	A P T I P Q G Q G K V T F N G T V V D A P C S I S Q K S A D Q S I
	PapA		V D N L T F R G K L I I P A C T T V S N T T V D W Q D
	PapE		D V Q I N I R G N V Y I P P C T I N N G Q N I V V D
	PapF		G W H N V M F Y A F N D Y L T T N A G N V K V I D Q
	PapG		
Type 1	Type 1A	K12	A A T T V N G — G T V H F K G E V V N A A C A
	Type 1A	J96	A A T T V N G — G T V H F K G E V V N A A C A
	Type 1A	C1214	A A T T V N G — G T V H F K G E V V N A A X A
	Type 1B	C1214	A T T V N G — G T V H F K G E V V
	Type 1B	C1023	V T T V N G — G T V H F K G E V V D

UTI = Urenary tract infective

[a] References from which the N-terminal amino acid sequences have been obtained: AFA-I (WALZ et al. 1985; LABIGNE-ROUSSEL et al. 1985), AFA-V (Ruffing and Schmidt 1988, unpublished work), F7 (KLEMM 1985), F7₂ (RHEN et al., 1985; VAN DIE and BERGMANS 1984), F9 (KLEMM 1985), F11 (VAN DIE et al. 1986), F12 (KLEMM et al. 1986), F13 (BAGA et al. 1984), PapE, F (LINDBERG et al. 1986), PapG (LUND et al. 1987), Type 1A, B, C (KLEMM 1985; RHEN et al. 1985)

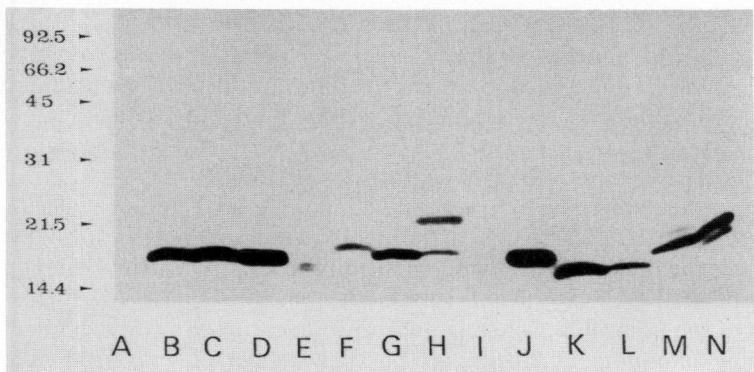

Fig. 2. Western blot of P-specific *E. coli* strains or purified P-specific pili (fimbriae) probed with anti-R5–12 antiserum diluted 1:200. Bound antibody was detected with ^{125}I-labeled protein A after an 8-h autoradiographic exposure. *Lane A*, nonpiliated K-12 strain HB101; *lane B*, strain J96 (F13); *lane C*, pap5 (HB 101) (F13); *lane D*, purified pili from recombinant strain HU849 (F13); *lane E*, C1979 (F12); *lane F*, AM 1727 carrying plasmid pPIL110-75 (expresses $F7_1$ of AD 110); *lane G*, AM 1727 carrying pPIL110-35 (expresses $F7_2$ of AD 110); *lane H*, C1212 (AD 110) ($F7_1$ and $F7_2$); *lane I*, 3669 (F9); *lane J*, pilin purified from pap5 (HB 101); *lanes K–N*, pyelonephritis strains P21, A42, A50, and A8. Positions and sizes (kd) of standard proteins run in parallel are shown at *left*. (SCHMIDT et al. 1988).

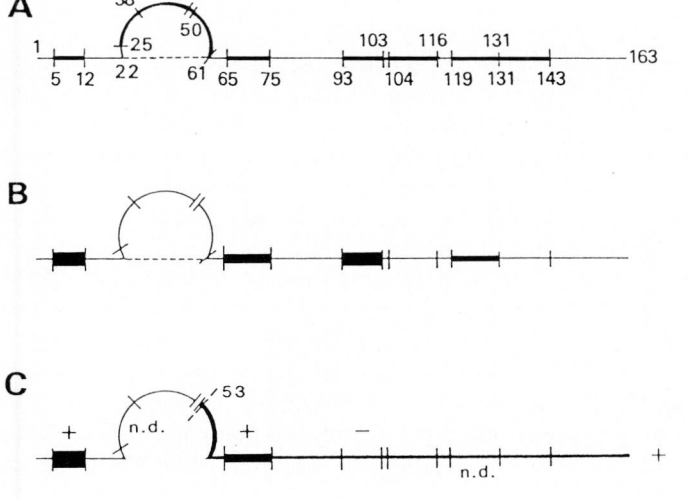

CNBr II RC53 (53–163)

Fig. 3. Linear antigenic and protective epitopes of the HU 849 pilin. The linear depiction of the pilin shows the location of the peptides synthesized (*A*), the identified linear antigenic determinants (*B*), and the synthetic peptides corresponding to protective epitopes (*C*) in the experimental pyelonephritis model

protective against heterologous challenge. This study demonstrated that a small selected peptide corresponding to a linear conserved immunorecessive epitope (R5–12) is able to confer protection on a mucosal surface in a relevant animal model against homologous bacterial infection equal to that obtained with the intact protein.

The molecular basis for the efficacy of peptides R5–12 and R65–75 as vaccines, however, was not addressed. In sera of immunized animals only fimbriae-specific IgG antibodies could be detected. In light of the identification and localization of the adhesin complex at the tip of the pilus filament (LINDBERG et al. 1987; HOSCHÜTZKY et al. 1989), a model for the efficacy of antipeptide antibodies directed against the N-terminal region of the fimbrillin protein is difficult to envisage. However, preliminary results using immunogold labeling of bacteria incubated with antibodies directed against the N-terminal sequence of the HU 849 pilus suggest a selective binding at the tip of the filament, thus offering a possible explanation for the protective effect (M. A. SCHMIDT, unpublished data). It remains to be seen from further experiments whether the protection demonstrated with *fimbrillin*-based synthetic peptides against homologous infections in an animal model can be extended to urinary tract infections due to heterologous strains. Although the employed animal model exhibits properties very close to the human situation, a real test of efficacy for any candidate vaccine can only be conducted in human trials.

7 Conclusion and Outlook

The possibility of employing fimbriae-based synthetic peptides for the development of synthetic vaccines has so far been investigated for fimbriae of *N. gonorrhoeae* and of uropathogenic P-specific *E. coli*. In both cases synthetic peptides were used to characterize the linear antigenicity of the immunodominant structural fimbrillin as a prerequisite for functional studies. Synthetic peptides corresponding to conserved antigenic determinants were assayed for their capacity to induce antibodies able to interfere with binding of the respective bacteria.

Synthetic peptides corresponding to a proposed conserved region of MS11 fimbriae induced cross-reactive antibodies which reduced the number of MS11 and R10 gonococi adhering to endometerial cells used in a tissue cultue adherence assay to about 10%. Studies in human volunteers, however, showed that the extensive antigenic variation found in gonococcal fimbriae is even more pronounced in the in vivo situation and furthermore always affected the proposed conserved region. Thus, the induction of protective immunity in humans by those synthetic peptides characterized so far seems rather unlikely.

In P-specific fimbriae of *E. coli* a conserved highly cross-reactive linear epitope has been identified. It could be shown that this determinant induces protective immunity towards homologous infection in an animal model relevant to the situation in humans. Though it remains to be seen whether these results can be extended to infections with heterologous strains, these studies showed that a small synthetic peptide corresponding to a linear antigenic fimbrial determinant is able to confer protection on a mucosal surface against bacterial infection. A similar approach is currently being applied to proteins more directly involved in the adhesive complex of P-specific fimbriae.

In considering possible applications of synthetic peptides as vaccines, one has to keep in mind that so far the theoretical basis for the selection of linear domains for synthesis is still rather crude. Detection and selection of peptide sequences likely to be employed as mimotopes are still largely based on trial and error. With the rapidly increasing knowledge of the molecular geometry of antigen-presenting MHC antigens as well as the corresponding T cell receptors, a more knowledge-based selection of antigenic domains might be possible, perhaps even leading to a predictive analysis of possible mimotopes. The incorporation of synthetic T_h and B cell sites in one synthetic antigen (CELADA and SERCARZ 1988) might also be a possible way of overcoming genetic restrictions (GOOD et al. 1987; FRANCIS et al. 1987). Furthermore, synthetic peptides can now be synthesized with inbuilt conformational restrictions (SATTERTHWAIT et al. 1988) so that the probability of eliciting cross-reactive antibodies will be highly increased. One of the more practical advantages of synthetic vaccines in general is that they can be obtained free of reactogenic contaminants which often present problems in vaccines derived from biologic sources. Synthetic peptides enjoy excellent prospects of being incorporated into semisynthetic or fully synthetic vaccines in the near future, including also the development of synthetic peptide vaccines on the basis of adhesion-linked proteins of pathogenic bacteria.

Acknowledgments. I like to thank Ms. Johanna Grosser for secretarial assistance and Dr. Inga Benz for the critical reading of the manuscript.

References

Abe C, Schmitz S, Moser I, Boulnois G, High NJ, Ørskov I, Jann B, Jann K (1987) Monoclonal antibodies with fimbrial FIC, F12, F13 and F14 specificities obtained with fimbriae from *E. coli* 04:K12:H⁻. Microb Pathogen 2: 71–77

Acres SD, Isaacson RE, Babiuk LA, Kapitany RA (1979) Immunization of calves against enterotoxigenic colibacillosis by vaccinating dams with purified K99 antigen and whole cell bacterins. Infect Immun 25: 121–126

Ada GL (1988) What to expect of a good vaccine and how to achieve it. Vaccine 6: 77–79

Allen PM (1987) Antigen processing at the molecular level. Immunol Today 8: 270–273

Allen PM, Matsueda GR, Evans RJ, Dunbar JB, Marshall GR, Unanue ER (1987) Identification of the T-cell and Ia contact residues of a T-cell antigenic epitope. Nature 327: 713–715

Andriole VT (1987) Urinary tract infections: recent developments. J Infect Dis 156: 865–869

Arnon R (1986a) Peptides as immunogens—prospects for a synthetic vaccine. In: Koprowski H, Melchers F (eds) Peptides as immunogens. Springer, Berlin Heidelberg New York, pp 1–12 (Current topics in microbiology and immunology, vol 130)

Arnon R (1986b) Synthetic peptides as the basis for future vaccines. Trends Biochem Sci (TIBS) 11: 521–524

Atassi MZ (1984) Antigenic structures of proteins. Eur J Biochem 145: 1–20

Babbitt BP, Allen PM, Matsueda G, Haber E, Unanue ER (1985) Binding of immunogenic peptides to Ia histocompatibility molecules. Nature 317: 359–361

Baga M, Normark S, Hardy J, O'Hanley P, Lark D, Olsson O, Schoolnik G, Falkow S (1984) Nucleotide sequence of the papA gene encoding the Pap pilus subunit of human uropathogenic *Escherichia coli*. J Bacteriol 157: 330–333

Barlow DJ, Edwards MS, Thornton JM (1986) Continuous and discontinuous protein antigenic determinants. Nature 322: 747–748

Beachey EH (1981) Bacterial adherence: adhesin-receptor interactions mediating the attachment of bacteria to mucosal surfaces. J Infect Dis 143: 325–345

Berzofsky JA (1988) Features of T-cell recognition and antigen structure useful in the design of vaccines to elicit T-cell immunity. Vaccine 6: 89–93

Bjorkman PJ, Saper MA, Samraoui B, Bennett WS, Strominger JL, Wiley DC (1987a) Structure of the human class I histocompatibility antigen, HLA-A2. Nature 329: 506–512

Bjorkman PJ, Saper MA, Samraoui B, Bennett WS, Strominger JL, Wiley DC (1987b) The foreign antigen binding site and T cell recognition regions of class I histocompatibility antigens. Nature 329: 512–518

Briand JP, Muller S, Van Regenmortel MHV (1985) Synthetic peptides as antigens: pitfalls of conjugation methods. J Immunol Methods 78: 59–69

Brinton CC, Bryan J, Dillon JA, Guerina N, Jacobson LJ, Kraus S, Labik A, Lee S, Levene A, Lim S, McMichael J, Polen S, Rogers K, To ACC, To SCM (1978) Uses of pili in gonorrhea control. Role of pili disease, purification and properties of gonococcal pili, and progress in the development of a gonococcal pilus vaccine for gonorrhea. In: Brooks GF, Gotschlich EC, Holmes KK, Sawyer WD, Young FE (eds) Immunology of *Neisseria gonorrhea*. American Society for Microbiology, Washington, pp 155–178

Britigan EB, Cohen MS, Sparling PF (1985) Gonococcal infection: a model of molecular pathogenesis. N Engl J Med 312: 1683–1694

Brown JH, Jardetzky T, Saper MA, Samraoui B, Bjorkman PJ, Wiley DC (1988) A hypothetical model of the foreign antigen binding site of class II histocompatibility molecules. Nature 332: 845–850

Buus S, Sette A, Colon SM, Jenis DM, Grey HM (1986) Isolation and characterization of antigen-Ia complexes involved in T-cell recognition. Cell 47: 1071–1077

Cease KB, Margalit H, Cornette JL, Putney SD, Robey WG, Ouyang C, Streicher HZ, Fischinger PJ, Gallo RC, Delisi C, Berzofsky JA (1987) Helper T cell antigenic site identification in the AIDS virus gp 120 envelope protein and induction of immunity in mice to the native protein using a 16-residue synthetic peptide. Proc Natl Acad Sci USA 84: 4249–4253

Celada F, Sercarz EE (1988) Preferential pairing of T-B specificities in the same antigen: the concept of directional help. Vaccine 6: 94–98

Chothia C, Levitt M, Richardson D (1977) Structures of proteins: packing of α-helices and β-pleated sheets. Proc Natl Acad Sci USA 74: 4130–4134

Chou PY, Fasman GD (1978) Empirical predictions of protein conformation. Annu Rev Biochem 47: 251–276

Contrepois M, Girardeau JP (1985) Additive protective effects of colostral antipili antibodies in calves experimentally infected with enterotoxigenic *Escherichia coli*. Infect Immun 50: 847–849

Cravioto A, Scotland SM, Rowe B (1982) Hemagglutination activity and colonization factor antigens I and II in enterotoxigenic and non-enterotoxigenic strains of *Escherichia coli* isolated from humans. Infect Immun 36: 189–197

Cresswell P (1987) Antigen recognition by T lymphocytes. Immunol Today 8: 67–69

de Graaf FK, Roorda I (1982) Production, purification and characterization of the fimbrial adhesive antigen F41 isolated from calf enteropathogenic *Escherichia coli* strain B41M. Infect Immun 36: 751–758

de la Cabada FJ, Evans DG, Evans DJ (1981) Immunoprotection against enterotoxigenic *Escherichia coli* diarrhea in rabbits by peroral administration of purified colonization factor antigen I(CFA/I). FEMS Microbiol Lett 11: 303–7

DeLisi C, Berzofsky JA (1985) T-cell antigenic sites tend to be amphipathic structures. Proc. Natl Acad Sci USA 82: 7048–7052

Dyrberg T, Oldstone MBA (1986) Peptides as antigens—importance of orientation. J Exp Med 164: 1344–1349

Dyson HJ, Cross KJ, Houghten RA, Wilson IA, Wright PE, Lerner RA (1985) The immunodominant site of a synthetic immunogen has a conformational preference in water for a type-II reverse turn. Nature 318: 480–483

Edwards, McDade RL, Schoolnik G, Rothbard JB, Gotschlich EC (1984) Antigenic analysis of gonococcal pili using monoclonal antibodies. J Exp Med 160: 1782–1791

Evans DG, Evans DJ (1978) New surface associated heat-labile colonization factor antigen (CFA/II) produced by enterotoxigenic *Escherichia coli* of serogroups O6 and O8. Infect Immun 21: 638–647

Evans DG, Evans DJ, Tjoa WS, DuPont HL (1978) Detection and characterization of the colonization factor antigen of enterotoxigenic *Escherichia coli* isolated from adults with diarrhea. Infect Immun 19: 727–736

Evans DG, Evans DJ, Clegg S, Pauley JA (1979) Purification and characterisation of the CFA/I antigen of enterotoxigenic *Escherichia coli*. Infect Immun 25: 738–748

Evans DG, de la Cabada FJ, Evans DJ (1982) Correlation between intestinal immune response to colonization factor antigen/I and acquired resistance to enterotoxigenic *Escherichia coli* diarrhea in an adult rabbit model. Eur J Clin Microbiol 1: 178–85

Evans DG, Graham BY, Evans DJ, Opekun A (1984) Administration of purified colonization factor antigens (CFA/I, CFA/II) of enterotoxigenic *Escherichia coli* to volunteers. Gastroenterology 87: 934–940

Francis MJ, Hastings GZ, Syred AD, McGinn B, Brown F, Rowlands DJ (1987) Non–responsiveness to foot-and-mouth disease virus peptide overcome by addition of foreign helper T-cell determinants. Nature 300: 168–170

Freter R (1981) Mechanism of association of bacteria with mucosal surfaces. Ciba Found Symp 80: 36–55

Gaastra W, Klemm P, de Graaf FK (1982) Prediction of antigenic determinants and secondary structures of the K88 and CFA/I fimbrial proteins from enteropathogenic *Eschericia coli*. Infect Immun 38: 41–45

Geysen HM, Rodda SJ, Mason TJ (1985) The delineation of peptides able to mimic assembled epitopes. Ciba Found Symp 119: 130–149

Geysen HM, Rodda SJ, Mason TJ (1986) *A priori* delineation of a peptide which mimics a discontinuous antigenic determinant. Mol Immunol 23: 709–715

Gibbons RJ (1977) Adherence of bacteria to host tissue. In: Schlessinger D (ed) Microbiology. American Society for Microbiology, Washington, pp 395–406

Good MF, Maloy WL, Lunde MN, Margalit H, Cornette JL, Smith GL, Moss B, Miller LH, Berzofsky JA (1987) Construction of a synthetic immunogen: use of a new T-helper epitope on malaria circumsporozite protein. Science 235: 1059–1062

Gotschlich EC (1984) Development of a gonorrhoea vaccine: prospects, strategies and tactics. Bull WHO 62: 671–680

Gross RJ, Rowe B (1985) *Escherichia coli* diarrhoea. J Hyg 95: 531–550

Hagblom P, Segal E, Billyard E, So M (1985) Intragenic recombination leads to pilus antigenic variation in *Neisseria gonorrhoeae*. Nature 315: 156–158

Hopp TP, Woods KR (1981) Prediction of protein antigenic determinants from amino acid sequences. Proc Natl Acad Sci USA 78: 3824–3828

Hoschützky H, Lottspeich F, Jann K (1989) Isolation and characterization of the α-galactosyl-1,4-β-galactosyl specific adhesin (P adhesin) from fimbrial *Escherichia coli*. Infect Immun 57: 76–81

Isaacson RE (1977) K99 surface antigen of *Escherichia coli*: purification and partial characterization. Infect Immun 15: 272–279

Isaacson RE, Dean EA, Morgan RL, Moon HW (1980) Immunization of suckling pigs against enterotoxigenic *Escherichia coli*-induced diarrheal disease by vaccinating dams with purified K99 or 987P pili: antibody production in response to vaccination. Infect Immun 29: 824–826

Janeway CA (1988) Frontiers of the immune system. Nature 333: 804–806

Jemmerson R (1987) Antigenicity and native structure of globular proteins: low frequency of peptide reactive antibodies. Proc. Natl Acad Sci USA 84: 9180–9184

Jerne NK (1960) Immunological speculations Annu Rev Microbiol 14: 341–358

Kaack MB, Roberts JA, Baskin G, Patterson GM (1988) Maternal immunization with P fimbriae for the prevention of neonatal pyelonephritis. Infect Immun 56: 1–6

Kaper JB, Levine MM (1988) Progress towards a vaccine against enterotoxigenic *Escherichia coli*. Vaccine 6: 197–199

Klemm P (1985) Fimbrial adhesins of *Escherichia coli*. Rev Infect Dis 7: 321–340

Klemm P, Mikkelsen L (1982) Prediction of antigenic determinants and secondary structures of the K88 and CFA/I fimbrial proteins from enteropathogenic *Escherichia coli*. Infect Immun 38: 41–45

Klemm P, Ørskov I, Ørskov F (1983) Isolation and characterization of F12 adhesive fimbrial antigen from uropathogenic *Escherichia coli* strains. Infect Immun 40: 91–96

Kyte J, Doolittle RF (1982) A simple method for displaying the hydropathic character of a protein. J Mol Biol 157: 105–132

Labigne-Roussel A, Schmidt MA, Walz W, Falkow S (1985) Genetic organization of the afimbrial adhesin operon and nucleotide sequence from a uropathogenic *Escherichia coli* gene encoding an afimbrial adhesin. J Bacteriol 162(3): 1285–1292

Lamb JR, Ivanyi J, Rees ADM, Rothbard JB, Howland K, Young RA, Young DB (1987) Mapping of T cell epitopes using recombinant antigens and synthetic peptides. EMBO J 6: 1245–1249

Lerner LA (1982) Tapping the immunological repertoire to produce antibodies of predetermined specificity. Nature 299: 592–596

Leszczynski JF, Rose GD (1986) Loops in globular proteins: a novel category of secondary structure. Science 234: 849–855

Levine MM, Kaper JB, Black RE, Clements ML (1983) New knowledge on pathogenesis of bacterial enteric infections as applied to vaccine development. Microbiol Rev 47: 510–550

Levine MM, Ristaino P, Marley G, Smyth C, Knutton S, Boedecker E, Blanck R, Young C, Clements ML, Cheney C, Patnaik R (1984) Coli surface antigens 1 and 3 of colonization factor antigen II-positive enterotoxigenic *Escherichia coli*: morphology, purification, and immune responses in humans. Infect Immun 44; 409–420

Lindberg F, Lund B, Normark S (1986) Gene products specifying adhesion of uropathogenic *Escherichia coli* are minor components of pili. Proc Natl Acad Sci USA 83: 1891–1895

Lindberg F, Lund B, Johansson L, Normark S (1987) Localization of the receptor-binding protein adhesin at the tip of the bacterial pilus. Nature 328: 84–87

Lund B, Lindberg F, Marklund BI, Normark S (1987) The PapG protein is the α-D-galactopyranosyl-(1-4)-β-D-galactopyranose-binding adhesin of uropathogenic *Escherichia coli*. Proc Natl Acad Sci USA 84: 5898–5902

Lund B, Lindberg F, Normark S (1988) Structure and antigenic properties of the tip-located P pilus proteins of uropathogenic *Escherichia coli*. J Bacteriol 170: 1887–1894

Male D, Champion B, Cooke A (1987) Advanced immunology. Gower Medical, London

Mimion FC, Abraham SN, Beachey EH, Gogen JD (1986) The genetic determinant of adhesive function in type 1 fimbriae of *Escherichia coli* is distinct from the gene encoding the fimbrial subunit. J Bacteriol 165: 1033–1036

Moch T, Hoschützky H, Hacker J, Kröncke KD, Jann K (1987) Isolation and characterization of the α-sialyl-β-2,3-galactosyl-specific adhesin from fimbriated *Escherichia coli*. Proc Natl Acad Sci USA 84: 3462–3466

Morgan RL, Isaacson RE, Moon HW, Brinton CC, To CC (1978) Immunization of suckling pigs against enterotoxigenic *Escherichia coli*-induced diarrheal disease by vaccinating dams with purified 987 or K99 pili: protection correlates with pilus homology of vaccine and challenge. Infect Immun 22; 771–777

Morris JA, Thorns C, Scott AC, Sojka WC, Wells GA (1982) Adhesion in vitro and in vivo associated with an adhesive antigen (F41) produced by a K99 mutant of the reference strain *Escherichia coli* B41. Infect Immun 36: 1146–1153

Nagy B, Moon HW, Isaacson RE, To CC, Brinton CC (1978) Immunization of suckling pigs against enteric enterotoxigenic *Escherichia coli* infection by vaccinating dams with purified pili. Infect Immun 21: 269–274

Niman HL, Houghten RA, Walker LE, Reisfeld RA, Wilson IA, Hogle JM, Lerner RA (1983) Generation of protein-reactive antibodies by short peptides is an event of high frequency: implications for the structural basis of immune recognition. Proc Natl Acad Sci USA 80: 4949–4953

Normark S, Båga M, Göranson M, Lindberg FP, Lund B, Norgen M, Uhlin BE (1986) Genetics and biogenesis of *Escherichia coli* adhesins. In: Mirelman D (ed) Microbial lectins and agglutinins (properties and biological activity). Wiley, New York, pp 113–145

Novotny J, Handschumacher M, Bruccoleri RE (1987) Protein antigenicity: a static surface property. Immunol Today 8: 26–31

O'Hanley P, Lark D, Falkow S, Schoolnik G (1985) Molecular basis of *Escherichia coli* colonization of the upper urinary tract in BALB/c mice—Gal-Gal pili immunization prevents *Escherichia coli* pyelonephritis in the Balb/c mouse model of human pyelonephritis. J Clin Invest 75: 347–360

Ørskov I, Ørskov F (1983) Serology of *Escherichia coli* fimbriae. Prog Allergy 33: 80–105

Parham P (1988) Presentation and processing of antigens in Paris. Immunol Today 9: 65–68

Pierce SK, Margoliash (1988) Antigen processing: an interim report. Trends Biochem Sci (TIBS) 13: 27–29

Rhen M, van Die I, Rhen V, Bergmans H (1985) Comparison of the nucleotide sequences of the genes encoding the KS71A and F7$_1$ fimbrial antigens of uropathogenic *Escherichia coli*. Eur J Biochem 151: 573–577

Richardson JS (1981) The anatomy and taxonomy of protein structure. Adv Protein Chem 34: 167–339

Robbins JB (1978) Disease control for gonorrhea by vaccine immunoprophylaxis: the next step? In: Brooks GF, Gotschlich EC, Holmes KK, Sawyer WD, Loung FE (eds) Immunobiology of Neisseria gonorrhoeae. American Society Microbiology, Washington, pp 391–394

Roberts JA, Kaack B, Korhonen TK, Källenius G, Svenson SB, Winberg J (1982) Protection against pyelonephritis by immunization with bacterial fimbriae of *Escherichia coli* in the primate model (Abstract no 243). In: Proceedings of the 7th international congress on infections and parasitic disease, Stockholm

Roberts JA, Kaack B, Kallenius G, Mollby R, Winberg J, Svenson SB (1984a) Receptors for pyelonephritogenic *Escherichia coli* in primates. J Urol 131: 163–168

Roberts JA, Hardaway K, Kaack B, Fussell EL, Baskin G (1984b) Prevention of pyelonephritis by immunization with P-fimbriae. J Urol 131: 602–607

Robson B, Suzuki E (1976) Conformational properties of amino acid residues in globular proteins. J Mol Biol 107: 327–356

Rose GD, Gierasch LM, Smith JA (1985) Turns in peptides and proteins. Adv Protein Chem 37: 1–109

Rothbard JB, Fernandez R, Schoolnik GK (1984) Strain-specific and common epitopes of gonococcal pili. J Exp Med 160: 208–221

Rothbard JB, Fernandez R, Wang L, Teng NNH, Schoolnik GK (1985) Antibodies to peptides corresponding to a conserved sequence of gonococcal pilins block bacterial adhesion. Proc Natl Acad Sci USA 82: 915–919

Rothbard JB, Taylor WR (1988a) A sequence pattern common to T cell epitopes. EMBO J 7: 93–100

Rothbard JB, Lechler RI, Howland K, Bal V, Eckels DD, Sekaly R, Long EO, Taylor WR, Lamb JR (1988b) Structural model of HLA-DR1 restricted T cell antigen recognition. Cell 52: 515–523

Rowlands DJ (1986) Vaccines—the synthetic antigen approach. In: Silver S, Silver S, Broda P, Chakrabarty AM, Collins J, Davies JE, Hopwood DA, Knowles CJ, Starlinger P, van Montagu M (eds) Biotechnology: potentials and limitations. Report of the Dahlem workshop 1985. Springer, Berlin Heidelberg New York, pp 139–154 (Dahlem workshop reports, vol 35)

Runnels PL, Moseley SL, Moon HW (1987) F41 pili as protective antigens of enterotoxigenic *Escherichia coli* that produce F41, K99 or both pilus antigens. Infect Immun 55: 555–558

Rutter JM, Jones GW (1973) Protection against enteric disease caused by *Escherichia coli*—a model for vaccination with a virulence determinant? Nature 242: 531–532

Sack RB (1980) Enterotoxigenic *Escherichia coli*: identification and characterization. J Infect Dis 142: 279–286

Satterthwait AC, Arrhenius T, Hagopian RA, Zavala F, Nussenzweig V, Lerner RA (1988) Conformational restriction of peptidyl immunogens with covalent replacements for the hydrogen bond. Vaccine 6: 99–103

Schmidt MA, O'Hanley P, Schoolnik GK (1984) Gal-Gal pyelonephritis *Escherichia coli* pili linear immunogenic and antigenic epitopes. J Exp Med 161: 705–715

Schmidt MA, O'Hanley P, Lark D, Schoolnik PA (1988) Synthetic peptides corresponding to protective epitopes of *Escherichia coli* digalactoside-binding pilin prevent infection in a murine pyelonephritis model. Proc. Natl Acad Sci USA 85: 1247–1251

Schoolnik GK, Tai JY, Gotschlich EC (1983) A pilus peptide vaccine for the prevention of gonorrhea. Prog Allergy 33: 314–331

Schoolnik GK, Fernandez R, Tai JY, Rothbard JB, Gotschlich EC (1984) Gonococcal pili—primary structure and receptor binding domain. J Exp Med 159: 1351–1370

Segal E, Billyard E, So M, Storzbach S, Meyer TF (1985) Role of chromosomal rearrangement in *N. gonorrhoeae* pilus phase variation. Cell 40: 293–300

Sela M (1960) Antigenicity: some molecular aspects. Science 166: 1365–1374

Settle A, Buus S, Colon S, Smith JA, Miles C, Grey HM (1987) Structural characteristics of an antigen required for its interaction with Ia and recognition by T-cells. Nature 328: 395–399

Shinnik TM, Sutcliffe JG, Green N, Lerner RA (1983) Synthetic peptide immunogens as vaccines. Annu Rev Microbiol 37: 425–446

Sojka WJ, Wray C, Morris JA (1978) Passive protection of lambs against experimental enteric colibacillosis by colostral transfer of antibodies from K99-vaccinated ewes. J Med Microbiol 11: 493–499

Steward MW, Howard CT (1987) Synthetic peptides: a next generation of vaccines? Immunol Today 8: 57–58

Stewart JM, Young JD (1984) Solid phase peptide synthesis. 2nd edn. Pierce Chemical, Rockford, Illinois

Svennerholm AM, Lopez Vidal Y, Holmgren J, McConnel MM, Rowe B (1988) Role of PCF8775 antigen and its coli surface subcomponents for colonization, disease, and protective immunogenicity of enterotoxigenic *Escherichia coli* in rabbits. Infect Immun 56(2): 523–528

Swanson J, Robbins K, Barrera O, Corwin D, Boslego J, Ciak J, Blake M, Koomey JM (1987) Gonococcal pilin variants in experimental gonorrhea. J Exp Med 165: 1344–1357

Tainer JA, Getzoff ED, Alexander H, Houghten RA, Olson AJ, Lerner RA, Hendrickson WA (1984) The reactivity of antipeptide antibodies is a function of the atomic mobility of sites in a protein. Nature 312: 127–134

Tainer JA, Getzoff ED, Paterson YA, Olson AJ, Lerner RA (1985) The atomic mobility component of protein antigenicity. Annu Rev Immunol 3: 501–535

Thomas LV, Cravioto A, Scotland SM, Rowe B (1982) New fimbrial antigenic type (E8775) which may represent a colonization factor in enterotoxigenic *Escherichia coli* in humans. Infect Immun 35: 119–1124

Thomas LV, McConnell MM, Rowe B, Field AM (1985) The possession of three novel coli surface antigens by enterotoxigenic *Escherichia coli* strains positive for the putative colonization factor PCF8775. J Gen Microbiol 131: 2319–2326

Thomas LV, Rowe B, McConnell MM (1987) In strains of *Escherichia coli* O167 a single plasmid encodes for the coli surface antigens CS5 and CS6 of putative colonization factor PCF8775, heat-stable enterotoxin and colicin Ia. Infect Immun 55: 1929–1931

Unanue ER (1984) Antigen-presenting function of the macro-phage. Annu Rev Immunol 2: 395–428

Van Die I, Bergmans H (1984) Nucleotide sequence of the gene encoding the F7$_2$ fimbrial subunit of a uropathogenic *Escherichia coli* strain. Gene 32: 83–90

Van Die I, Zuidweg E, Hockstra W, Bergmans H (1986) The role of fimbriae of uropathogenic *Escherichia coli* as carriers of the adhesin involved in mannose resistant hemagglutination. Micro Pathogen 1: 51–56

Van Regenmortel MHV (1986) Which structural features determine protein antigenicity? Trends Biochem Sci (TIBS) 11: 36–39

Van Regenmortel MHV (1987) Antigenic cross-reactivity between proteins and peptides: new insights and applications. Trends Biochem Sci (TIBS) 12: 237–240

Watts C, Davidson HW (1988) Endocytosis and recycling of specific antigen by human B cell lines. EMBO J 7: 1937–1945

Wilson IA, Niman HL, Houghten RA, Cherenson AR, Conolly ML, Lerner RA (1984) The structure of an antigenic determinant in a protein. Cell 37: 767–778

Wilson IA, Haft DA, Getzoff ED, Tainer JA, Lerner RA, Brenner S (1985) Identical short peptide sequences in unrelated proteins can have different conformations: a testing ground for theories of immune recognition. Proc Natl Acad Sci (USA) 82: 5255–5259

Subject Index